WINEMAKING

The Chapman & Hall Enology Library

Principles and Practices of Winemaking by Roger B. Boulton, Vernon L. Singleton, Linda F. Bisson, and Ralph E. Kunkee

Wine Microbiology by Kenneth C. Fugelsang

Winery Utilities
Planning, Design and Operation by David R. Storm

Winemaking
From Grape Growing to Marketplace by Richard P. Vine, Ellen M. Harkness, Theresa Browning and Cheri Wagner

Wine Analysis and Production by Bruce W. Zoecklein, Kenneth C. Fugelsang, Barry H. Gump, and Fred S. Nury

WINEMAKING
From Grape Growing to Marketplace

Richard P. Vine
Ellen M. Harkness
Purdue University
and
Theresa Browning
Cheri Wagner
Indiana Wine Grape Council

CHAPMAN & HALL

I(T)P® International Thomson Publishing
New York • Albany • Bonn • Boston • Cincinnati • Detroit • London • Madrid • Melbourne
Mexico City • Pacific Grove • Paris • San Francisco • Singapore • Tokyo • Toronto • Washington

JOIN US ON THE INTERNET
WWW: http://www.thomson.com
EMAIL: findit@kiosk.thomson.com

thomson.com is the on-line portal for the products, services and resources available from International Thomson Publishing (ITP). This Internet kiosk gives users immediate access to more than 34 ITP publishers and over 20,000 products. Through *thomson.com* Internet users can search catalogs, examine subject-specific resource centers and subscribe to electronic discussion lists. You can purchase ITP products from your local bookseller, or directly through *thomson.com*.

Visit Chapman & Hall's Internet Resource Center for information on our new publications, links to useful sites on the World Wide Web and an opportunity to join our e-mail mailing list. Point your browser to: http://www.chaphall.com or http://www.thomson.com/chaphall/foodsci.html for Food Science

A service of I(T)P®

Cover design: Saïd Sayrafiezadeh, emDASH inc.
Art Direction: Andrea Meyer
Copyright © 1997 by Chapman & Hall
Printed in the United States of America

Chapman & Hall
115 Fifth Avenue
New York, NY 10003

Chapman & Hall
2-6 Boundary Row
London SE1 8HN
England

Thomas Nelson Australia
102 Dodds Street
South Melbourne, 3205
Victoria, Australia

Chapman & Hall GmbH
Postfach 100 263
D-69442 Weinheim
Germany

International Thomson Editores
Campos Eliseos 385, Piso 7
Col. Polanco
11560 Mexico D.F
Mexico

International Thomson Publishing–Japan
Hirakawacho-cho Kyowa Building, 3F
1-2-1 Hirakawacho-cho
Chiyoda-ku, 102 Tokyo
Japan

International Thomson Publishing Asia
221 Henderson Road #05-10
Henderson Building
Singapore 0315

All rights reserved. No part of this book covered by the copyright hereon may be reproduced or used in any form or by any means—graphic, electronic, or mechanical, including photocopying, recording, taping, or information storage and retrieval systems—without the written permission of the publisher.

1 2 3 4 5 6 7 8 9 10 XXX 01 00 99 98 97

Library of Congress Cataloging-in-Publication Data
Winemaking : from grape growing to marketplace / Richard P. Vine...
   [et al.].
      p.   cm. – (Chapman & Hall enology library)
   Includes bibliographical references and index.
   ISBN 0-412-12221-9
   1. Wine and wine making.   I. Vine, Richard P.   II. Series.
TP548.W773   1997
641.2'2–dc20   96-41930
                                                         CIP

British Library Cataloguing in Publication Data available

To order this or any other Chapman & Hall book, please contact **International Thomson Publishing, 7625 Empire Drive, Florence, KY 41042.** Phone: (606) 525-6600 or 1-800-842-3636. Fax: (606) 525-7778. e-mail: order@chaphall.com.

For a complete listing of Chapman & Hall titles, send your request to **Chapman & Hall, Dept. BC, 115 Fifth Avenue, New York, NY 10003.**

Contents

PREFACE		xi
ACKNOWLEDGMENTS		xiii
INTRODUCTION		xv
1	**HISTORY OF WINE IN AMERICA**	**1**
	Eastern America	2
	Western America	9
	National Prohibition	14
	American Wine in the 20th Century	17
	The American Wine Booms	20
2	**VITICULTURE (GRAPE GROWING)**	**24**
	Site Selection	26
	Climate	26
	Soils	26
	The Vine	27
	Cultivar Selection	30
	Vitis vinifera	30
	Major *Vitis vinifera* White Varieties	32
	Major *Vitis vinifera* Red Varieties	34

Vitis labrusca	37
Major *Vitis labrusca* White Varieties	38
Major *Vitis labrusca* Red Varieties	38
Vitis riparia	39
Vitis aestivalis	40
Vitis rupestris	40
Vitis berlandieri	41
Vitis rotundifolia	41
Hybrids and Grafting	41
French-American Hybrids	42
Major Hybrid White Wine Cultivars	43
Major Hybrid Red Wine Cultivars	44
Vineyard Establishment	44
Cost and Value	44
Start-Up	47
Labor Costs	47
Equipment Costs	47
Variable Costs	47
Fixed Costs	47
Budget	48
Site Preparation	58
Vineyard Layout	58
Planting	59
Vine Training	60
Trellising	60
Training Systems	60
Machine Harvesting	61
Weed Control	63
Herbicide Injury	64
Vineyard Management	64
Pruning	64
Pest Control	65
Diseases	65
Insects	70
3 WINE MICROBIOLOGY	**73**
Yeasts	73
Bacteria	80
Molds	83
The Microbiological Workplace	83
Equipment	85
Factors Affecting Microbial Growth	87
Commercial Starter Cultures	87
Microbial Spoilage in Juice and Must	88
Wine Preservation	88
Sterile Filtration	88
Pasteurization	89
Chemical Agents	89

	Microbial Spoilage of Juice Concentrates	89
	Microbial Spoilage of Bulk Wine	89
	Detection	90
	Control	90
	Microbial Spoilage of Bottled Wine	90
	Detection	90
	Control	93
	Winery Sanitation	93
	Sulfite and Citric Acid Sanitizing	94
4	**ENOLOGY (WINEMAKING)**	**95**
	Grape and Wine Components	96
	Pectic Enzymes	96
	Sugars and Sweetness	98
	Late-Harvest Grapes	99
	Acids, Acidity, and pH	101
	Oxygen and Oxidation	102
	Sulfur Dioxide	103
	Phenols, Phenolics, and Polyphenols	105
	Color	106
	Nitrogenous Compounds—Proteins	107
	Malolactic Fermentation	108
	Grape Flavors	108
	Wine Blending	111
	Blend Expression Rationale	111
	Key ATF Blending Regulations	112
	Key ATF Blending Regulations for Labeling	115
	Meritage Blending	117
	Proprietary Blending	118
	Blending Cost Rationale	119
	Fining	119
	Detartration	119
	Barrel Aging	120
	Cost Implications	121
	American Oak Versus French Oak	126
	Conditioning	126
	New Barrels	128
	Maintenance of Empty Barrels—Short Term	129
	Maintenance of Empty Barrels—Long Term	129
	Soda Ash Treatment	129
	Chlorine Treatment	130
	Citric Acid Treatment	130
	Filtration	131
	Microfiltration and Ultrafiltration	133
	Unfiltered Wines	133
	Packaging	134
	Bottles	134
	Closures	135

	Capsules	136
	Label Rationale	137
	Label Shapes	138
	Label Colors	138
	Label Humor	139
	Label Experiments	139
	Key ATF Labeling Regulations	140
	Winemaking Perils and Pitfalls	145
5	**WINE CLASSIFICATION**	**148**
	Table Wines	148
	Sparkling Wines	150
	Dessert Wines	151
	Aperitif Wines	152
	Pop Wines	152
6	**WINERY DESIGN**	**153**
	Management	154
	Feasibility and Finance	157
	Insurance	159
	Location	160
	Scale and Size	162
	Design Motif	163
	Case Studies	163
	Remodeled Small Barn	163
	Simplistic Custom Design	165
7	**REQUIREMENTS, RESTRICTIONS, AND REGULATIONS**	**169**
	Application for Basic Permit	170
	Application to Establish and Operate Wine Premises	170
	Bonds	171
	Environmental Impact	172
	Personnel Questionnaire	172
	Signature Authority	173
	Label Approvals	173
	Formula Wines	173
	State ABC Regulations	174
8	**GETTING STARTED**	**176**
	Record Keeping	176
	Sanitation	183
	Quality Assurance	183
	Analytical Instrumentation	184
	pH Meter	184
	Brix and Balling	185
	Residual Sugar	186
	Total Acidity	186
	Volatile Acidity	187
	Sulfur Dioxide	187

Sensory Evaluation	187
Laboratory Refrigerator	188
Winery Equipment	188
Scales	188
Crusher–Destemmer–Must Pump	189
Press	190
Pumps	192
Hose	192
Tanks, Barrels, and Other Bulk Wine Vessels	194
Key ATF Tank Regulations	195
Fermentation Locks	196
Refrigeration	197
Filters	198
Fillers	200
Corkers	201
Labelers	201
Materials	202
Potassium Meta-bisulfite	203
Pectic Enzymes	204
Yeast	204
Fining Agents	204
Ascorbic Acid	205
Citric Acid	205
Sorbic Acid	206
Bottles	206
Corks	208
Capsules	209
Labels	209
Cases	210
9 WHITE TABLE WINES	**211**
Grape Varieties	211
Increasing White Wine Fruit Flavor Intensity	212
Harvesting	214
Crushing–Destemming	214
Pressing	215
Brix (Sugar) Adjustments	215
Acid Additions	219
Yeast Inoculation	219
Fermentation	220
Stuck Fermentation	220
First Racking	221
Second Racking	222
Blending	222
Fining and Stabilization	223
Aging	228
Balancing and Preservation	228

	Filtration	229
	Bottling	231
	Packaging	231
	Records	235
10	**RED TABLE WINES**	**236**
	Grape Varieties	236
	Increasing Red Wine Fruit Flavor Intensity	237
	Harvesting	238
	Crushing–Destemming	238
	Brix (Sugar) Adjustments	238
	Acid Additions	240
	Yeast Inoculation	240
	Fermentation	240
	Pressing	244
	Maceration	244
	Free-Run and Press Wine	245
	Racking	245
	Blending	246
	Malolactic Fermentation	247
	Fining and Stabilization	247
	Barrel Aging	248
	Balancing and Preservation	249
	Filtration	249
	Bottling and Corking	249
	Bottle Aging	250
	Packaging	252
	Records	254
11	**BLUSH TABLE WINES**	**255**
	Grape Varieties	255
	Crushing–Destemming, Sugar and Acidity Adjustments, and Yeast Inoculation	256
	Pressing and Fermentation	257
	Rackings	258
	Blending	258
	Clarification and Stabilization	259
	Aging, Balancing, and Preservation	259
	Filtration	259
	Bottling, Corking, and Recording	260
12	**FRUIT AND BERRY WINES**	**261**
	Apple Wine	262
	Bramble Berry Wine	263
	Blueberry Wine	264
	Cherry Wine	265
	Peach Wine	266
	Strawberry Wine	267

	Dried Fruit Wines	268
	Mead	269
13	**MARKETING**	**270**
	Rationale	270
	Marketing Plan	271
	Marketing Budget	273
	Public Relations	273
	Tourism	274
	Traveler Motivation	274
	Winery Video	275
	Motorcoach and Group Tours	275
	Wine Trails	276
	Networking	277
	Education	277
	Media Relations	278
	Journalism Basics	278
	News Releases	279
	Television	280
	Media Formats—Timing and Deadlines	280
	Interviews	281
	Interviewing Techniques	282
	Visuals	282
	Media Kits	283
	Special Events	284
	Newsletters	288
	Advertising	289
	Brochures	290
	Direct Mail	291
	Database Marketing	292
	Billboards	292
	Internet	293
	Word of Mouth	294
	Sales Promotion	294
	Sales	295
	Sales Personnel	295
	Niche Sales	296
	Winery Tour, Tasting Room, and Retail Shop	296
	Customer Service	298
	Gift Shop	298
	Sales to the Wholesale Channel	300
	Sales to the Retail Channel	301
	Sales to Consumers	301
	Market Feedback	302
	Reselling	302
	Keeping in Touch	303

xii   Contents

	Testimonials	303
	Referrals	304
APPENDIX A	**SOURCES**	**305**
	Microbiology Laboratory Material and Equipment	305
	New Equipment	305
	New Tankage	306
	Used Equipment and Tankage	306
	Grapes, Must, and Juice	306
	Laboratory Ware and Computer Software	307
	Materials	307
	Periodicals	307
APPENDIX B	**ANALYTICAL PROCEDURES**	**309**
	Microscopy	309
	Gram Stain Procedure	312
	Apparatus and Reagents	312
	Procedure	313
	Malolactic Fermentation Determination	
	by Paper Chromatography	314
	Apparatus and Reagents	314
	Procedure	314
	Bottle Sterility	315
	Winery Sanitation	317
	Swab Test	317
	pH Meter	319
	Apparatus and Reagents	319
	Procedure	319
	Brix–Balling by Hydrometer	320
	Analysis	320
	Procedure	320
	Brix by Refractometer	321
	Apparatus	321
	Procedure	322
	Alcohol by Ebulliometer	336
	Apparatus and Reagents	336
	Procedure	337
	Extract by Nomograph	338
	Total Acidity	338
	Apparatus and Reagents	338
	Procedure	340
	Volatile Acidity by Cash Still	342
	Apparatus and Reagents	342
	Procedure	343
	Free Sulfur Dioxide	344
	Apparatus and Reagents	345
	Procedure	345

	Total Sulfur Dioxide	347
	Apparatus and Reagents	347
	Procedure	347
	Sensory Evaluation	348
	Apparatus and Conditions	350
	Procedure	350
	Visual Mode	350
	Olfactory Mode	353
	Gustatory Mode	356
	Overall Impression	361
	Cardinal Scales	361
APPENDIX C	**CHARTS AND TABLES**	**362**
APPENDIX D	**GLOSSARY**	**414**
BIBLIOGRAPHY		**425**
INDEX		**427**

Preface

During the past several decades considerable interest has developed in the United States for the wines that are produced in small wineries across our nation. This interest continues to intensify, especially for the truly good wines that are reasonably priced. Consumers are unforgiving. Second-class wines will not be acceptable just because a vintner may be newly established. The functions that must take place in the small estate-type wine cellar and the controls that can be realistically exercised by winemasters are essential in the creation of superior products.

Although wine can be a comparatively simple food to produce, it is a very vast topic. Perhaps much the same as with other art forms, it is the infinite variability of factors at the root of the subject that renders it so complex. There are hundreds of different vine varieties cultivated around the world, and doubtlessly an even greater number of fruit and berry cultivars. Combined with such factors as soils, climates and mesoclimates (which may change with each vintage season), cultivation techniques, harvesting criteria, and overall operational philosophy, one can easily understand the enormous breadth and depth of variation which exits. This diversity, along with more than 5 years of enological development, generates a number of different wine possibilities that can only be conceived as something vastly exponential.

In larger wineries, one can usually find chemists, enologists, record-keepers, viticulturists, and other well-trained professionals who respond and supervise within their respective fields of expertise. The comprehension and responsibility of estate-type winery production dynamics may weigh heavily, often entirely, on the winemaster. It is not necessary for the small winery vintner to have advanced degrees in the physical sciences, but a solid academic background is highly desired. Even so, some people are more adept than others at grasping the analytical and operational functions which are necessary in learning the elements of commercial winegrowing. More importantly, we, as authors, have been sensitive to the challenge of good questions from talented beginners and the benefit of good comments from accomplished veterans—inputs repeated over and over again throughout our experience. This book is constructed with these valuable inputs as focal points.

One should fully understand the integral involvement of the Bureau of Alcohol, Tobacco, and Firearms (ATF) in the permits and regulations which must be complied with in operating a winery. *Part 240 of Title 27, Code of Federal Regulations* is a source which is quoted herein for outlining the application, documentation, and reporting forms and procedures necessary for vintners in the maintenance of a legal and sound bonded winery operation. Each state has supplemental requirements which vintners should ensure are fulfilled.

There are hundreds of volumes already published that provide a wealth of information about commercial winemaking. Some that come to mind are the works of Doulton et al. in *Principles and Practices of Winemaking* and Zoecklein et al. in *Wine Analysis and Production,* Winkler et al. in *General Viticulture,* Weaver in *Grape Growing,* as well as Gump et al. in *Beer and Wine Production,* to list only five of the best. These authors have made outstanding contributions in providing up-to-date technology which can and should be applied in modern-day wineries.

This book is designed to be a foundation text from which both would-be and established vintners can find the elements of viticulture, enology, and marketing wine—a focal point from which the technology mentioned in the above paragraph can be interpreted and implemented. *Winemaking* should serve professional marketers as a reference from which to understand the rationale of methodology employed by grape-growers and vintners. This book should be of similar value to students searching for an initial overview grasp of the pathway of contemporary viniculture. Advanced amateur winemakers will also find this volume a handy guide in their cellars. Similarly, wine connoisseurs can also find value in this book for a deeper understanding of crafting wine. Overall, this work should bridge the ever-widening gap between the art and science of winegrowing.

This book should be owned by everyone who has more than a casual interest in wine. The investment of its price should be repaid many times over with an easy rationale for beginners to understand the manner of wine made by experts—and experts to explain these ways to beginners.

Acknowledgments

The authors would like to thank Purdue University for its generous support in the preparation of this book. In particular, Dean of Agriculture, Victor Lechtenberg, and Associate Deans of Agriculture, Bill Baumgardt and Hank Wadsworth have given of their personal time and energies to this project. Dr. Eldon Ortman, Dr. Doug Curry, and Mr. John Trott have also answered our calls to Purdue administration for help.

Dr. Phil Nelson, Dr. Randy Woodson, and Dr. Dave King, Purdue department heads of Food Science, Horticulture and Agriculture Communications, respectively, have also given us sound advice, along with considerable resources—without which this work would have not been possible. As with most every book authored amid an academic community, much is owed to all the enthusiastic students who contribute far more than is generally thought about. Thank goodness for their boundless energy.

We are grateful to the Indiana Wine Grape Council for its many contributions towards creating *Winemaking: From Grape Growing to Marketplace*. Council President, Mr. Bill Oliver, Jr., along with former President, Jim Butler, and all members of the Council provided both material and inspirational influences. Special words of gratitude go to Purdue Extension Specialist Dr. Bruce Bordelon and Indiana vintner Dave Gahimer for their advice and contributions in key sections of this book.

Similarly, the Indiana Wine Guild, from which the initial need for this book was first identified, must be recognized. Special thanks go to Bill Oliver, Sr., and Ben Sparks for their leadership in the rebirth of commercial winegrowing in midwestern America. Indiana Wine Guild President, Steve Thomas, continues this guidance and has been an essential resource for us.

Most of all, we would like to thank our families and friends who excused us from considerable personal quality time in order to get this manuscript to Chapman & Hall before the deadline.

To all of you, we are deeply grateful.

Introduction

Man made wine long before the dawn of civilization. Paleontologists infer that crude fruit wines were familiar to early human beings more than 100, years ago.

According to Genesis 9:20–21 in the Bible, Noah landed his ark on Mount Ararat across the northern border of Turkey and grew grapes for wine. Documents affirm that pre-Islamic Arabs drank wine in early Mesopotamia. Hammurabi wrote strict laws for wine commerce in the 18th century B.C. The great Egyptian pharaohs enjoyed a variety of wines grown from irrigated vineyards along the Nile, and Homer wrote of fine wines made in ancient Greece. In July of 1996, The Associated Press reported that King Herod's 2-year-old wine jug was discovered near Jerusalem.

It was several centuries later, during the rise of the Roman Empire, that wine was introduced to Europe, but it was Christianity which developed the wine industry as we know it today. When Christ offered the cup at the Last Supper, He willed that wine be the everlasting symbol of His blood—an event which was the motivation for monks to establish vineyards wherever they settled in the Middle Ages and the Renaissance. It continued in the New World—the first wines in New York and California, among other states, were influenced by the Church.

Wine continues to replace fouled water supplies, as it has throughout the history of Western civilization. For thousands of years wine was our most important medicine and our only anaesthetic. The "magic" of wine fascinated the ancients—that by simply crushing grapes wine would literally make itself. It was not until Pasteur's discoveries in 1860 that the phenomenon of fermentation could be explained—whereby microscopic one-celled plants, called yeasts, convert simple sugars into ethyl alcohol and carbon dioxide gas.

In many countries, winemaking is conducted in much the same regard as we would preserve fruits and vegetables from the garden. They endure little fanfare about alcohol content, as it serves as a natural wine preservative, it enhances digestion, and adds to carbohydrates in the diet.

The wine industry is vast—one person in every 100 around the world is employed either in a vineyard, a winery, or in the marketing of wine. Wine is, however, far from global. Some nations find difficulty in cultivating the vine because of harsh environs and other countries are populated with cultures which restrict wine drinking or are prohibitionist.

Grapevines grow abundantly in the United States. In nearly every state, one type of vine or another can be grown. Of the 60-odd species and thousands of varieties known, a greater selection of vine types can be grown in the United States than in any other country.

Wine consumption in France and Italy surpasses 16 gallons per person per year, which far outpaces the average of just two gallons annually per capita across the United States. Despite this comparatively small demand, total wine consumption in the United States creates more than 530,000 jobs and in excess of $8 billion in wages, and $2.6 billion in taxes. Winegrowing in the United States is regulated nationally by the U.S. Bureau of Alcohol, Tobacco, and Firearms (ATF) and by individual state Alcohol Beverage Control (ABC) commissions.

Consistent winemaking success requires a systematic approach and the design herein is to help organize thoughts and materials into an efficient step-by-step methodology. It is fundamental but still offers plenty of room for variations which may better suit individual tastes. Although one should expect good results from following the recommendations made in this book, it cannot, however, be considered as any form of assurance or guarantee that any resulting product will be acceptable or superior.

The text is written in the "language" of the wine industry so that, as one becomes proficient, the commercial wine literature will serve to expand upon the same fundamental precepts.

WINEMAKING

CHAPTER 1

HISTORY OF WINE IN AMERICA

In comparison to the 6000 years of winegrowing history in Europe, the 400-year span of American vines and wines seems minuscule. But it should be remembered that most of the major development in the Old World took place during these same four centuries. Few modern-day European wine producers can boast continuous ownership of estate vineyards and cellars since before the 17th century.

The first wines in America were made more than 500 years ago by the Papago Indians who fermented the juice of the Saguaro cactus growing in what is now Arizona. Cactus wine is still made there, but demand for it falls far short of wine made from grapes.

This profusion of vines is doubtlessly what Leif Erikson found when he first visited the North American continent in the year 1000. The following passage comes from *The Discovery of America in the Tenth Century,* written several centuries ago by Charles C. Prasta:

> . . . A delightful country it seemed, full of game, and birds of beautiful plumage; and when they went ashore, they could not resist the temptation to explore it. When they returned, after several hours, Tyrker alone was missing. After waiting some time for his return, Leif, with twelve of his men, went in search of him. But they had not gone far, when they met him, laden down with grapes. Upon their enquiry, where he had stayed

so long, he answered: "I did not go far, when I found the trees all covered with grapes; and as I was born in the country, whose hills are covered with vineyards, it seemed so much like home to me, that I stayed a while and gathered them." . . . And Leif gave name to the country, and called it Vinland, or Wineland.

An enchanting story, and convincing, too, as wild grapevines are still prolific along the coastline of eastern America.

EASTERN AMERICA

A ship's log kept by navigator Giovanni da Verrazano dated in 1524 made note of vines in what is now North Carolina—and the possibility of making wine from them.

French Huguenots settled in Florida during the mid-1500s and made America's first wines from some of the native grapes found there. This fruit was from the species *Vitis rotundifolia,* known today as the "Muscadine" varieties. These New World grapes must have seemed strange to the Huguenot winemasters, as Muscadines ripen their fruit by individual berries in small clumps, rather than in large clusters. Further, the flavor of Muscadines is very exotic when compared to the more subtle nuances found in the Old World *Vitis vinifera* grapes.

In 1565, Sir John Hawkins, a British admiral, reported on winemaking from the native Florida grapes during his visits there. It seems likely that the Florida wines met with little favor among the early settlers.

Further north, Captain John Smith wrote of his observations in Virginia during the early 17th century:

> Of vines great abundance in many parts that climbe the toppes of highest trees in some places, but these beare but few grapes.
>
> There is another sort of grape neere as great as a Cherry, they [probably Indian natives] call *Messamins,* they be fatte, and juyce thicke. Neither doth the taste so well please when they are made in wine.

Beauchamp Plantagenet, in a London-published text which described the colonies, made these observations of the native grapes:

> *Thoulouse Muscat, Sweet Scented, Great Fox* and *Thick Grape;* the first two, after five months, being boiled and salted and well fined [clarified], make a strong red Xeres [Sherry]: the third, a light claret; the fourth, a white grape which creeps on the land, makes a pure, gold colored wine.

The "Great Fox" which Plantagenet mentioned was probably a variety of *Vitis labrusca,* a species that bears its heavily scented fruit in moderately sized clusters. Even in modern times, *foxy* is a rather loose organoleptic term used to describe the aroma and flavor which is typical of the *labrusca* species—the Catawba and Concord being prime examples.

Even further north, more native vines were discovered in the early 17th century by the Frenchman Samuel de Champlain, as he explored the islands in the St. Lawrence River. One of these islands was so overgrown with vines, doubtlessly *Vitis labrusca,* that Champlain named it "Bacchus Isle."

The colonists often built their community inn or tavern near their church. Between services, the worshippers would gather around the tavern's log fire, enjoying food, drink, and good fellowship. Vineyards were mostly private plantings and the wines homemade.

Disappointment with the native American vines, and the unfamiliar wines they made, no doubt prompted the plantings of European *Vitis vinifera* varieties in the Colonies during the 1600s. The London Company of Virginia may have made the first attempt at growing Old World grapes in the New World. Lord Delaware brought French viticulturists and vines to establish Virginia vineyards in the 1620s. The vines died, however, and colonists blamed the failure on the French vigneron's mishandling of the project.

As the 17th century progressed, hope continued that the European vines could be successfully grown in the New World. Viticulture was encouraged through economic incentives and even required by law in some places. In New York, wine could be made and sold without taxation. Virginia promulgated a law whereby

> all workers upon corne and tobacco shall plant five vynes. . upon penaltie to forfeit one barrell of corne.

Governor John Winthrop of Massachusetts planted vines on Governor's Island and, in Maryland, Lord Baltimore established vineyards. William Penn had a vineyard plot near Philadelphia where he tried to grow classic European vines. One by one, all the Old World plantings died. In Georgia, at Savannah, the enthusiastic Abraham de Lyon, from Portugal, planted a large-scale *vinifera* vineyard in 1730. Despite encouragement throughout the community, his vines died.

King Charles II ordered vines to be planted in Rhode Island, which turned out to be yet another failed attempt to grow the classic Old World *Vitis vinifera.* John Mason offered to trade the British monarch all of what is now New Hampshire in return for 300 tons of French wine, but Charles II rejected the proposal.

The major inspiration for the frustrated hopes of early colonial winegrowing pioneers was, of course, the profusion of wild vines; if the native vines grew so luxuriantly, why should not the cultivated European varieties do all the better?

The failures were, of course, not the fault of anyone—rather the then unknown *Phylloxera vastatrix* root louse, plus a lack of hardiness against the colder winter temperatures. These adverse conditions thwarted eastern American winegrowing for more than two centuries.

Thomas Jefferson maintained a very keen interest in American vineyards but continued to import wine from Europe—where, as Ambassador to France, his appreciation for fine was well known by his peers. He walked the entire breadth of France visiting Burgundy, Bordeaux, and other winegrowing regions. As U.S. Secretary of State, Jefferson chose the wines for President Washington. When Jefferson became president, he selected and paid for his wines personally—keeping careful inventory records. He also assisted in stocking the presidential cellars of John Adams, James Madison, and James Monroe. Some of his own personal wine bottles and casks remain in the Monticello cellars as illustrated in Figure 1–1.

The first commercial cultivation of native American *Vitis labrusca* vines is credited to John Alexander, a Pennsylvanian whose viticultural research in the latter 1700s brought forth the aptly named "Alexander" grape. The Alexander became famous for its heavy production of fruit and resistance to disease and cold winters. The wine from Alexander, however, still fell short of the prized vines from mother Europe.

Jean Jacques Dufour left his native Vevey, in Switzerland, during the late 1700s, eventually settling in southeastern Indiana along the Ohio River. By 1804, Dufour's first vines were planted near the settlement village of Vevay, near Madison, in what is now Switzerland County. Others followed his lead and the region soon became known as the "Little Rheinland," owing to the ascending riverbank terraces upon which hundreds of acres of the Alexander vine were planted. Indiana became one of the top 10 states in volume of commercial wine production.

Fig. 1–1 Thomas Jefferson's wine cellar at Monticello.
Courtesy: Vinifera Wine Grower's Association.

The Alexander grape variety interested the likes of Thomas Jefferson and Henry Clay, who also provided encouragement for Dufour to write *The American Vine Dresser's Guide,* a grape-grower's manual first published in 1826.

In 1809, following some of his own unsuccessful *Vitis vinifera* experiments at Monticello, Thomas Jefferson stated that

> it will be well to push the culture of this grape [Alexander] without losing time and effort in the search of foreign vines which it will take centuries to adapt of our soil and climate.

But Jefferson was not the only early hopeful winegrower with good foresight. William Penn was also of the opinion that the proper course was to move toward cultivating native *labrusca* vines.

John Adlum, from Georgetown, District of Columbia, introduced the *Vitis labrusca* variety, Catawba, named for the Catawba River in North Carolina, where the vine was discovered in the wild. The comparatively delicate fruit flavor of Catawba brought it quickly to the attention of winegrowers. Without modesty, Adlum informed the world that his providing the world with the Catawba grape had given a greater service to America than if he "had paid the national debt."

Thomas Jefferson never lost his optimism that native American winegrowing would eventually be achieved on a commercial scale—despite his fondness for the fine wines from Europe. He foresaw in 1808 that

> we could in the United States make as great a variety of wines as are made in Europe, not exactly of the same kinds, but doubtless as good.

And later,

> wine being among the earliest luxuries in which we indulge ourselves, it is desirable it should be made here and we have every soil, aspect & climate of the best of wine countries . . . these South West mountains, having a S.E. aspect, and abundance of lean & meager spots of stony & red soil, without sand, resembling extremely the Cote of Burgundy from Chambertin to Montrachet, where the famous wines of Burgundy are made.

Jefferson kept a *Garden Book* at Monticello from 1766 to 1826 in which he kept the data of his vine plantings and experiments. Among his notes are recorded the planting of

> 30 plants of vines from Burgundy and Champagne with roots, 30 plants of vines of Bordeaux with roots.

Several years later was another entry reporting the planting of several famous Italian vines, which included *Trebbiano* and *Aleatico,* among others. The *Garden Book* does not, however, show any significant success at the Monticello vineyard in testing any of the European selections.

The first commercial American winery may have been the Pennsylvania Vine Company, situated near Philadelphia, and founded by Pierre Legaux in 1793. Legaux was a Frenchman determined to succeed in cultivating French vines, but as had all previous attempts to grow the vaunted *vinifera,* this was yet another failure.

Located in Washingtonville, New York, about 50 miles northwest of New York City, the Brotherhood Winery has caves which are reminiscent of many in Europe. Founded in 1816 by a French immigrant shoemaker, Jean Jacques, Brotherhood remains the oldest operating winery in the United States.

Nicholas Longworth, a real estate magnate, built a majestic winery near Cincinnati—extending the "Little Rheinland" approximately 50 miles northeastward along the Ohio River from Vevay, Indiana. Longworth was so highly impressed by Adlum's Catawba grape that he planted a large acreage of the variety near his winery in the early 1820s. Longworth made the first American "Champagne"—labeling his product with the honest name of "Sparkling Catawba." Longworth had great pride for his wine, declaring that it was even superior to the sparkling wine from Champagne in France.

Henry Wadsworth Longfellow, a noted wine critic and connoisseur, was intrigued by the lofty claims of Nicholas Longworth and paid a visit to the Ohio winery in the mid-1850s to taste Longworth's sparkling wine. Impressed and convinced, Longfellow wrote a poem in 1854, entitled *Ode to Catawba Wine,* which proclaims

> Very good in its way
> Is the Verzenay
> Or the Sillery soft and creamy;
> But Catawba wine
> Has a taste more divine,
> More dulcet, delicious, and dreamy.

Longworth's wine must have been truly remarkable, as Longellow's testimonial ranking it superior to the French Champagne from Verzenay is, indeed, very high praise.

The Church grew very wealthy from winegrowing during the Renaissance in Europe. Christianity was influential in New World winegrowing, too—this time from Protestant denominations existing in eastern America. Elijah Fay, a Baptist deacon, planted the first vineyard in western New York State in 1818.

The vineyards of central New York State are nestled along the banks of the Finger Lakes. (See Fig. 1–2.) Local legend has it that when God finished making the Heavens and Earth, He laid His hand upon that particular site to bless his creation and, hence, the great handlike impression which identifies the beautiful Finger Lakes map. Geologists point out, however, that the glens and hills of the Finger Lakes region resulted from the clawing effects of the Ice Age.

In either event, St. James Episcopal Church still stands today as a landmark in

Fig. 1–2   Finger Lakes vineyards in Upstate New York.

Hammondsport, New York. This village, situated at the southern tip of Lake Keuka, the "thumb" of the Finger Lakes, is where the New York State wine industry began. Father William Bostwick propagated cuttings of Catawba and Isabella, another native *labrusca* variety, into roots and planted them behind the church in 1829 in order to make wine for his Anglican Eucharist.

The success of Bostwick's work became well known and attracted the interest of several prospective commercial vintners. In 1860, Charles Champlin, heading an investment consortium, founded the Pleasant Valley Wine Company just south of Hammondsport in Rheims, New York. The original cave was lined with stone walls more than 6 feet thick—a remarkable cellar which was to become the first federally licensed winery in America.

Champlin's obsession was to make the finest "Champagne" in the New World—certainly to share the fame generated in Ohio 30 years earlier. He hired Jules Masson, a French winemaster, to ensure that the very finest New York State "Champagne" was made. In 1871, a tasting of some of the first wine from Pleasant Valley was held at the Parker House in Boston. Among the prestigious critics was Marshall Wilder, then president of the American Horticultural Society, who proclaimed that Champlin's sparkling wine was "great" wine from the new "western" world. The famous "Great Western" name was thus born and became the trademark for the Pleasant Valley Wine Company.

The early 1860s marked the beginning of a promising era for American winegrowing. The success of Champlin and Longworth, among others, awakened new commercial winegrowing interests in the east and prompted considerable vineyard expansion—in turn, bringing about a fierce competition in the marketplace for the consumer's scarce dollars. In no less than 20 states and territories, significant vineyard and winery investments were taking place. Commercial American wines were existing, if not thriving. Mary Todd Lincoln was the first to serve American-grown wines in the White House.

New market potential for barrels, casks, and vats enticed cooper Walter Taylor, descendant of a sailing ship captain, to set up a cooperage shop on Bully Hill—at the foot of which the village of Hammondsport is situated. Taylor's farm was in an area known for large stands of white oak timber and, high on Bully Hill, overlooked miles and miles of the young but dynamic New York State wine industry. Taylor also dabbled in winemaking, but his expertise laid in the crafting of fine cooperage—much of which was in service at the local wineries under credit agreements.

The wine production capacity of the Finger Lakes grew faster than the market's propensity to consume, and the competition proved to be too much for many of the financially weaker vintners. As this worsened, Walter Taylor was ultimately forced to act on his delinquent accounts receivable—often taking bulk wine inventories for payment. Taylor and his three sons, Fred, Clarence, and Greyton, promptly found themselves deep in the wine business. During the ensuing 50 years, the Taylor family parlayed this somewhat impromptu start into one of the single largest wineries of its time.

During the mid-1800s, the Catawba and Isabella grape varieties found their way to Missouri and were adapted to yield large bounties of fruit. This success attracted widespread plantings there, much the same as in New York.

Many of the Missouri growers were inexperienced with grapes, however, and did not recognize infestations of the dreaded Black Rot fungus. The attack of this disease throughout the heartland of America caused discouragement even among dedicated believers. A large number of the Old Swiss growers in the "Little Rheinland" of Indiana moved to northern Ohio where they joined German immigrants in establishing vineyards near Sandusky and out on the Bass Islands in Lake Erie.

Ephraim Bull, from Concord, Massachusetts, introduced eastern vinegrowers to the "Concord" variety. It came from the long labors of Bull's research with vine hybrids and was released in 1854 after 7 years of testing. The Concord was bred from disease-resistant native *Vitis labrusca* varieties and, despite the then astronomical cost of $5 per vine, it was of interest to the sparse wine industry remaining in Missouri.

Along with another new disease-resistant and high-producing variety called "Norton's Virginia," Concord rekindled enthusiasm in Missouri once more. By the mid-1860s, vineyards had expanded to the largest acreage ever—making Missouri the leading grape-producing state in the Union. It was during this time that the first vine rootstock materials were taken from American vineyards to combat the *Phylloxera* root louse in Europe. Nurserymen Bush and Meissner of St. Louis published their initial catalog in 1869 and, by 1876, a French translation was being used by European winegrowers for ordering the rootstocks necessary for the great grafting project in the Old World. Yet today, classic vines in France and other European countries are grafted on American roots.

The finest years for Missouri viticulture proved to be the 1860s, as production again grew to far oversupply demand. New York City newspaper editor, Horace Greeley, advised, "Go west, young man, go west." People were doing exactly that—the Gold Rush, the land grants, and the transcontinental railroad were among many fascinations for opportunity in the west. Viticulture Professor Husmann moved from Missouri to California, where he proclaimed that "this [California] was the true home of the grape."

In 1870, California took the lead as the largest winegrowing state in America—a position that the Golden State has never relinquished.

WESTERN AMERICA

Winegrowing in the western states was pioneered, as in Europe and in eastern America, by the Church.

The "island" of California was first recognized by Pope Alexander VI in 1493. Spain made expeditions to explore the California coastline in 1542 and 1602. The Conquistadors found new frontiers to explore along the western shores of the Americas, and vineyards were established in their wake.

Unlike the eastern states, the European *Vitis vinifera* vines flourished in the western New World. In fact, vines grew so well that in 1595 the Spanish king decreed new vineyard plantings be terminated in the provinces. The edict was intended, of course, to maintain sales of Spanish wine to the west, and it remained in effect for nearly 150 years. This decree offers an interesting historical parallel to a similar abatement made by Rome some 1500 years earlier to curtail vineyard plantings in its colonial European provinces.

Despite the order from Madrid, Franciscan missions continued planting vines in the Americas. The first vineyard in the *Baja,* or lower California, was established at Mission Xavier by Father Juan Ugarte in the late 1690s.

Gaspar de Portola, explorer-captain of the Eighth Company of Dragoons, arrived in Mexico from Spain in 1767. The following year he was ordered by General Jose de Galvez to establish missionary settlements in *Alta* (upper) California to prohibit Russian fur trappers and British explorers from taking over the Spanish territorial claims.

De Portola embarked upon his mission project in March of 1769 with a small force of soldiers and a devout missionary priest, Padre Junipero Serra. (See Fig.

Fig. 1–3 Serra's personal quarters in Mission San Diego.

1–3.) It was Serra's responsibility to teach the native American Indians the ways of Christianity in order to subdue any pagan violence which may be encountered. The procession arrived in California in July of the same year and Serra raised the cross that marked the first mission—called "San Diego." Because Padre Serra knew he would need wine, he brought vines and planted the very first vineyard in California.

More missions were established northward, each a hard day's horseback ride apart, along what became known as the "El Camino Real." By 1771, the vineyard at the San Gabriel Mission had become a large plot—called the *vina madre,* or "mother vineyard," because of its superior soil and weather conditions. By the end of the 18th century, Serra had carried his vines to five outposts in southern California: San Diego, San Juan Capistrano, San Buenaventura, San Gabriel, and Santa Barbara. His "Mission" vines flourished in the California sun, and the wines they produced, although perhaps not of the finest quality, were abundant. In all, there were 12 missions constructed as the El Camino Real progressed to San Luis Obispo, Santa Cruz, San Jose, and yet further north—the last begun at Sonoma in 1823 by Father Jose Altimira.

As merchant ships began to stop more frequently at ports along the Alta California coast, the missions of the El Camino Real came increasingly into contact with the outside world which, in turn, brought about increased demand for wine there during the early 1800s.

This growth was short-lived, however, as the Mexican revolt against Spain resulted in the curtailment of all agricultural endeavors by the missions. The situation worsened, and some of the early Church vineyards in California were abandoned—others destroyed by the enraged monks themselves. Consequently, although the birth of California wine must be credited to the Church, the huge development of the California wine industry was made by the commercial laity.

Joseph Chapman planted the first commercial vineyard in California on the Los Angeles pueblo land in 1824. Chapman was originally an easterner who had gained experience in viticulture by working at Junipero Serra's Santa Ynez and San Gabriel missions.

Junipero Serra's Mission vines proved to be hardy, vigorous, and exceptionally well adapted to California. Some specimens eventually grew to become truly magnificent plants. The U.S. Department of Agriculture reported that a single vine, the Carpinteria vine, had a trunk nearly 3 feet in diameter and once yielded a crop of 8 tons of grapes! According to Ruth Teiser in her book *Winemaking in California,* the Carpinteria vine was dug up and sent off to the 1876 World Fair Exposition in St. Louis.

With the concentrated efforts of the monks, winegrowing had developed faster in southern than northern California. With the Mexican loss of California to the United States in 1846 and the Gold Rush of 1849, that situation was quickly reversed. Many European winegrowers emigrated to the San Francisco area—seeking refuge from political oppression in Europe and a mother lode of gold. California welcomed this new populace and encouraged land investments for vineyards. The State Agricultural Society promised that "capital put into vineyards would bring greater returns than when outlayed in fluming rivers for golden treasures."

The face of California winegrowing, indeed, in all of America, would be changed forever by Agoston Haraszthy, an exiled Hungarian count. Haraszthy initially emigrated to Wisconsin in the early 1800s, planting vines and building a beautiful stone winery near what is now the village of Prairie du Sac. His vines died, either from *Phylloxera* or the Wisconsin winters, or both. In any event, about one year prior to the Gold Rush, Haraszthy moved to the Sonoma Valley, near the last mission of the El Camino Real.

Count Haraszthy's optimism over the possibilities of quality winegrowing became infectious in northern California. He developed the Buena Vista Vinicultural Society and was determined that the wine industry there should be built on a sound foundation. His speeches and articles prescribed that only the finest varieties of *Vitis vinifera* should be planted in only the choicest land and that only the latest viticultural methods be used. (See Fig. 1–4.)

In 1861, Governor Downey delegated Haraszthy to make a trip to Europe to select the choicest *vinifera* vine stocks there and bring back cuttings from them to California. About 300 different varieties were shipped, but difficulties in handling

Fig. 1–4 Original Haraszthy vineyard in Sonoma, California.

and labeling the plant materials resulted in some of the varieties becoming either mislabeled or lost altogether. It took years to untangle the puzzle of identifying varieties. Some of which, such as the now famous "Zinfandel," have yet to be traced with absolute surety.

The estate and winery at Buena Vista were lavish—their cost eventually leading to Haraszthy's bankruptcy. With further financial support denied, Haraszthy left California quite mysteriously. His classic Old World vines remained and have since become a perpetual monument to his dedication.

Ten miles to the east of Sonoma is the Napa Valley—felt by many to be the single finest winegrowing district in America. A number of choice European varieties were planted there in the 1860s—some via Haraszthy's Sonoma project and others who had found their own sources for Old World *vinifera* vines.

The reputation of the wines being grown in the Napa Valley spread rapidly. Robert Louis Stevenson commented in the 1880s that the Napa locale was "where the soil sublimated under sun and stars to something finer, and the wine is bottled poetry." Stevenson went on with his lauds, comparing Napa wines with those from the great growths of Bordeaux and Burgundy—namely Chateau Lafite and the Clos de Vougeot.

The future seemed very bright indeed, but Sonoma and Napa vines were soon attacked by the *Phylloxera* root louse in much the same manner as in Europe. Having studied winegrowing in Michigan and Mississippi, Eugene Waldemar Hilgard, who had published in *Hilgardia* magazine, was named the first professor at the grape and wine laboratory at the University of California. Through the viticultural research work of Hilgard and the ex-Missourian Hussmann, the *Phylloxera* was thought to be finally held in check.

With the root louse seemingly out of the way, a general opinion emerged that the potential of the California wine industry was limitless. As in New York and Missouri earlier in the century, wine production in California was developed far in excess of demand. This situation was aggravated further by get-rich-quick vintners who marketed cheap, poor-quality wines. Their schemes crashed in the late 1860s, bringing about a depressed market for California wines. Industry leaders learned that more technical and economic knowledge was required before California wines could become the American standard. Professor Hilgard wrote in 1879:

> As the depression was, beyond doubt, attributable chiefly to the hasty putting upon the market of immature and indifferently made wines, so the return of prosperity has been, in great measure, the result of steady improvement in the quality of the wines marketed—such improvement being partly due to the introduction of grape varieties better adapted than the Mission grape to the production of wines suited to the taste of wine-drinking nations.

Two colossal personalities led the recovery of California winegrowing in the 1880s—George Hearst and Leland Stanford. Hearst made a fortune in mining, became a U.S. Senator, and helped launch the monumental publishing career of his son, William Randolph Hearst. George Hearst also purchased a 400-acre tract in Sonoma County, but *Phylloxera* promptly killed his vines. Undaunted, Hearst ordered new grafted vines, replanted, and ultimately attained production capacity of nearly 250,000 gallons of wine each year. Stanford, railroad magnate and patron of Stanford University, built a winery amidst the 350 acres of vineyard he owned near San Jose. By 1888, his output exceeded one million gallons.

Because of projects such as these, California vineyards once again expanded to the point of overproduction. A subsequent recession was even worse than that which took place in the previous decade. Grapes were sold for less than $10 per ton and wines for less than eight cents per gallon. To make matters worse, another mysterious disease was found in vineyards, principally in the Los Angeles vicinity. Vines rather suddenly became defoliated and died. This condition, which bewildered plant pathologists for decades, was finally identified as a bacterium introduced to vines by a leafhopper insect. It was called "Pierce's disease" and, to date, no cure has been found. Pierce's disease is generally found in subtropical climates and was doubtlessly involved in many of the *vinifera* failures experienced in the South.

But most winegrowers forged ahead with little concern for economic and biological threats—perhaps encouraged by French winegrowers who admitted that

California could grow Old World grapes well enough to be "capable of entering serious competition with the wines of Europe."

The gauntlet was dropped. American vintners were invited by the French to enter their wines in competition at the Paris World Exposition of 1889. The results stunned the international wine scene—U.S. vintners won 42 medals awarded by French wine judges! Among the California winners were Beringer, Krug, and Schramsberg. Stone Hill was a Missouri medalist, as was Great Western from New York State.

NATIONAL PROHIBITION

It had been building for some 80 years—a disaster worse than any disease or depression that had yet been known to American vintners. It was a reflection of profound and perhaps excessively rapid change in the fabric of American life. As wine had been the symbol of the Church throughout the entire history of western civilization, the Church had become, ironically, the primary influence used to destroy winegrowing.

The end of the Civil War did not end America's differences. Dynamic advances in industrial technology, a huge influx of immigrants coupled with a population explosion, increased urbanization, lingering influences from the Victorian era, and Protestant sectarianism—all were an aftermath which contributed to the idea that alcohol, in any form or measure, was sinful and a detriment to the welfare of mankind.

It formally began with the 1846 passage of prohibition in Maine, with Massachussets, Rhode Island, and Vermont following suit in 1852. More notable, however, was the 1880 enactment of prohibition in Kansas. The moving force behind this action was the infamous Carrie Nation, who assumed, at one time or another, the roles of judge, jury, and executioner in taking action against any facility which may have had anything to do with an alcoholic beverage. (See Fig. 1–5.) Her disregard for private property and human rights spread quickly. As a result, much of our early American wine history was lost, as books, diaries, and journals were burned by the prohibitionists.

The first signal of National Prohibition came with the "Dry Movement." Proponents demanded that any mention of wine be struck from textbooks and from the *U.S. Pharmacopoeia*. The "Drys" published books claiming that the word *wine* in the Bible was really grape juice—at the same time advocating the ban of classical Greek and Roman literature which mentioned wine. By the time World War I broke out in Europe, a total of 33 states had voted to go dry. By 1919, the Drys had succeeded in gaining wartime prohibition. The 18th Amendment to the U.S. Constitution, in conjunction with the Volstead Act, brought National Prohibition across all of America in 1920.

History of Wine in America 15

Fig. 1–5 Carrie Nation ready for battle.
Courtesy: The Kansas State Historical Society.

Vintners who had foreseen the reality of Prohibition taking shape geared up their production and marketing forces for medicinal, sacramental, and salted cooking wines which remained legal under the Volstead Act. Some winemasters went into the grape juice business, but many were forced to close their doors altogether.

Consumers who were determined to keep drinking wine found ways around Prohibition. Fake churches and synagogues were founded for the purpose of dispensing legal sacramental wines. Wine was often prescribed as "medication" for all sorts of human conditions. Of all the schemes, perhaps the most blatant was the sale of an array of grape juice products with a "pill" of dehydrated dormant yeast attached for home winemakers. The complete package was usually labeled with a printed warning not to add the pill "because if you do, this will commence fermenting and will turn into wine, which would be illegal."

Home winemaking soon became legal, with the head of each American household allowed to make up to 200 gallons per year for personal and family consumption. Some growers converted their vineyards from wine grapes to juice grapes. Concord growers knew that their grapes made high-quality grape juice. In fact, the founder of Welch's grape juice company was a devout prohibitionist.

The owners of Stone Hill Wine Cellars at Hermann, Missouri had the novel idea of remodeling their cellars into caves for mushroom farming. The Glen Winery in Hammondsport, New York was transformed into a storehouse for Glenn Curtiss' airplane parts.

A number of vintners desperately tried to continue their operations in secret, illegally, becoming "bootleggers." For the most part, though, such efforts were ultimately discovered and destroyed by U.S. Treasury officers, or "T-men," who, standing amid the rubble of a winery just destroyed, would strike triumphant poses for news photographers. Ironically, most of the bootleggers who prospered were criminals who had access to the gangster network of underground labor and resources which could circumvent or corrupt the enforcement efforts of the T-men. It was a conflict in which the names Al Capone and Eliot Ness will be etched forever.

California vineyard land actually grew during Prohibition, totaling more than 640,000 acres. Most of the grapes grown were of the variety "Alicante Bouchet," which resisted spoilage in storage and transport. More important was that this variety is so dark and dense that one gallon of crushed grapes could be ameliorated with another gallon or more of water and sugar. The resulting wine, often referred to as "Dago Red," gave the bootleggers a beverage they could market cheaply to lower income people.

Despite the tenacity of the T-men, the nation was well supplied with alcoholic beverages in countless ways—perhaps the memorable "speakeasy" private clubs being the most colorful. Bootlegger greed eventually made Prohibition wine expensive—and made the gangsters rich. Desperate alcoholics who could not afford the contraband prices resorted to drinking rubbing alcohol, aftershave lotions, colognes, and the like, which were made from butanol, methanol, and other toxic alcohols.

It did not take long for most people in America to realize that the "great experiment" of National Prohibition was a huge mistake. Many attempts were made for the reversal of the Volstead Act, as well as some proposals for modification of the prohibitionist regulations, but such efforts were fragmented with egoistic disunity. Thomas Jefferson's statement, a century earlier, rang vividly appropriate: "No nation is drunken where wine is cheap."

Finally, recognizing that Prohibition was destroying America, the nation commenced to rally around the demand for National Repeal. The impetus was, of course, not from the underworld, for they had built an empire with "Dago Red" and "Bathtub Gin." Nor did the repeal movement originate with the legal vintners. They were, except for the remaining medicinal and sacramental wine producers, long since out of business. The support for repeal came from a force perhaps best described as a national conscience. Early in December 1933 repeal took effect, bringing to an end 14 years of devastation to the entire country—and particularly the American wine industry.

AMERICAN WINE IN THE 20TH CENTURY

Repeal came during the midst of the Great Depression, which was a major hurdle for commercial winegrowing recovery. Consequently, few winegrowers in 1933 believed that American vines and wines could ever be resurrected to become a viable industry again, much less ever offer serious competition to the wines of Europe, which made great advances during the American Prohibition.

Apart from the Italian-Americans who continued to drink Dago Red, the new generation of young Americans maturing during Prohibition found that dry table wines were an unfamiliar, "sour," foreign-type of drink. In addition, many states voted to remain prohibitionist, especially in the Deep South—Mississippi remaining so until 1966. Other states voted in rigorous tax laws and government control; still others heavily restricted marketing activities.

Perhaps the only positive thing borne out of Prohibition was the need to start over again from scratch. The Prohibition years effectively ended many of the traditional vineyards, wineries, and practices which had been detrimental to the advancement of the wine art. With the slate swept clean, there was an opportunity to formulate a truly American wine industry, rather than a poor imitation of the European one. In any event, American winegrowers, presented with the ruins of abandoned vineyards, decayed cellars, and broken lives, began their work anew.

Unfortunately, a plethora of articles and books written by self-proclaimed American wine experts disseminated wine fallacy and snobbery. The notions that only certain wines should accompany specific entrees, and then at precise serving temperatures and in specific types of stemware, and other nonsense confused and intimidated potential wine consumers. Nevertheless, the U.S. wine industry slowly reemerged. Franklin Delano Roosevelt saw to it that American wines were once again served in the White House. Leon Adams, a young San Francisco news reporter and wine enthusiast, was a key figure in founding the Wine Institute. Adams (see Figure 1–6) spent his life as a wine educator.

It was during the late 1930s that several foresighted pioneers were to make a permanent mark on the American wine industry—each with very different philosophies of winegrowing. Philip Wagner was a Baltimore newspaper editor

Fig. 1–6   The late Leon Adams, Co-founder of The Wine Institute. *Courtesy:* Vineyard and Winery Management magazine.

and an amateur winemaker. After visiting Europe, Wagner had become deeply interested in some of the little-known French-American hybrid vines. These had been developed as an alternative to save Europe from the *Phylloxera* root louse. The idea had been to breed new cultivars of grapes from parentages of *vinifera,* with *labrusca, riparia, rupestris,* and other species—endeavoring to combine the high wine quality of the Old World vines with the disease resistance and high productivity of the native American selections. From thousands of crosses, several hundred promising new hybrids were developed—each identified by number codes instead of names. Those which captured Wagner's attention were imported and propagated in his Maryland nursery.

Wagner befriended many winegrowing people in the east, particularly in New York State. Greyton Taylor, youngest son of the Taylor Wine Company founder, planted experimental plots of Wagner's imported hybrids. Success begged that the number codes be changed to marketable names—the best whites now known as *Seyval Blanc, Vidal Blanc,* and *Vignoles,* along with reds, *Chambourcin, Chelois,* and *Marechal Foch.*

The French-American hybrids were instantly rejected by *vinifera* purists and *labrusca* traditionalists. True enough, it took more than 50 years to perfect the viticultural and vinification techniques necessary to result in competitive wines. Today, they compete very well, indeed. In 1992, a Missouri Vignoles was determined by more than 30 premier wine judges at the New World Wine Competition in California to be the "Champion White Table Wine"—besting more than 700 competing whites of every specie. Hybrid grape wines continue to win awards in prestigious competitions. Nevertheless, considerable controversy still surrounds the French-American cultivars. According to some, the hybrids are a viable alternative to native *labrusca* grapes and good competition to the Old World *vinifera*. To others, the hybrids are bastard children not worth planting on the most marginal of vineyard land. There are some who still think only wines from Europe are worthy of serious consideration.

Another prominent post-Prohibition figure was Charles Fournier (Fig. 1–7),

Fig. 1–7 The late Charles Fournier.
Courtesy: Vineyard and Winery Management magazine.

who emigrated in 1934 from the Champagne province in France to Gold Seal Vineyards in Hammondsport. With 200 years of failure in the eastern United States, Fournier was fully forewarned that *Vitis vinifera* would not survive the disease and harsh winters of upstate New York, but he experimented anyway.

A decade or so later, Fournier met Dr. Konstantin Frank, a Russian winegrower refugee. Fournier was convinced by Frank that it was possible to grow Old World vines in colder climates by employing highly selective cultivation techniques learned in mother Russia. Dr. Frank was hired by Fournier to work in the Gold Seal vineyards and they experimented together with the tender *vinifera*. During the late 1950s, they succeeded in harvesting the first commercial quantities of Old World grapes grown in eastern America. Although devastating winters still bring caution and fear among eastern winegrowers, the development of Old World wines continues to advance upon the breakthroughs by Frank and Fournier.

Brothers Ernest and Julio Gallo had their roots in Asti, in the Piedmont region of Italy. The Gallos first grew grapes at Antioch, California, where their parents had immigrated from Italy via South America. During Prohibition, the Gallo family shipped grapes for home winemaking to Chicago and New York. Instead of the Alicante Bouschet variety, however, they marketed the higher-quality Zinfandel. Following repeal, the Gallos became interested in the commercial wine business and leased a warehouse in Modesto from the Santa Fe railroad in which to commence their initial operations. Demand for their bulk wines grew rapidly, and by World War II, Gallo wines were also competing for consumers in the retail marketplace. The marketing genius of Ernest and the production expertise of Julio combined in a management team of unequaled drive and success. In less than 40 years, Gallo became the largest single winery in the world. Its success heralded the greatest growth in the U.S. wine industry to date.

THE AMERICAN WINE BOOMS

The first major U.S. consumer rush to buy wine commenced during the late 1940s—perhaps nurtured by returning American servicemen who brought home a newly acquired appreciation for wine from Europe. Subsequently, a burgeoning commercial airline industry gave business and pleasure travelers an entirely new exposure to European wine consumption.

The quest for wine technology was again rekindled. The American Society of Enologists was founded in 1950 at the University of California at Davis, where wine researchers followed up the early work of Hilgard. Funding was made available in other states for commencing or renewing viticulture and enology programs of study. As both European and American winemasters conducted intensive research programs, the traditions of Old World methodology commenced to give way to innovations based upon New World technology.

The forward steps were by no means confined only to production. There were many new ideas for wine marketing, too. New and more efficient distribution channels opened up better access to retail shelf space and restaurant wine lists. People began to tour the wineries—learning the jargon and sampling wines. Vintners printed sales promotion pamphlets replete with wine cookery recipes and down-to-earth wine serving tips. Slowly, the shrouds of wine snobbery and intimidation began to fall.

By far the most important contribution of the 1950s was a simple commonsense idea expanded upon by Frank Schoonmaker, a devoted and highly respected American wine authority. Schoonmaker's concept was that the labels of American-made wines should carry the name of the grape variety, rather than continuing to borrow the traditional European geographic names. A wine made from Pinot Noir grapes in a style similar to that of French Burgundy would no longer be labeled "California Burgundy," rather "California Pinot Noir." The great German *vinifera* vine varieties, Johannisberg Riesling and Sylvaner, were much more honest and descriptive terms on labels than, say, "California Rhine Wine."

The conversion proved to be rather easy in California, especially in the Napa, Sonoma, and other north coast districts where many of the classic Old World varieties had been grown since the Haraszthy influence a century earlier.

In the east, however, conversion to varietal labeling proved to be more difficult. The native American grape varieties, such as Catawba and Concord, were not internationally known, and the French-American hybrid cultivars were not yet widely planted. Consequently, many eastern vintners retained the practice of using European geographic names, such as "New York State Burgundy," "Ohio Rhine Wine," and "Michigan Chablis," among many others. Whereas some of these wines were tasty, the practice was nonetheless false labeling, as native and hybrid grapes cannot be vinified to *vinifera* flavor. Consequently, commercial eastern winegrowing in the east became suspect to consumers during the 1950s.

Clever eastern vintners devised many efforts to conserve their market share. One innovation was for "Pop Wines"—simple table wines flavored with synthetic essences and given hip names. A mixture of "Champagne" and "Sparkling Burgundy" made a bubbly called "Cold Duck," which was an instant sensation. A traditional Spanish drink called "Sangria," a mixture of fruit and light red wine, achieved sales in the millions of cases. An even greater success was a rush for fruit wines from California—particularly Gallo's "Boone's Farm" brand. The success of these created a second wine boom which inspired both European and Californian competition—creating yet another glut of production.

The third wine boom in the United States gathered momentum during the late 1970s and early 1980s via a totally different operational philosophy—the establishment of more than 1000 small "boutique" wineries. The typical estate, as illustrated in Figure 1–8, was a few acres of prime vineyard land planted to several select *vinifera* varieties which was centered by a state-of-the-art production facil-

Fig. 1–8 Silverado Vineyards winery—The Walt Disney estate.

ity housed in an architectural masterpiece. Although most of this development occurred in the north coast counties of California, boutique wineries emerged in most every state except the colder extreme upper central region of the United States.

While these new vintners prided themselves on individual wine character and style, they were unified by the common expression of producing wines of high quality—wines which could challenge the classical Old World vineyards.

Small wine contests became comprehensively structured wine competitions. It was in the early 1990s that U.S. wine consumption first surpassed all distilled spirits combined. New consumer wine periodicals keep Americans abreast of news from the vineyards. Chapters of national and international wine societies and brotherhoods continue to grow.

Much of this success is credited to the technological advances in winegrowing across the United States. As previously mentioned, the University of California at Davis has pioneered viticulture and enology research, but other institutions have contributed significantly, too. Among these are the Universities of Arkansas, Florida, Missouri, and Pennsylvania, along with the California State University at Fresno, Cornell University, plus Michigan State, New Mexico State, North Carolina State, Ohio State, Oregon State, Purdue, Texas A&M, Virginia Tech, and Washington State.

Out of all this has come a national interest in fine wine such as America had never seen before. It is common to find large metropolitan wholesale distributors carrying portfolios of several thousand different selections for retail wine shops and restaurant wine lists.

Annual wine production of approximately 450 million gallons now ranks the United States fifth in wine production among nations—behind France (2 billion gallons), Italy, Spain, and Argentina, respectively.

CHAPTER 2

VITICULTURE (GRAPE GROWING)

Viticulture is that part of horticulture which deals solely with the growing of *Vitis*, the grapevine. A commercial grape grower is a viticulturist.

The genus *Vitis* is divided into two subgenera, *Euvitis*, the bunch grape, and *Muscadinia*, the muscadine grape. They are separated based on various morphological characters and somatic chromosome number. The term *bunch grape* refers to the fact that the berries of *Euvitis* cultivars are borne in clusters, and all berries ripen over a concentrated period so that the entire cluster is harvested as a unit. In contrast, cultivars of the *Muscadinia* subgenus bear fruit in small clusters and the berries ripen individually over an extended period, so they are harvested individually rather than as a whole cluster unit.

More than 20 million acres of vines are cultivated in vineyards around the world with more than 60 different species identified. The most widely grown species for wine is the *Vitis vinifera*, or the "Old World" vine a native to Europe. *Vinifera* is also the principal species for wine in California, Oregon, Washington, and other states having moderate climates. In general, *V. vinifera* cultivars require a long growing season, relatively high summer temperatures, low atmospheric humidity, a ripening season free from rain, and mild winter temperatures.

Some species are native to the United States, such as *Vitis labrusca*, which are more disease-resistant and endure colder climates than the *vinifera*. Labrusca is found in all of the midwestern and northeastern states, as well as the lower reaches of Canada. Hybrid cultivars are also found in most every winegrowing region. Grape plants are confined primarily to the temperate zones of the world, although they are grown in subtropical and tropical climates. Commercial grape production occurs primarily between 20 and 51°N latitude and between 20 and 40°S latitude.

There can be an extra level of satisfaction gained in growing one's own grapes for winemaking. Wines grown from vineyards owned by commercial vintners are often labeled with the statement "estate bottled." (See Fig. 2–1.)

It may be the fifth or sixth year before a full vintage of grape production is taken. All things considered, it is common to find these years of vineyard establishment costs totaling more than $7000 per acre—not including the cost of land.

Fig. 2–1 Northern California estate vineyard and winery.

best places to begin the search for information relating to growing finding sources for grapes is with the Cooperative Extension Service and regional winegrowing associations.

SITE SELECTION

Climate

The major limiting factors to grape production in most temperate climate locales are cold-temperature injury from winter cold and spring frosts—along with diseases encouraged by hot, humid weather and frequent rainfall during the growing season. Selection of a site with a desirable climate helps reduce the occurrence of these problems. The best sites for grapes are those with full sun exposure, mild winter temperatures, freedom from frost, and good soil drainage. Cultivar selection is obviously essential in determining the suitability of a given site.

Two important considerations in site selection are minimum winter temperatures and freedom from frost. Cultivar selection is often dictated by the minimum temperature expected for an area. Very hardy cultivars are capable of withstanding $-15°F$ with little injury, whereas tender cultivars will suffer significant injury at temperatures slightly below $0°F$. Varietal selection, preconditioning of the vine, its state of acclimation, along with the rate and extent of temperature change can dramatically affect the amount of cold injury sustained.

Freedom from spring and autumn frost events is another important characteristic of vineyard sites. Frost-free sites are usually areas with gentle slopes which are elevated above surrounding areas. Cold air drains from higher sites to lower levels—reducing the risk of damaging frost at the elevated sites. Higher elevations also have better air drainage throughout the growing season, which promotes rapid drying of foliage following dew or rain—greatly reducing the incidence of disease organisms.

The length of the growing season is another important factor for grape production. It is determined by the dates of first and last frosts of the year. The length of the growing season (frost-free days) ranges from 150–160 days in cooler regions to 200 days or more in warmer locales. A frost-free period of at least 150 days is required for the earliest ripening cultivars. The length of the growing season is often extended within a region on sites near large bodies of water which temper drastic changes in temperatures.

Soils

Grapevines are adapted to a wide range of soil types but perform best when they have healthy, well-developed root systems. Soil conditions favorable to root growth include good aeration, loose texture, and moderate fertility, as well as good surface and internal drainage. Proper soil drainage cannot be overemphasized as a necessity for successful vine culture. Soils which are consistently wet during the growing season due to an impervious subsoil, high water table, or other drainage problem should be avoided. Root growth in poorly drained soils is usually limited to the top 2 feet or less, whereas in deep, well-drained soils, roots may penetrate 6 feet or more. When root growth is restricted because of poor drainage, vine growth and fruit yields are generally low and plant survival is limited to only a few years.

THE VINE

The grapevine is a woody perennial which can live for many years, generally several decades of productive growth, with proper care. The leaves, stems, and roots of the grapevine classify it biologically as a higher-form plant. Such plants are able to use sunlight to make sugars from carbon dioxide and water through the process of photosynthesis. Some of these sugars are used in the metabolic processes of the plant's life system. Another portion of these precious sugar compounds is stored in the grapevines's woody tissues until *veraison*—a French term referring to that time when grapes commence to ripen and sugars are transported to berries.

Taxonomically, the grapevine is classified as follows:

Group	Spermtophyta
Division	Tracheophyta
Subdivision	Pteropsida
Class	Angiosperm
Subclass	Dicotyledoneae
Order	Ramnales
Family	Vitaceae (Ampelidaceae)
Genus	*Vitis*
Subgenera	*Euvitis* and *Muscadinia*

There are more than 60 species of *Vitis* known, and many of these are indistinguishable to all but highly skilled viticulturists. Origins of the vine are largely limited to the Northern Hemisphere; the North American continent is particularly abundant in native species. The subgenus *Euvitis* includes all the world's grapevines which produce grapes in clusters—more than 60 species. The most important species of *Euvitis* is *Vitis vinifera*, vines which originated in Asia Minor and Europe.

Viticulture (Grape Growing)

Vitis is a deciduous plant which climbs by grasping supporting objects with outgrowths of very special leaf-type organs called *tendrils*. In the wild, vines may reach the tops of tall trees, but in culturing the vine, the natural growth is controlled by annual pruning. Typically, vines in vineyards are trained onto post-and-wire trellis. (See Fig. 2–2). These supports are designed to optimize both the qual-

Fig. 2–2 Diagram of vine and trellis.

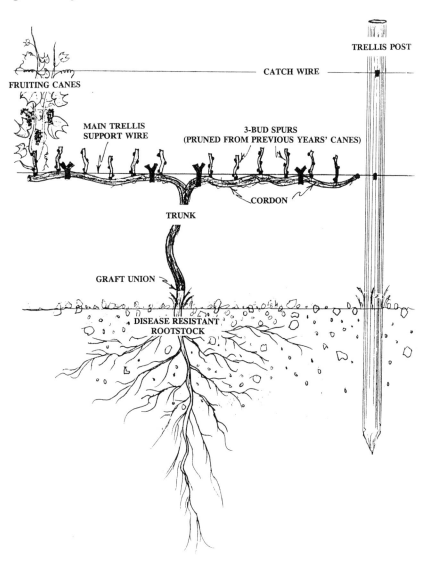

ity and the quantity of the plant. Exposure of grape leaves to sunlight increases the production of grape sugars by allowing photosynthesis to transform carbon dioxide, which eventually will result in grapes of varying levels of sweetness.

The following is a chemical equation for photosynthesis:

$$6CO_2 + 6H_2O \xrightarrow{\text{sunlight}} C_6H_{12}O_6 + 6O_2$$

Carbon dioxide is the substrate from which the reaction is fueled, catalyzed by sunshine—with sugar as the product, and oxygen as a by-product.

Traditionally, grape quality is closely linked to sugar content and grape growers monitor sugar levels carefully. Pruning is often practiced for balanced production of optimal quality fruit. Lesser pruning is generally associated with a larger quantity of lower-quality fruit, and severe pruning is generally associated with smaller crops of higher quality.

Grape flowers and fruit are borne only on new shoots which arise from dormant buds formed on the previous season's growth. As a shoot matures and drops its leaves, it is known as a "cane". It is from canes that the next year's fruiting wood is selected at pruning time.

Grape buds are classified as compound buds because they contain a primary bud and one to four separate, smaller buds. When growth starts in the spring, the primary (center) bud breaks dormancy and produces the fruiting shoot. Low midwinter temperatures may kill primary buds, and spring frosts may occasionally kill the tender primary shoots during early stages of growth. If this happens, one of the smaller buds usually develops into a shoot. These shoots are typically less vigorous and produce fewer, smaller fruit clusters. However, as these shoots grow, new compound buds are formed which will provide the potential for a full crop the following season.

Flower clusters develop on shoots at nodes opposite a leaf. At the perimeter of the flower several male *stamens* produce pollen which, through wind, insect activity, or other mechanical means, finds its way to the central female *pistil*, where fertilization takes place and formation of grape berries commences.

At some nodes where fruit clusters are not formed, a tendril will develop. The tendrils aid the plant in attaching to supports so that the leaves will be exposed to as much sunlight as possible. The number of clusters which develop per shoot is mostly a genetically controlled characteristic of the cultivar, with little influence by the environment. The number of berries per cluster, however, can be greatly influenced by the environment as well as the genetics of the cultivar. The health and vigor of the vine, environmental conditions during bloom, and other factors can affect the number of berries that form. Proper care of the vine is essential in meeting full fruiting quality and quantity potential. Bunches of grapes may be comprised of less than 10 berries each for some Muscadine cultivars, to more than 300 berries each on some bunch grape cultivars. The grape is morphologically a true berry, as it is a simple fruit which has a pulpy *pericarp* consisting of four skin layers: the epidermis, hypodermis, outer wall, and inner wall. (See Fig. 2–3.) The color, flavor, and aromatic compounds in the berry are located in the skin.

Cluster shapes range from long, cylindrical forms (with or without *shoulders*) to shorter conical structures. (See Fig. 2–4.)

The grapevine can adapt to some rather difficult climate and soil conditions. Vines often form *clones*, which are mutations that result in genetic changes which can affect certain characteristics of the plant and/or its fruit. These occur as slight variations in character responding to natural selection processes. As compared to the classic type, some clones may be more resistant to cold winter temperatures, whereas others may require a shorter growing season, and still others produce heavier crops, and so forth. Some varieties have been cataloged with hundreds of different clones from which selections are made to best meet the vintner constraints and designs.

Young vines are commercially propagated asexually, usually from cuttings or graftings grown in a nursery for a year or so. A newly planted vineyard may be allowed to produce a token amount of grapes during its third year of growth; other clusters are pinched off shortly after bloom.

CULTIVAR SELECTION

Selection of the proper vine varieties and cultivars are obviously elemental to the product line which will become an estate winery's product portfolio. Selection of the appropriate vine selections is equally elemental in successful grape growing. In that the desired cultivars may not be ideal for the vineyard site selected, choices may be difficult to make.

Wine enthusiasts are sensitive to the pedigree of vine varieties upon wine labels; some estate vintners accept high risk in planting high-profile varieties on sites which are marginal. It is recommended that supplemental cultivars be considered in order to make alternative wines and reduce that risk. Cultivar selection should be based on the intended use of the fruit and its versatility—along with, of course, relative cold hardiness and disease resistance.

VITIS VINIFERA

Hundreds of varieties of *Vitis vinifera* are cultivated around the world today as the "Old World" vine. More than 90% of the wine grown globally is from this genus and species.

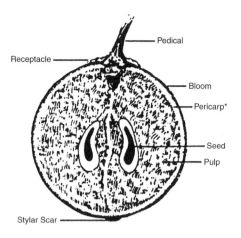

*Pericarp – consists of epidermis, hypodermis, outer wall and inner wall

Fig. 2–3   Diagram of a grape berry.

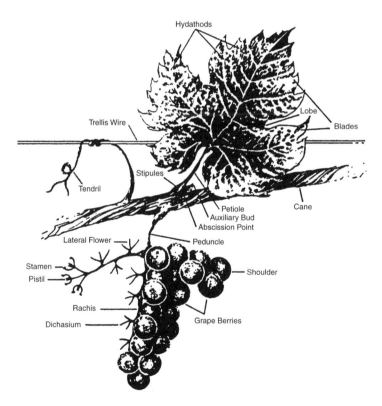

Fig. 2–4   Diagram of a grape cluster.

Major Vitis vinifera White Varieties

Chardonnay (shar-doe-NAY)

This classic variety is native to France but is grown around the world. It ripens late mid-season to small compact cone-shaped clusters of light green berries (Fig. 2–5). Chardonnay produces sparse crops—the very best quality from struggling vines in marginal soils and climates. It is famous for complex apple–citrus, honey–olive flavors, from the noble vineyards in the Chablis, Meursault, and Montrachet districts in Burgundy. Chardonnay is also the principal variety grown in the Cote de Blancs district of Champagne. Chardonnay is widely planted in California and several other states.

Fig. 2–5 Chardonnay cluster.
Courtesy: Food and Wines From France.

Sauvignon Blanc (so-veen-YOHN BLAWN)

The white Sauvignon is another classic variety native to France and also very popular in many other countries. The name is derived from the French *sauvage* or "wild." It is a vigorous vine with long, loosely clustered yellowish green berries maturing early mid-season. From the Pouilly-Fume vineyards in the upper reaches of the Loire Valley region, Sauvignon Blanc wines are typically intense in blossom–melon aroma, or "fume" in flavor character—a style sometimes labeled as "Fume Blanc" in the New World. In Bordeaux and in much of northern California, the wine has a pale greenish tint and a bold grassy–pineapple flavor.

Johannisberg Riesling (yo-HANNES-berg REEZ-ling)

This vine is native to Germany and is the principal variety grown there for that country's very best wines. The fruit takes the form of short cylindrical clusters of yellowish gray berries dotted with russet sports. The prized Riesling, sometimes labeled as "White Riesling" or just "Riesling" in the United States and other countries, can result in a golden wine of intense floral–apricot flavor. It is a variety increasingly planted in the Great Lakes and northeastern United States, as well as in Oregon—less so in California.

Semillon (SEM-mee-yohn)

This variety is most closely identified with the famed sweet wines made from purposely over-ripened grapes in the Sauternes district of Bordeaux. Clusters of Semillon are comparatively small, with tightly compact yellowish golden berries. Although Sauternes has a rich golden amber hue and profound honey–peach flavor, dry (not sweet) wine made from Semillon has a shy yellow color with a delicate pear–melon character. It is finding growing popularity in the warmer climes of California and Washington State.

Gewürztraminer (ge-VOORTS-trah-meener)

Native to northern Italy, Gewürztraminer is generally considered the major variety of the Alsace winegrowing region situated in northeastern France. It yields moderately heavy crops of small cylindrical bunches tightly packed with small berries curiously yellowish pink in color. the German "gewürz," or "spicy," refers to its flavor—a rather ginger–cinnamon aroma. The acreage planted to Gewürztraminer has been slowly on the rise in the cooler climate locales of Oregon, the Great Lakes, and the northeastern states.

Muscat de Frontignan (MOOSkat dah frawn-teen-YAWN)

One of many Muscat grape varieties that are prized for their bold, powerful fig–guava flavor. It is widely cultivated in southern France and a close relative,

Moscato, is grown in northern Italy for the famed Asti Spumante sparkling wines. Muscat de Frontignan produces elliptically shaped berries on compact medium-sized clusters. Only token vineyards of this variety are cultivated in the United States—although this and other varieties of Muscat are increasingly planted for wines which are blended to enhance the fruit aroma in other white wines.

Pinot Blanc (PEE-noh BLAWN)

This variety is closely identified with the lower districts in the Burgundy region of France. Pinot Blanc gets its name from its "pine" cone cluster shape, packed tightly with small green berries. Its wine is very shy in flavor, recalling very light nuances of fig, melon, and raisins. Despite this, it is gaining some popularity among vintners in California and other locales in the United States.

Sylvaner (sil-VAHN-ner)

Sylvaner (sometimes spelled "Silvaner") is native to Germany and can generally be found in the warmer, less rigorous soils and climates of the southern wine-growing districts. It matures to small, cylindrical, compact clusters of spherical yellow-green berries. Sylvaner generally yields heavier crops, but with less flavor intensity, than its sibling, the noble Johannisberg Riesling. The Alsace region in France cultivates many Sylvaner vineyards. Interest in the variety has never been widespread in the United States.

Pinto Grigio

This variety is produced in northern Italy and is the same as Rulander in Germany and Pinot Gris in the Balkan countries. It produces light, but distinctively fruity, white table wines and is gaining increasing popularity in the United States.

Major Vitis vinifera Red Varieties

Cabernet Sauvignon (cab-bair-NAY so-veen-YOHN)

The quintessential red Sauvignon is perhaps the most famous grape in the world. There are diverse opinions as to its origin, but it remains the principal variety grown for the renowned chateaux estates of Bordeaux in France. The vine is vigorous and makes long, loose, conical clusters of small spherical blue/black berries (Fig. 2–6). The deeply lobed leaves of Cabernet Sauvignon are often used as the models for viniculturally inspired artwork. The wine is very dark ruby red with a aroma of green bell pepper and cedar with flavors of black currants. Cabernet Sauvignon is widely grown in California and Washington State—with several other states adding to this acreage.

Merlot (MAIR-loh)

Next to Cabernet Sauvignon, Merlot is the most important grape variety grown in Bordeaux. Merlot clusters are long and conical with compact medium-sized,

Fig. 2–6 Cabernet Sauvignon cluster.
Courtesy: Food and Wines From France.

spherical, blue-violet berries. Wine from Merlot is typically a bit lighter in color density and texture than Cabernet Sauvignon—with similar flavor components, but markedly more fruit-like in overall character. In recent years, the demand for Merlot has achieved remarkable growth—with a commensurate acreage of new vineyards planted in California, Washington State, and other milder-climate states.

Pinot Noir (PEEN-oh nwah)

This is the regal variety of the magnificent "Slope of Gold" known as the grand "Cote d'Or" of northern Burgundy in east central France. It is called Spätburgunder in Germany. Pinot Noir yields its best quality fruit in cooler climates and poorer soils epitomized in Burgundy. It yields small blue berries tightly packed on cone-shaped clusters. Red wine from Pinot is very light in color hue and density with distinctive coffee-like aroma and bramble-berry flavor. Some districts in California and Oregon are building fine reputations for the great Pinot, as are certain locales in upper New York and other eastern States.

Gamay (gam-AAY)

Gamay is the premier grape variety of the famous Beaujolais district of the Burgundy wine region in France. Confusingly, the variety Gamay Beaujolais is neither Gamay nor Beaujolais, but rather a clone of Pinot Noir. Gamay is a vigorous vine producing abundant medium-sized conical, compact clusters of large, slightly elliptical, blue berries. Young Gamay wine has a medium color density with a regal purplish hue, a black-cherry aroma, and plum flavor. There are comparatively few U.S. vineyards planted to Gamay.

Sangiovese (san-gee-oh-VAAYZ-ah)

This is the revered vine of the Tuscany region in north-central Italy. A clone called Sangioveto is grown for the famous Chianti wines of Tuscany. Sangiovese requires a long, warm growing season in order to produce its best fruit. Clusters are moderately large with oval black berries filled with an abundance of large seeds. Young Sangiovese wine is very dark with a rich scarlet-ruby hue with flavors recalling truffles and blackberries. Although there are still very small acreages of this variety planted in the United States, interest seems to be increasing among northern California vintners.

Nebbiolo (neb-BYO-loh)

Along with Sangiovese, Nebbiolo is the other aristocrat among red wine grapes in Italy. The grand vineyards of Barolo and Barbaresco in the Piedmont region are testimony to the famous Nebbiolo. It ripens late to long, conical clusters which are often shouldered. Medium-sized berries are spherical and a "foggy" (*nebbia* means "foggy" in Italian) gray-blue in color. Young Nebbiolo wine is dense in purplish garnet color and it typified by an aroma of violets with prune-like flavors. There are only several commercial Nebbiolo vineyards in the United States—although interest grows slowly.

Zinfandel (ZIN-fun-dell)

This variety is closely identified with California vineyards. Controversy continues as to the origin of Zinfandel; the most widely accepted notion being that it is perhaps the very same vine as the Primitivo di Gioia in southern Italy. Zinfandel is a productive vine, yielding large cylindrical clusters, tightly packed, and often heavily shouldered. Berries are spherical and have a small brown apical scar. Young

Zinfandel red wine has rich plum–berry flavors. The popular "White Zinfandel" is only partially fermented in contact with berry skins so that a faint "blush" color is extracted.

Syrah (sear-RAH)

Syrah was brought to the Rhone Valley in southern France from Asia Minor by returning Crusaders more than 700 years ago. It was called Shiraz then; that name is still used in Australia and other countries. Syrah ripens rather early with moderate production of compact clusters of slightly oval shaped berries and is often confused with Petite Sirah, a totally different variety. Syrah has a dense blue-black hue and truffle/tar-like aroma, but its great intensity of color and flavor are its real character.

Cabernet Franc (cab-bair-NAY FRAHNK)

Another of the grand Bordeaux varieties still used primarily for blending but now finding an increasing popularity as an individual—particularly in cooler climates across the United States and Australia. Cabernet Franc berries are a bit larger, with less density in color, acidity, and tannin as compared to its Sauvignon cousin. The aroma of Cabernet Franc is distinctively tobacco leaf with herbal and grassy nuances; the young wine is characterized by a fresh, but complex earthy-berry flavor.

Mourvedre (moor-VEHD-rah)

This vine is perhaps the same, if not a close relative, of the Spanish Mataro vine. Cultivated primarily for blending in the Rhone Valley of France, Mourvedre has new popularity building as a varietal red table wine in the United States. Clusters of Mourvedre are rather large, often double shouldered, with spherical dark blue berries.

Petite Sirah

This variety is from Northern California and is the same as Duriff in the Rhone Valley of France. It makes dense, dark, full-bodied, often tannic red table wine.

VITIS LABRUSCA

This species is the often called "fox" grape. Some people contend that the character of *labrusca* varieties is related to one or another odors associated with foxes. This takes a bit of imaginative stretch in the mind of most anyone who has ever been close to such little animals. More plausible legends relate to foxes being attracted by the intense fruit flavor character often associated with "grape" soft drinks, candies, and so forth.

Major Vitis labrusca White Varieties

Catawba (kah-TAW-bah)

A vine found growing in the wild near the Catawba River in North Carolina circa 1802. Catawba was found to be disease resistant, while productive and hardy. It was widely planted in southwestern Ohio during the middle 1800s and made America's first commercial sparkling wine. It ripens late, with medium-sized, conical bunches of large, pink-skinned spherical berries. The Catawba remains heavily planted in eastern U.S. vineyards, but interest is gradually diminishing.

Delaware

This native American variety was discovered in Delaware County, Ohio during the early 1800s. It is widely cultivated in eastern U.S. vineyards, remaining primarily because of a once huge demand for "New York State Champagne"—most of which has long since been replaced by California sparklers. Delaware ripens rather early to small, cylindrical clusters of tiny spherical, pink-skinned berries (Fig. 2–7). In the hands of a talented winemaster, grapes from this variety can make wines which compare favorably with wine types found in Germany and Austria.

Niagara

This is a vine developed by researchers in Niagara County, New York during the 1060s. In preferred locales, Niagara is a vigorous and productive vine; its rich, heavily scented flavor is the standard for robust, fruity, native American white wines. Most vineyards of Niagara today are harvested for fresh table grapes and for white grape juice production.

Major Vitis labrusca Red Varieties

Concord

The quintessential *labrusca* grape. Concord, a cultivar developed by E. W. Bull in Concord, Massachusetts during the 1840s is better known for its juice, jelly, and table grape production than for wine. Taking all its various uses together, it is unquestionably the most widely grown cultivar in the United States. Concord yields large crops of large spherical blue-black berries loosely arranged on moderate-sized clusters (Fig. 2–8).

Ives

This cultivar was developed by Henry Ives in Cincinnati, Ohio in the 1840s. It has immense color and flavor. It was first planted along the banks of the Ohio River during the 1850s, and to a greater extent in upstate New York later. The grand New York State "Port" wines, very popular after the repeal of Prohibition through the

Fig. 2–7 Delaware cluster.

1960s, were dependent on Ives. It is a rather temperamental vine, not as productive as Concord, nor as cold and disease resistant—which accounts for why it has long since been replaced with specially selected hybrid grape cultivars.

VITIS RIPARIA

This species is referred to as the "post-oak" or "River Bank" grape by some viticulturists. It is widely adapted to almost all of temperate North America, but is found mostly east of the Rocky Mountains from Canada to the Gulf Coast. Because of its hardiness in cold weather climates, phylloxera resistance, and disease resistance, it is often used for grafting as a rootstock or in grape breeding by researchers. Derived cultivars include 'Beta,' 'Baco Noir,' 'Marechal Foch,' 'C3309,' '5BB,' and 'S04.'

Other wild American species have also been important in grape culture, either

Fig. 2–8 Concord clusters.

as producers or as hybrids developed for rootstock purposes. They are particularly valuable where own-rooted (ungrafted) vines have failed because of the grape *phylloxera* root louse or where adverse soil conditions exist. Several species are of significance to the wine industry which should be noted.

VITIS AESTIVALIS

This is often called the "summer grape," or "pigeon grape." It is a species that has been used primarily for disease resistance. The most important derived cultivar is "Cynthiana," which is also known as "Norton."

VITIS RUPESTRIS

This is commonly known as the "rock grape," or the "mountain grape." It is a species which has been used for *phylloxera* resistance, as well as for disease and drought resistance. One derived cultivar, AXR 1, once relied upon widely as a rootstock in the United States, has been overcome by the *phylloxera* root louse. This has led to a massive replant of vineyards, particularly in the North Coast counties region of California. Another *rupestris* cultivar is "St. George."

VITIS BERLANDIERI

Also called the "Spanish grape," or "winter grape," this species has been used primarily for tolerance to high lime content soils. Derived cultivars include "41 B" and "99 R".

VITIS ROTUNDIFOLIA

The subgenus *Muscadinia* is more commonly known in America's Deep South as "muscadine"—generally small, loose clusters of large berries typified by a thick skin and dense pulp consistency and literally bursting with jammy fruit flavors. The bronze-skinned varieties are collectively referred to as "Scuppernongs," with the red known as "muscadines." Most are actually hybrid cultivars in the *Vitis rotundifolia* species, with the most commercially important of these being Carlos, Magnolia, Noble, and Tarheel.

HYBRIDS AND GRAFTING

Hybrids are created by man, instead of nature. The process is one whereby pollen is gathered from the male stamens of selected grape flowers and used to fertilize female pistils on other specific grape flowers. (See Fig. 2–9.) As a result of *meiosis*, a genealogical combination takes place in seeds which reflect parental attributes. Crosses are often planned to compliment the characters of each parent. That was the purpose of crossing male vines of wild American species that had resistance to phylloxera with female vines of *V. vinifera* that had excellent fruit quality. The resultant seedlings displayed various combinations of phylloxera resistance and high fruit quality. Breeders selected the best among each progeny for release or for further breeding.

Fig. 2–9 Illustration of cross-pollination.

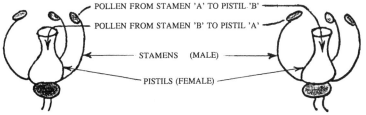

The rationale for hybridization is to improve resistance to disease and adverse climate/soil conditions—to increase yields and advance fruit quality. Unfortunately, it is common to find negative characteristics associated with dominate genes. History indicates that it may take hundreds of crosses to find just one cultivar worthy of commercial cultivation. University breeding programs continue, but state-of-the-art research looks to the advent of biotechnology which may serve to make classic varieties increasingly resistant to adverse climates and disease, with greater productivity of better quality fruit.

Hybridizing should not be confused with *grafting*. In avoiding one or another adverse disease or soil conditions, viticulturists will often graft one variety upon another. The most common reason for grafting is to resist the *Phylloxera* root louse. Great care must be taken to ensure that the *xylem* and *phloem* sap-flow tissue matches perfectly between the *scion* (upper fruit-bearing variety) and the *rootstock* (root-support variety). Vines can be "bench grafted" (Fig. 2–10) and propagated asexually in a nursery, or "top grafted" to change a mature vine from one variety to another in a vineyard.

FRENCH-AMERICAN HYBRIDS

The term "French-American hybrid" evolved from efforts to develop vines which were resistant to the great *Phylloxera* root louse blight which ravaged most of the

Fig. 2–10 Photograph showing bench grafting.

world winegrowing community during the latter 19th century. French *vinifera* vines, well-known for fine wine quality, were bred with disease-resistant *riparia* and other species of American vines.

Major Hybrid White Wine Cultivars

Seyval Blanc (SAY-vahl BLAWN)

This cultivar is considered by many to be one of the finest of the French-American hybrids. It ripens mid-season to large, compact, conical bunches of slightly elliptical greenish yellow berries. (See Fig. 2–11.) Well-made Seyval Blanc can compete favorably with some dry white wines made from Chardonnay and Pinot Blanc.

Vidal Blanc (VEE-doll BLAWN)

In some environs, this cultivar matures fruit with a rather distinctive Teutonic character. Vidal Blanc has very long cylindrical clusters (almost always shouldered)

Fig. 2–11 Seyval Blanc cluster.

tightly packed with small greenish white berries with Riesling-like russet spots. It ripens late, often left on the vine for "late harvest"—a natural dehydration process which concentrates sugars and flavors. Late-harvest Vidal Blanc can compare to similar wines made from Sylvaner.

Vignoles (VEEN-yole)

A Missouri-grown Vignoles shocked the world in 1992 when it was selected as the finest white table wine in a prestigious international wine competition held in California. Vignoles is a shy producer with small yellowish white berries tightly packed on rather small clusters. Its flavor has a delicate flower-blossom character.

Major Hybrid Red Wine Cultivars

Chambourcin (SHAM-burr-sahn)

A cultivar which has garnered considerable attention among eastern U.S. winegrowers during the past decade or so—and may now be the most widely planted red hybrid cultivar. Chambourcin usually yields good crops of medium–sized blue berries loosely clustered on long bunches which ripen rather late during the vintage season. (See Fig. 2–12.) The young wine has a moderate garnet hue with abundant berry–cherry flavors.

Chancellor

Chancellor is known for its rich plum–cedar aroma. If frequent downy mildew attacks can be controlled in the vineyard, Chancellor can be a bountiful producer. Clusters are long and loose with medium-sized berries which ripen mid-season. Chancellor wine is dense in ruby–violet color, heavy bodied, and can be quite astringent unless tannins are tamed in oak.

Marechal Foch (MARR-shawl FOHSH)

This is a cultivar which ripens early with small berries tightly compact on small clusters. From early-harvest fruit, young Foch wine can exhibit a moderate crimson hue and delicious cranberry–currant flavors.

VINEYARD ESTABLISHMENT

Cost and Value

Establishment and operating costs can vary significantly within a region due to differences in cost for land, labor, machinery, and materials. Costs will also be af-

Fig. 2–12 Chambourcin cluster.

fected by the vineyard site, grape cultivar, vine spacing, (See Fig. 2–13.) training system, pest management program, and other cultural practices selected.

The major costs during vineyard establishment are land, land preparation, labor, grapevines, trellis materials, pest management materials, and opportunity cost (interest paid on debt or return lost on capital). Additionally, irrigation (if needed), animal fencing, and major land preparation such as clearing, leveling, rock removal, and drainage tiling can greatly increase initial costs. The added expenses of land improvement, however, will usually result in earlier and more consistent production of superior fruit.

Operating costs vary in much the same way as establishment costs according to vineyard site, grape cultivar, vine spacing, training system, pest management program, and other factors. In particular, a new grower should carefully consider the advantages and disadvantages of site selection and cultivar selection because of the profound impact these decisions have on the profitability of the planting.

Fig. 2–13 Young vineyard planting.

Knowing the establishment and production costs is still not enough to make a sound decision about entering into grape production. Potential marketability and profitability must also be considered. Some varieties of grapes are more marketable, both as fruit and as value-added wine, than others and are thus more valuable. Appellation (location expressed on a wine label) of production is also an important characteristic for some vintners which can greatly increase grape value. Yield is a function of site, soils, vine variety, and vineyard management, as well as the highly unpredictable effects of weather. Grape values and yields obviously have a combined direct effect on profitability. Table 2–1 illustrates typical grape prices and their impact of various sizes of wine production units.

Table 2–1   Value of grapes in wine at various prices per ton.

	$500/ton	$600/ton	$700/ton	$800/ton	$900/ton	$1000/ton
$/bottle	0.74	0.89	1.04	1.19	1.33	1.48
$/gallon	3.70	4.44	5.19	5.93	6.67	7.41
$/case	8.93	10.71	12.50	14.29	16.07	17.86

Note: Value base on 150 gallons of juice per ton minus 10% loss in processing for a total finished wine volume of 135 gallons per ton of grapes.

Each additional $100 per ton paid for grapes adds 0.15 to the cost per bottle of wine.

Start-Up

Labor Costs

Labor represents 30% or more of the cost of vineyard establishment, depending on the amount of mechanization available, management skills, and quality of the labor pool. It is highly recommended that prospective grape growers seek discussion with experienced growers and state university Cooperative Extension Service personnel for advice and assistance.

Equipment Costs

Equipment ownership costs depend on the particular circumstances of each vineyard operation. A diversified fruit farm which has all the necessary equipment available should have relatively low equipment costs for vineyard establishment. However, a grower who is starting totally anew will experience comparatively high capital requirements for a truck, tractor, sprayer (Fig. 2–14), and other major equipment needs.

Major factors to consider in calculating equipment costs are initial costs, salvage value, years of life, annual use and maintenance, repair costs, insurance, interest, and operating expenses such as fuel and lubricants. As accounting for the total cost for each piece of equipment is usually allocated over units of production, it may be difficult to justify starting a commercial vineyard of less than 5 acres. Sometimes it is possible for growers to find economies in one form or another by sharing equipment needs. Table 2–2 illustrates typical equipment costs a prospective grower may expect in starting a vineyard.

Variable Costs

Variable costs are those which change directly with increases or decreases in acreage. Examples are trellis posts and wires, pesticides, fertilizer, wages, machinery fuel, and lubricants, and even the grapevines themselves. The enterprising grower can reduce vineyard establishment costs by identifying alternative sources—such as making his own trellis posts, seeking assistance from family and friends to reduce wage costs, and by propagating his own young vines. When a vineyard has been fully established and initial costs are recovered, labor costs will generally represent about 60% of the annual cost of producing grapes. In larger operations, particularly those large enough to justify a machine harvester, this percentage reduces significantly. Table 2–3 is provided in order to illustrate typical equipment variable costs a grower may expect in establishing a vineyard.

Fixed Costs

Fixed costs are often referred to as overhead and are those which to not change directly with increases or decreases in acreage. Overhead includes machinery depre-

Fig. 2–14 Vineyard tractor and sprayer.
Courtesy: Wine East *magazine*

ciation, opportunity cost, insurance, and taxes. In diversified farms or ranches, total overhead should, of course, be appropriately allocated to the vineyard operation.

Budget

A 7-year budget is provided in Tables 2–4 through 2–10 in order to illustrate typical costs a prospective grower may expect in starting a vineyard.

Table 2–2 Equipment purchase cost, operational cost, and anticipated annual hourly use.

Machine	Initial cost	Cost/hour	Hours/year 1	2	3	≥4
Tractor, 35–50 hp	20,000	4.05	85.2	14.0	2.17	26.7
Pickup truck (cost/mile & miles/yr)	18,500	0.29	500.0	500.0	500.0	500.0
Herbicide sprayer, 50 gal	2,125	1.27	2.0	2.0	1.0	1.0
Post driver	1,500	1.20	11.2			
Airblast sprayer, 100 gal	6,000	2.82	3.0	4.0	5.7	5.7
Mower/brush chopper, 6 ft	1,700	0.63	2.0	2.0	2.0	2.0
Fertilizer & seed spreader	1,800	1.44	0.5			
Auger, 12 in.	1,400	1.12	47.0			
Wagon/trailer, 5-ft × 12-ft flatbed	1,575	0.58	15.0		2.9	7.0
Total machinery purchase expense:	$54,600					

Machinery capital recovery:
 Machinery investment financed at 8% for 7 years: $54,000 × 0.1921^a = $10,489.
 Annual cost per acre (10-acre vineyard): $10,489/10 acres = $1,049.

^{a}0.1921 is the capital recovery factor for the financing terms specified above.

Table 2–3 Vineyard equipment variable costs per hour of operation and per acre.

Machine	Cost/hour	Cost/acre for first 8 years 1	2	3	4–8
Tractor, 35–50 hp	4.05	345.1	56.7	87.9	108.1
Pickup truck (cost/mile)	0.29	145.0	145.0	145.0	145.0
Herbicide sprayer, 50 gal	1.27	2.5	2.5	1.3	1.3
Post driver	1.20	13.4			
Airblast sprayer, 100 gal	2.82	8.5	11.3	16.1	16.1
Mower/brush chopper, 6 ft	0.63	1.3	1.3	1.3	1.3
Fertilizer & seed spreader	1.44	0.7			
Auger, 12 in.	1.12	52.6			
Wagon/trailer, 5-ft × 12-ft flatbed	0.58	8.7		2.9	7.0
Total machinery cash expense		577.8	216.8	254.4	278.7

Table 2–4 Cost of vineyard establishment per acre—year 1.

Operation	Hours/ acre	Unit cost	Units/ acre	Cost/ acre
Site Preparation				
Liming—3 tons/acre		16.00	3	48
Plow & disk—labor	3.0	7.50		23
Sow cover crop—labor	1.0	7.50		8
Cover crop seed		0.68	50	34
Vineyard Layout				
Mark vine & post locations	5.0	7.50		38
Planting				
Vines		2.00	545	1090
Auger vine holes (two men @ 3 min/hole)	54.5	7.50		409
Planting (2 min/vine)	18.2	7.50		136
Trellising Materials				
3-in. × 8-ft. CCA-treated line posts		3.50	187	654
6-in. × 9-ft. CCA-treated end post		10.00	22	220
12.5 ga. HT wire (4000 ft per roll)		60.00	2.5	150
End-post anchors (5.8-in. × 4-ft helical end)		4.00	22	88
In-line ratchet wire strainers for cordon wires		2.00	11	22
Wirevise wire strainers for trunk support wires		1.50	11	17
Staples (28 lbs.) and wire crimping sleeves (44)				31
Trellising Labor				
Distribute & drive line posts (2 men @ 3 min/post)	18.7	7.50		140
Auger and set end posts	6.0	7.50		45
Mark & drill end posts for wirevises	3.0	7.50		23
Mark post for wires	2.0	7.50		15
Install end-post anchors	5.0	7.50		38
String wire, staple to posts, & tighten	15.0	7.50		113
Weed Control				
Postemergent application prior to planting		44.04	0.17	7
Preemergent application after planting		64.00	0.33	21
Herbicide application labor	2.0	7.50		15
Mowing row middles (6 times)	4.0	7.50		30
Hoeing/weeding labor (2 times)	30.0	7.50		225
Fertilization				
Ammonium nitrate (0.25 lbs. per vine)		0.14	136	19
Hand application of fertilizer	2.0	7.50		15

Table 2–4 (*continued*)

Operation	Hours/ acre	Unit cost	Units/ acre	Cost/ acre
Canopy Management (training)				
Bamboo/wood stakes for trunk support		0.25	545	136
Shoot thinning & tying to stakes (twice)	20.0	7.50		150
Tie materials				5
Flower cluster removal	2.0			15
Disease and Insect Control				
Spray materials				109
Spray labor (6 sprays @ 0.5 h/spray)	3.0	7.50		23
Machinery				
Cash operating expenses only (See Table 2–2)				578
Operating Interest		½ yr @		
Interest charged on yearly cash expenses for ½ yr		8.00%	4690	188
Annual Cash Expenses—Year 1				**4878**

General assumptions made in developing this budget include the following:

1. Costs are based on a 10-acre vineyard planted on a good to excellent site to own-rooted (ungrafted) vines adapted to the region.
2. Vines are planted 8 feet apart in rows spaced 10 feet apart, for a total of 545 vines per acre.
3. Row length is approximately 400 feet. Shorter rows would cost slightly more, longer rows slightly less.
4. The trellis is a standard two-wire system for either cordon training or cane pruning. All trellis posts are CCA-treated timber. End posts are 6 in. × 9 ft; line posts are 3 in. × 8 ft. spaced 24 feet apart.
5. The vineyard floor is managed as grass row middles with herbicide strips under the row.
6. Labor cost is calculated at $7.50 per hour. Time required to complete the tasks will vary and the times listed are considered averages.
7. All machinery and equipment expenses reflect *operating costs only;* no charges have been included for ownership items such as insurance, taxes, or depreciation/capital recovery.
8. Fixed costs for ownership of land are not included in this budget.
9. Accumulated costs are finances at a 8% annual percentage rate. Interest on annual operating costs is computed at 8% APR for 6 months.

Table 2–5 Cost of vineyard establishment per acre—year 2.

Operation	Hours/ acre	Unit Cost	Units/ acre	Cost/ acre
Dormant Pruning				
Pruning & tying canes for trunks	10.0	7.50		75
Weed Control				
Preemergent application in spring		64.00	0.33	21
Spot treatment with postemergent		44.04	0.08	4
Total herbicide application labor	2.0	7.50		15
Mowing row middles (6 times)	4.0	7.50		30
Fertilization				
Ammonium nitrate (0.5 lbs. per vine)		0.14	250	35
Hand application of fertilizer	2.0	7.50		15
Replanting				
Plants (2% of initial planting)		2.00	11	22
Replanting labor	1.0	7.50		8
Canopy Management (training)				
Shoot thinning & tying to stakes (twice)	30.0	7.50		225
Flower cluster removal	8.0	7.50		60
Tie materials				5
Disease and Insect Control				
Spray materials				200
Spray labor (10 sprays @ 0.5 hrs/spray)	5.0	7.50		38
Machinery				
Cash operating expenses only (See Table 2–2)				217
Operating Interest		½ yr @		
Interest charged on yearly cash expenses for ½ yr		8.00%	970	39
Interest on Year 1 Accrued Cash Expense		8.00%	4878	390
Annual Cash Expenses—Year 2				1399
Total Accumulated Cash Expense (Years 1 and 2)				6277

Table 2–6 Cost of vineyard establishment per acre—year 3.

Operation	Hours/acre	Unit cost	Units/acre	Cost/acre
Dormant Pruning				
Spur pruning and brush pulling	20.0	7.50		150
Cordon training/tying	20.0	7.50		150
Weed Control				
Preemergent application in spring[a]				
(oryzalin)		64.00	0.33	21
(simazine)		2.85	1.0	3
Spot treatment with postemergence				
herbicide	2.0	44.04	0.08	4
Total herbicide application labor	4.0	7.50		15
Mowing row middles (6 mowings)		7.50		30
Fertilization[b]				
Ammonium nitrate (0.5 lbs. per vine)		0.14	250	35
Spreading fertilizer	0.5	7.50		4
Leaf petiole sampling	0.3	7.50		2
Tissue analysis				20
Canopy Management (training)				
Shoot and cluster thinning	15.0	7.50		113
Shoot positioning	10.0	7.50		75
Tie materials				5
Disease and Insect Control				
Spray materials				287
Spray Labor (12 sprays @ 0.5 hrs. per spray)	6.0	7.50		45
Harvest Costs[c]				
Picking costs ($1.25/25-lb. lug for 2.5 tons/acre)		1.25	200	250
Grape lugs		4.00	200	800
Machinery				
Cash operating expenses only (See Table 2–2)				254
Operating Interest		½ year @		
Interest charged on yearly cash expenses for ½ yr		8.00%	2263	91
Interest on Year 2 Accrued Cash Expense		8.00%	6277	502
Annual Cash Expense—Year 3				2856

Table 2–6 (*continued*)

Operation	Hours/ acre	Unit cost	Units/ acre	Cost/ acre
Total Accumulated Cash Expense				
Years 1–3				9133
Income[d]		500	2.5	1250
Net Investment at End of Year 3				7883

[a] Oryzalin and simazine may be used separately or in combination from year 3 on. Costs shown are for combination.

[b] The need for a particular nutrient fertilizer is best determined through visual assessment of vines and by routine leaf petiole sampling for nutrient analysis. Fertilizer costs will vary depending on the particular needs of the vines. A single tissue sample may suffice for an entire 10-acre planting.

[c] Harvest costs include purchase of one-half the total number of lugs needed to harvest the expected full crop yield of 5 tons/acre, and harvest costs of $1.25 per 25-lb. lug, or $100 per ton. Expected yield in year 3 is one-half the expected full crop yield.

[d] Income based on harvest of 2.5 tons of grapes and a price of $500/ton.

The cost of vineyard establishment and operation at various prices per ton for 5 tons per acre yields are given in Table 2–11 for a 24-year period.

Based on the budgets presented herein, the costs of establishing an acre of wine grape vineyard may be expected to range from $7,000 to $10,000—and far more in marginal vineyard sites.

It is essential that vine variety selections for a prospective vineyard be considered from several major viewpoints. The wine from some varieties has far more consumer demand than others. In most metropolitan markets, varietal wines grown from *Vitis vinifera* and prime French-American hybrid cultivars are preferred. In more suburban and countryside locales, wines made from the native *Vitis labrusca* varieties and fruit wines may be more popular.

These consumer demands must, of course, be weighed against the suitability of a given to site for the desired selection of vine variety types. A major concern is potential damage from extreme cold temperatures. Some varieties, especially *Vitis vinifera* varieties are so tender that they will frequently suffer bud and/or trunk damage during bitter winter temperatures. Measures such as hilling up in fall to protect the graft union and taking out in spring are very labor intensive and require special equipment; and still may fail to protect from a particularly severe winter. Untimely spring frosts can damage fruiting shoots on any vine.

There are considerations for each cultivar's individual susceptibility to diseases, pests, yields, and resulting fruit quality. Further, some vine types require more care and handling than others, such as shoot positioning, cluster thinning (green harvest), and leaf removal.

Table 2-7 Cost of vineyard establishment per acre—year 4.

Operation	Hours/acre	Unit cost	Units/acre	Cost/acre
Dormant Pruning				
Spur pruning and brush pulling	20.0	7.50		150
Cordon training/tying	5.0	7.50		38
Weed Control				
Preemergent application in spring[a]				
(oryzalin)		64.00	0.33	21
(simazine)		2.85	1.0	3
Spot treatment with postemergence				
herbicide	2.0	44.04	0.08	4
Total herbicide application labor	4.0	7.50		15
Mowing row middles (6 mowings)		7.50		30
Fertilization[b]				
Ammonium nitrate (0.5 lbs. per vine)		0.14	250	35
Spreading fertilizer	0.5	7.50		4
Leaf petiole sampling	0.3	7.50		2
Tissue analysis				20
Canopy Management (training)				
Shoot and cluster thinning	15.0	7.50		113
Shoot positioning	10.0	7.50		75
Tie materials				5
Disease and Insect Control				
Spray materials				287
Spray labor (12 sprays @ 0.5 h. per spray)	6.0	7.50		45
Harvest Costs[c]				
Picking costs ($1.25/25-lb. lug for 5 tons/acre)		1.25	400	500
Machinery				
Cash operating expenses only (See Table 2-2)				279
Operating Interest		½year @		
Interest charged on yearly cash expenses for ½ yr		8.00%	1626	65
Interest on Year 2 Accrued Cash Expense		8.00%	7883	631
Annual Cash Expense—Year 4				**2,322**
Total Accumulated Cash Expense Years 1-4				**10,205**
Income[d]		500	5	2,500
Net Investment at End of Year 4				**7,705**

Table 2–7 (*continued*)

[a]Oryzalin and simazine may be used separately or in combination from year 3 on. Costs shown are for combination.

[b]The need for a particular nutrient fertilizer is best determined through visual assessment of vines and by routine leaf petiole sampling for nutrient analysis. Fertilizer costs will vary depending on the particular needs of the vines. A single tissue sample may suffice for an entire 10-acre planting.

[c]Harvest costs of $1.25 per 25-lb. lug, or $100 per ton. Expected yield in year 4 and beyond is 5 tons/acre.

[d]Income based on harvest of 5 tons of grapes and a price of $500/ton.

Table 2–8 Cost of vineyard operation per acre—year 5.

Operation	Unit cost	Units/ acre	Cost/ acre
Management Expense—(Similar to Year 4, including interest on yearly cash expenses for $1/2$ yr)			1,691
Interest on Accrued Investment	8.00%	7,705	616
Annual Cash Expense—Year 5			2,307
Total Accumulated Cash Expense			10,012
Harvest Income 5 ton/acre yield	500	5	2,500
Net Investment at End of Year 5			7,512

Table 2–9 Cost of vineyard operation per acre—year 6.

Operation	Unit cost	Units/ acre	Cost/ acre
Management Expense Similar to Year 4			1,691
Interest on Accrued Investment	8.00%	7,512	601
Annual Cash Expense—Year 6			2,292
Total Accumulated Cash Expense			9,804
Harvest Income 5 ton/acre yield	500	5	2,500
Net Investment at End of Year 6			7,304

Table 2–10 Cost of vineyard operation per acre—year 7.

Operation	Unit cost	Units/ acre	Cost/ acre
Management Expense Similar to Year 4			1,691
Interest on Accrued Investment Minus Year 6 Crop Sale	8.00%	7,304	584
Annual Cash Expense—Year 7			2,275
Total Accumulated Cash Expense			9,579
Harvest Income 5 ton/acre yield	500	5	2,500
Net Investment at End of Year 7			7,079

Note: Cost analysis can be continued in this matter each year.

Table 2–11 Cost of vineyard establishment and operation at various prices per ton for 5 tons per acre yields.

	Total accumulated cash expense			
Year	$500/ton	$600/ton	$700/ton	$800/ton
1	4878	4878	4878	4878
2	6277	6277	6277	6277
3	7883	7629	7379	7129
4	7705	6930	6160	5390
5	7512	6175	4844	3512
6	7304	5360	3423	1484
7	7079	4480	1888	(706) 1st net
8	6836	3529	230	(2309) yearly net
9	6574	2502	(1561) 1st net	
10	6291	1393	(1809) yearly net	
11	5985	195		
12	5655	(1098) 1st net		
13	5298	(1309) yearly net		
14	4913			
15	4497			
16	4048			
17	3563			
18	3039			
19	2473			
20	1862			
21	1202			
22	489			
23	(281) 1st net			
24	(809) yearly net			

Note: Yearly cost of production from year 4 and on is $1691/acre at 5 tons/acre. Increased profits from grapes sold at higher prices greatly reduces the number of years before establishment costs are recovered.

The additional costs anticipated for such jeopardies must be figured into the cost of production in order to determine grape values. Years of reduced or no crops should also be calculated into the cost of production so that the value for grapes harvested during more successful years can accurately reflect the overall profitability on a sustainable basis.

Site Preparation

Thorough preparation of the planting site is essential to minimize weed problems and to increase the organic matter content of the soil. This should be done 1 year before planting. Problem weeds should be controlled with herbicides prior to soil preparation. Green manure crops such as rye add organic matter and help reduce weed problems.

Soil testing will provide information on fertility and liming needs for the first-year vineyard. If the soil pH is below 5.5, apply agricultural limestone to raise the pH to a more desirable level—in the 5.5–6.5 range. Lime applications should be made the year before planting, and the limestone should be incorporated deeply into the soil. Other nutrients, especially potassium, should be applied during site preparation if the soil test shows a deficiency.

Avoid poorly drained soils for grape production. If the soil has only fair internal drainage and drain tile is not installed, it may be advantageous to sub-soil (chisel plow) to a depth of at least 24 inches and establish a ridge on which the vines will be planted by mounding the soil from the edges of the row to the middle. The ridge should be about 10 inches high and 2 feet wide. This will provide improved surface and internal drainage, as well as allow deeper rooting of the vines. It also prevents ponding of water at the base of vines which can lead to disease problems.

Vineyard Layout

Grape vines are generally planted in straight rows running in a north–south direction for maximum sunlight interception. However, sites on sloping land may require contour planting or straight rows across the slope, regardless of compass direction. In some areas, it may be advantageous to orient rows parallel to the prevailing wind to aid in disease control. Row spacing depends on the training system and the equipment to be used in the vineyard. Nine to 10 feet between rows is commonly used, but it may be increased to accommodate larger equipment or on sites with steep slopes. Spacing between vines within the rows is usually 8 feet but may be closer for cultivars with low vigor. At maturity, the vines should completely fill the trellis space without competing with each other for sunlight. (See Fig. 2–15.)

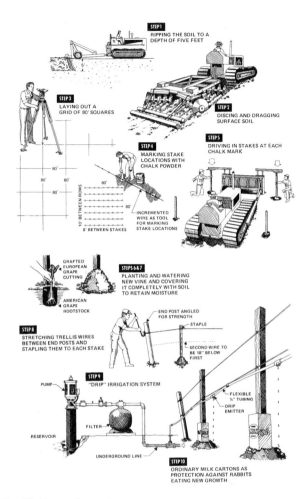

Fig. 2–15 Establishing a vineyard.
Courtesy: Sebastiani Vineyards.

Planting

Planting should take place in early spring—after the danger of hard freezing temperatures have passed. Vines are usually purchased as dormant bare-root plants and should be stored in a cool, moist place, preferably cold storage, until planted. Soak the roots in water for 24 hours prior to planting to prevent the root system from drying out during the handling and planting process. Prune off broken and damaged roots and shorten excessively long roots for convenience when planting.

The planting hole should be large enough to accommodate the root system with ease. Excessive pruning of the root system is not advised. However, it is better to prune a few roots rather than stuff the roots into a small planting hole. Spread the roots, cover with soil, and tamp well.

If soil moisture is low, water as needed until the plants have developed a root system large enough to support themselves during dry periods. Own-rooted (ungrafted) vines should be set to a depth where the lowest shoot of the dormant plant is just above the soil level. For grafted vines, the graft union should be situated at least 2–3 inches above the soil level to prevent scion rooting.

Vine Training

It is important to properly train vines during the first few years of growth to establish a vine form which will be easy to manage. After planting, but before growth begins, the top of the dormant plant should be pruned back to a single cane with two to five buds. After growth starts, all but the best two to four shoots should be removed. One or more of these shoots will become the trunks. Support should be provided for new shoots to kept them off the ground. This will greatly reduce disease problems and provide full sun exposure for maximum growth. The trellis should be established soon after planting to provide this support. String can be tied from a side shoot of the vine to the wires and the new shoots wrapped around the string. Never tie around the main trunk of the plant because the trunk will expand rapidly during the first growing season and can be girdled by the string. A stake can be driven next to each plant and the shoots tied to the stake instead of the trellis wire.

TRELLISING

Grape vines require support for ease of management and the heavy weight of ripening fruit. This can vary from a simple wire trellis to an elaborate arbor. Vine support should hold the vine up where it can be managed and cared for efficiently. The trellis should also expose a greater portion of the foliage to full sunlight, which promotes the production of highly fruitful buds. Trellis design features, such as the number and placement of wires, is dictated by the type of training system used. All training systems require a strong trellis assembly to support the weight of fruit and vines.

Training Systems

Grape vines are trellis trained by several dozen major configurations. The most efficient methods provide a well-spaced even distribution of fruiting wood along the trellis and promote full sun exposure for clusters and basal nodes. However, the selection of a training system usually is made after careful consideration of climate history, soil profile, and vigor of the variety.

Cane training systems are devised by selecting superior canes from the previous season of growth and arranging them in a manner which serves best. An Umbrella Kniffin (Fig. 2–16) may be employed for larger crops from vigorous vines in fertile soils situated in moderately cool environs. A Keuka Low Renewal system (Fig. 2–17) is typically used for grafted vines and in colder locales. The short trunk conserves vine metabolism generated for woody tissues and allows for snow insulation or burying during winter if needed. Fewer, shorter canes control crop size in order to optimize vine vigor. Cordon training systems function by extending trunks along the trellis wires with shoots and fruit developing from buds left on short 'spurs' after pruning. The Bilateral Cordon (Fig. 2–18) is often the choice in moderate climates and average soils. Vigorous vines cultivated in richer soils may do best upon a Geneva Double Curtain (Fig. 2–19) whereby the trellis is divided to provide additional exposure to sunlight for enhanced photosynthesis.

Machine Harvesting

The price tag on grape harvesters (Fig. 2–20) range from $50,000 to more than double that—depending on whether the device is pulled by a tractor or self-

Fig. 2–16 Umbrella Kniffin cane training system. (Courtesy of M.L. Hayden, Department of Horticulture, Purdue University)

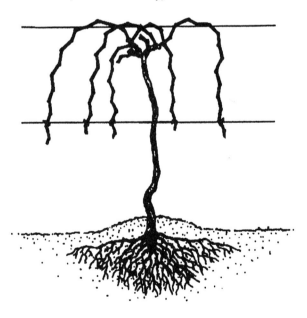

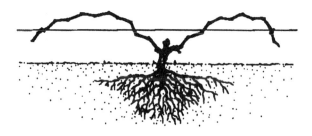

Fig. 2–17 Keuka Low Renewal cane training system. (Courtesy of M.L. Hayden, Department of Horticulture, Purdue University)

powered. That amount can be doubled again by the extent of ancillary equipment needed, such as vineyard trailers, bins, heavy-duty trucks, and winery receiving machinery. Consequently, it is rare that estate vintners can find economies justifying individually owned and operated machine harvesting equipment. On the other hand, there may be an opportunity to share such investments with other estate vintners in the area, or to hire custom machine harvesting services. In any case, harvesting by machine requires trellis systems specially designed to accommodate the type of device which will be employed.

Fig. 2–18 Bilateral Cordon training system. (Courtesy of M.L. Hayden, Department of Horticulture, Purdue University)

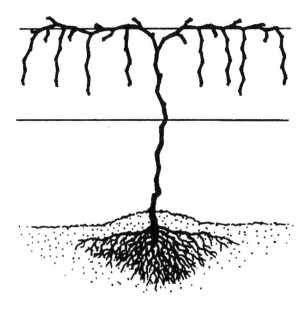

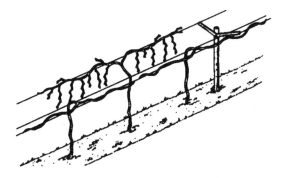

Fig. 2–19 Geneva Double Curtain Cordon training system. (Courtesy of M.L. Hayden, Department of Horticulture, Purdue University)

Weed Control

It is especially important to provide good weed control during the first several years of vineyard establishment. Young vines do not compete well with grasses and weeds for soil nutrients, moisture, and sunlight. Grasses and weeds can also harbor disease organisms which can harm developing vines. Plastic and

Fig. 2–20 Machine harvesting.
Courtesy: Sebastiani Vineyards.

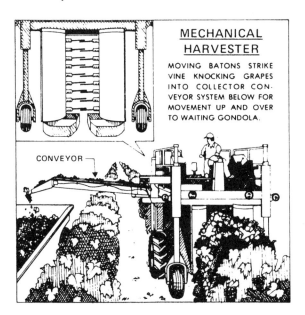

biodegradable mulches, as well as organic compost mulches, can be used to control weed growth. Mechanical weed control is effective as long as devices do not penetrate the soil deep enough to destroy vine feeder roots. For larger plantings, the use of herbicides may be necessary. Refer to Cooperative Extension Service small fruit spray guides for information relating to suitable herbicides.

Herbicide Injury

Injury from 2,4-D herbicide is frequently confused with disease or insect injury. Most cultivated grapes are highly susceptible to injury by this herbicide and related compounds. These herbicides are used in a variety of forms on many different crops and noncrop situations. Because of their extreme volatility, many 2,4-D compounds can injure vines at considerable distances from the point of application. The use of these compounds in field crop production, especially reduced tillage situations, may limit the production of grapes in certain areas.

Vintners should diplomatically inform the local community, especially neighboring farmers, that grape vines will be cultivated in the area—and that 2,4-D or related compounds are a threat to this endeavor. Requests should be made that they use only low-volatile formulations and to make applications with equipment designed to reduce drift. Vines are most sensitive to this type of chemical injury during the early part of the growing season, when shoots and leaves are expanding rapidly.

Symptoms of 2,4-D are easily diagnosed. The youngest terminal growth at the time of exposure will be stunted, misshapened, and have thick veins which tend to run parallel to each other rather than branching as in normal leaves. Growth will be stunted for several weeks following injury. Severely injured vines may die or not recover for several years.

VINEYARD MANAGEMENT

Pruning

Pruning is the most important cultural practice in the management of grapes. It is performed to select superior potential fruiting wood, to maintain vine shape and form for the best sun and wind exposure, and to regulate the number of buds retained per vine.

Grape vines require annual pruning to remain productive and manageable. It can be done anytime during the dormant season, but it is generally best to delay until late winter or early spring, as pruned vines are more susceptible to cold injury. In addition, delayed pruning allows for adjustments in the number of retained buds in the event of winter injury. It is, thus, essential to assess bud survival before final pruning.

An average mature grapevine will have 200–300 buds developed during the previous growing season. If all of these were retained, the result would be a large crop which would not ripen properly, reduced vine vigor, and poor cane maturation for succeeding years of production. To avoid this situation, researchers have developed a method of pruning to balance the fruit productivity and vegetative growth which will give maximum yields without reducing vine vigor or cane maturity. This procedure is commonly referred to as "balanced pruning." The number of buds retained is balanced to match the situation in which the vine grows—its soil, its climate, and its vigor. All of this is termed "vine size," which is measured by the weight of one year old canes taken from vines during pruning.

To balance prune a grapevine, estimate the vine size, then prune the vine, leaving enough extra buds to provide a margin of error, usually 70–100 total buds. Next, weigh the 1-year-old cane prunings with a small spring scale (See Figs. 2–21 through 2–24). Then use the pruning formula information in Table 2–12 to determine the number of buds to retain per vine.

For example, a moderately vigorous vine cultivated in a cool climate having been pruned to a total of 3 pounds of 1-year-old prunings would be afforded 20 buds for the first pound of pruning wood, and 10 buds for each of the additional 2 pounds. This is summed up to be a total of 40 fruiting buds left from which the upcoming crop of grapes would be borne. Note that even if there were 4 or 5 pounds of prunings, the maximum total number of buds recommended for a vine in this situation would still be 40. After determining the appropriate number of buds to retain, prune the extra buds off, taking care to space the fruiting buds evenly along the trellis. Once several vines in a given vineyard are measured, experienced pruners will be able to closely estimate vine size so as to dispense with the arduous task of weighing every vine. Pruning formulas developed for many cultivars have been devised and published by the Cooperative Extension Service in some states.

On some cultivars balanced pruning alone will not adequately balance the crop load with vegetative growth. Cluster thinning and/or shoot removal may be required on some cultivars, especially certain selections of hybrids which are highly productive. Cluster thinning typically involves the removal of all but the basal cluster, which is usually the largest. Shoot removal is performed by removing shoots which occur from secondary buds and latent buds from wood older than 1 year.

Pest Control

Diseases

Control of disease is necessary to maintain healthy leaves and protect the fruit from various types of deterioration. Most diseases of grapes are caused by fungi which thrive in hot, humid conditions.

The most common cluster and leaf diseases are black rot, bunch rot, downy

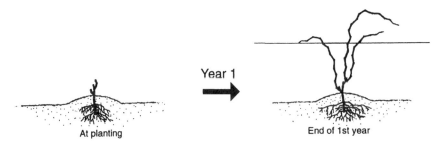

Fig. 2–21 Typical pruning and growth during first year.

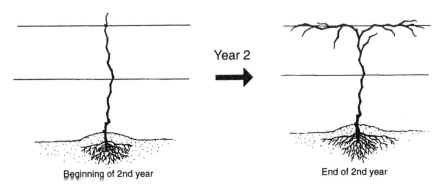

Fig. 2–22 Typical pruning and growth during second year.

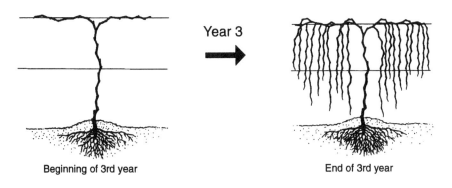

Fig. 2–23 Typical pruning and growth during third year.

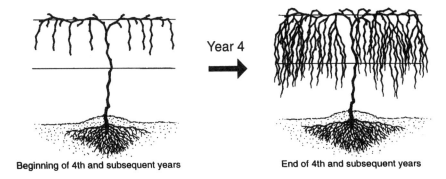

Fig. 2–24 Typical pruning and growth during fourth and mature years.

Table 2–12 Fundamental pruning formula for grapevines.

Vine condition	No. of buds left for first pound of prunings	No. of buds left for each additional pound	Max. total buds
Low vigor and/or cold climate	10	5	20
Medium vigor and/or cool climate	20	10	40
High vigor and/or moderate climate	30	20	70

mildew, and powdery mildew. (See Figs. 2–25 through 2–28.) Many factors affect the amount of disease which can occur, such as site selection, vine vigor and susceptibility, canopy density, and other factors. Most cultivars are prone to one or another of the disease infections. A few cultivars are only slightly susceptible to the major diseases and can be cultivated with little or no pesticide use. Normally, however, some pesticide use is necessary in order to product top quality fruit.

Pierce's Disease (See Fig. 2–29) is a rather "arteriosclerosis-like" terminal condition vectored by the Sharpshooter Leafhopper. This insect introduces Pierce's Disease bacteria into vines as it sucks sap from succulent shoot tips. The bacteria multiply upon xylem walls, causing eventual total restriction of sap flow, and vine mortality. The adult insects emerge each year during summer and feed on grape foliage. The damage can be severe if uncontrolled, especially on young or weak vines.

68 Viticulture (Grape Growing)

Fig. 2–25 Black rot.
Courtesy: Mississippi State University.

Fig. 2–26 Bunch rot.

Fig. 2–27 Downy mildew.
Courtesy: Garth A. Cahoon, Ohio Agricultural Research and Development Center.

Fig. 2–28 Powdery mildew.
Courtesy: Mississippi State University.

Fig. 2–29 Pierce's disease.

Crown Gall (See Fig. 2–30) is usually caused by freeze or mechanical injury whereby lesions resulting in the vine trunks allow for *Agrobacterium tumefaciens* to enter the xylem vessels. Growth of this bacteria forms large galls which kill the vine. The bacterium itself does not cause disease but vectors a plasmid, a circular piece of DNA, that enters the plant cells, becomes incorporated into the plant genome and causes the cells to divide abnormally, leading to the development of tumors or galls.

Responsible pesticide use requires a knowledge of pest identity and biology. Growers should frequently inspect their vineyards carefully and seek help in identifying problems. Most fungicides registered for use on vines are effective against only certain diseases. Correct identification of disease problems is necessary for effective and responsible disease control. Consult state Cooperative Extension Service small fruit spray guides for information relating to suitable fungicides and pesticides.

Insects

There are relatively few insect pests found in grapevine cultivation. Grape berry moth, Japanese beetle, and various leafhoppers are the most common.

Another important insect pest is the Grape Flea Beetle. These insects emerge early in the spring and feed on developing grape shoots. If left uncontrolled, these pests can completely destroy all primary buds and cause heavy losses in production.

Fig. 2–30 Crown gall.

Fig. 2–31 Phylloxera.

One of the most infamous of viticultural insects is phylloxera (*Daktulosphaira vitifoliiae*), (See Fig. 2–31) the root louse insect which devastated millions of acres of vineyards in Europe during the latter 1800s and recently invaded vineyards in northern California again.

Insecticide control of destructive pests on vines should be made only if the infestation is present and the amount of damage is severe enough to warrant the cost of chemical control.

CHAPTER 3

WINE MICROBIOLOGY

Microbiologically speaking, unknown wine is much safer than unknown water. Of the thousands of microorganisms now classified, only relatively few have been found growing in wine. This is due to the nearly antiseptic combination of alcohol and acids which occur in wine. Most microscopic organisms taken from contaminated water actually die if introduced into wine. It is from this phenomenon that wine has served in human health throughout Western civilization.

This chapter will consider the major microscopic organisms which are involved in wine. It should be understood that the entire realm of wine microbiology embraces hundreds of yeasts, bacteria, and mold organisms.

The procedures for microbiological laboratory analysis are found in Appendix B.

YEASTS

There are millions of natural, or 'wild', microscopic yeast cells in the *cutin* of a grape berry. The cutin, often called the "bloom," is a waxy coating on grape skins.

Fermentation is a natural process in which the grape sugars (mostly glucose and fructose) are transformed into ethyl alcohol and carbon dioxide gas. In 1810, the French chemist and physicist Joseph-Louis Gay-Lussac described this process in the following chemical equation:

$$C_6H_{12}O_6 \longrightarrow 2C_2H_5OH + 2CO_2$$
(grape sugars) $\longrightarrow$ (ethyl alcohol + carbon dioxide gas)

Today, we know that wine fermentation is carried out by certain enzymes (protein compounds produced by yeast cells) which act as catalysts in facilitating the sugar to alcohol/CO_2 conversion. Students of the fermentation process will learn that it very closely parallels the glycolysis cycle which is elementary to modern biochemistry and that heat energy is also a product of the transformation. Thus, the original Gay-Lussac equation is modified as follows:

$$C_6H_{12}O_6 \xrightarrow{\text{yeast enzymes}} 2C_2H_5OH + 2CO_2 + 56 \text{ kcal of energy}$$

Note that this reaction is the opposite of that given in the Chapter 2 discussion which related to the photosynthesis in grapevine leaves. Thus, photosynthesis and fermentation comprise a natural cycle.

It was Theodor Schwann who first discovered that this transformation was caused by a "sugar fungus" (i.e., *Saccharomyces* organism; (see Fig. 3–1). Louis Pasteur published a paper in 1857 which revealed an observation of a similar microscopic organism which was responsible for wine fermentation. Prior to this, fermentation was thought to have been due to a phenomena known as "spontaneous generation"—loosely related to various forms of "magic." The Greeks called fermentation *zestos,* or "boiling," from which the term *yeast* was fashioned. Thus, the discipline of microbiology was born in France, largely in response to Pasteur's curiosity about Gay-Lussac's equation.

Yeasts are egg-shaped cells about 1/300,000 of an inch in length. Because they lack stems, leaves, roots, and chlorophyll, yeasts are referred to as lower-form plants. (See Fig. 3–2.) In tumultuous fermentation, they can reproduce once every half-hour or so—a rate which can generate a huge population in a relatively short period of time.

Wild yeasts may, or may not, perform the fermentation function adequately. *Saccharomyces apiculata* are, as the name suggests, apiculated with irregular shapes pointed at one or both ends. These and other natural yeasts can lose viability well before fermentation is completed—the resulting wine being residually sweet and unbalanced. An additional hazard exists with spoilage yeast strains which can foul the wine with a vinegar (acetic acid) or paint thinner (ethyl acetate) flavor. Despite the risks, some winemakers still prefer to employ natural yeasts.

Wine Microbiology 75

Fig. 3–1 Electron microscope photo of *Saccharomyces cerevisiae*.
Courtesy: Michael Sullivan, SEM, Mississippi State University.

Yeasts are taxonomically classified as follows:

Phylum	Thallophyta
Subphylum	Fungi
Class	Eumycetes
Subclass	Ascomycetes
Order	Endomycetales
Family	Saccharomycetaceae
Subfamily	Saccharomycoideae
Genera	Saccharomyces

Saccharomyces cerevisiae (Fig. 3–3) is the species within which the most important cultured wine fermentation yeasts are found. The variety *ellipsoideus* has many strains available for commercial vintners. It is common to find such strains as *Champagne, Epernay, Montrachet, Prisse de Mousse,* and *Steinberg* offered in dehydrated pellets which are stirred in warm water by winemakers just prior to inoculating in new grape juice or must for fermentation. Another common choice

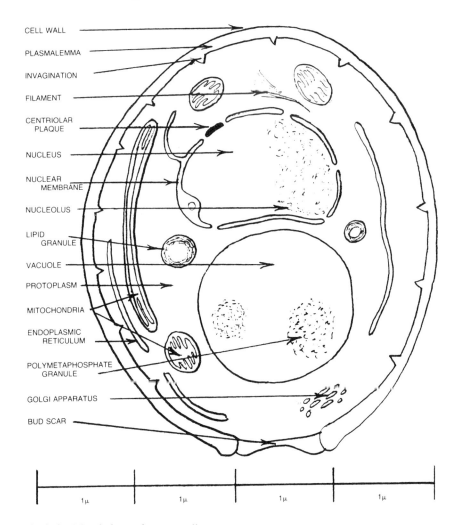

Fig. 3–2 Morphology of a yeast cell.

nowadays are yeasts which have a "killer factor," allowing them to inhibit many wild spoilage yeasts.

Alcohol is, of course, a by-product of fermentation. Yeasts have difficulty remaining viable in levels of alcohol exceeding 14% by volume.

Most wine yeasts grow in suspension throughout the entire volume of fermenting juice. There are, however, also surface-growing yeasts classified as *Saccharomyces beticus,* more often referred to as *flor* yeasts. These microorganisms serve to convert alcohol into aldehyde and furfural compounds which are some-

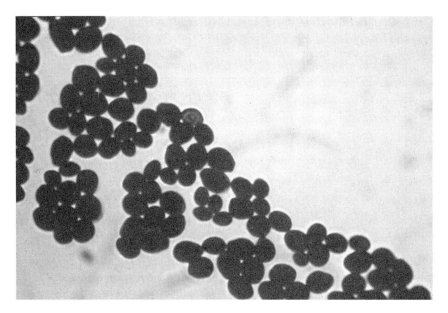

Fig. 3–3 Light microscope illustration of *Saccharomyces cerevisae.*

what caramel and nut-like in flavor. This is a desired character in "madeirized" wines, such as those grown on the island of Madeira—namesake to similar wines made in the Sherry region of Spain and Marsala in Italy.

Other surface-growing yeasts such as *Brettanomyces* (Fig. 3–4), *Candida* (Fig. 3–5), *Kloeckera* (Fig. 3–6), and *Pichia* (Fig. 3–7) are spoilage types which can cause dirty, musty, off-flavors. A chalky, wrinkled, white film appears on the surface of wines infected with these yeasts.

Yeasts which do not require oxygen to grow are classified as *fermentative*. All of the strains of *Saccharomyces cerevisae* which are cultured for primary wine fermentations are fermentative yeasts. *Brettanomyces* is also fermentative and can produce large amounts of acetic, isobutyric, and isovaleric acids, resulting in "barnyard" odors and flavors. Some winemakers actually prefer a trace of this type of character for added complexity. *Kloekera,* with its growth primarily associated with higher levels of acetic acid and aldehyde compounds, is also a fermentative yeast.

Yeasts requiring oxygen for growth are called *oxidative* and are generally undesirable for table wine fermentations. These microorganisms utilize alcohol and organic acids as carbon sources to produce volatile esters and aldehydes. Again, some winemakers prefer trace levels of these compounds in their wines to gain added complexity.

78 *Wine Microbiology*

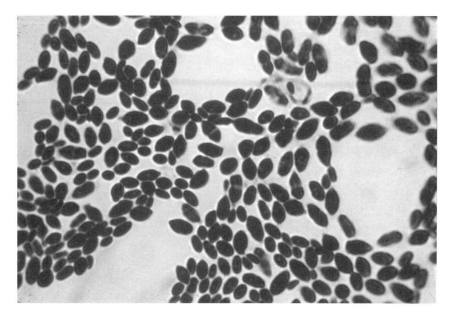

Fig. 3–4 Light microscope illustration of *Brettanomyces*.

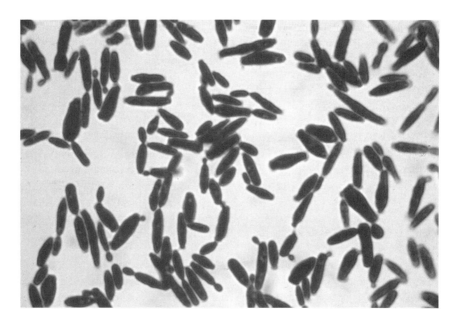

Fig. 3–5 Light microscope illustration of *Candida*.

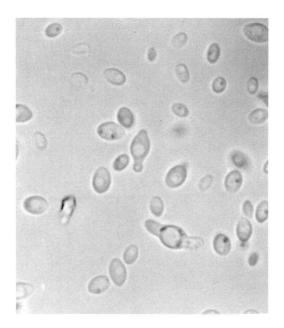

Fig. 3–6 Light microscope illustration of *Kloekera*.
Courtesy: A. Dumont, Lallemand, Inc.

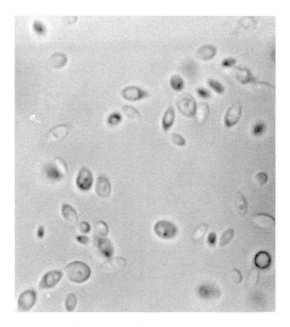

Fig. 3–7 Light microscope illustration of *Pichia*.
Courtesy: A. Dumont, Lallemand, Inc.

BACTERIA

The microbes most feared by wine masters are the vinegar bacteria—more properly known as *Acetobacter,* or acetic acid bacteria. Fortunately, we know that these one-celled organisms require oxygen to grow—a condition which can be controlled in well-equipped wineries. The most common strain of vinegar bacteria is *Acetobacter aceti* (Fig. 3–8), although *Acetobacter pasteurianus* and *Acetobacter peroxydans* are also identified.

Another type of wine bacteria can be beneficial to some wines and detrimental to others. These are the well-known "lactic acid" bacteria which, as their name suggests, transform malic acid to lactic acid along with carbon dioxide gas. Malic acid is an organic acid which has a rather apple-like flavor; lactic acid has somewhat of a cheesy character. *Leuconostoc oenos* (Fig. 3–9) is the low pH widely cultured strain of lactic acid bacteria and is often employed to reduce acidity and add complexity to red wines and heavier white wines, as well.

Fig. 3–8 Electron microscope illustration of *Acetobacter aceti.*
Courtesy: Michael Sullivan, SEM, Mississippi State University.

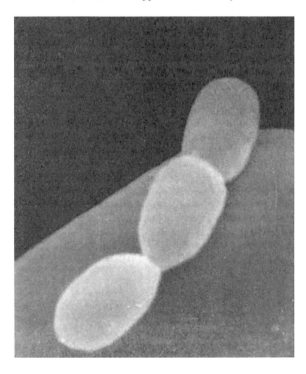

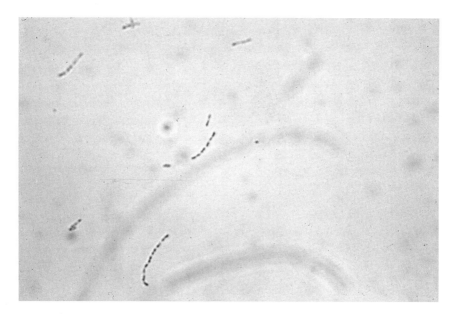

Fig. 3–9 Light microscope illustration of *Leuconostoc oenos.*

Lactobacillus brevis (Fig. 3–10) is another of several lactic acid bacteria species which can severely reduce total acidity, leaving a wine insipid and lifeless.

Natural malolactic fermentations may occur in virtually all viticultural areas, although cultured lactic acid bacteria inoculums are generally preferred by vintners. The chemical reaction for malolactic fermentation is as follows:

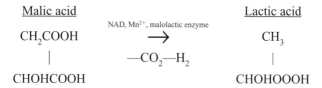

This is a stoichiometric reaction in which very little energy is released. Whether natural or cultured bacteria perform the malolactic fermentation, time duration is not easily predicted. The reaction may commence immediately or lag for months. Low wine pH, cool temperature, and free SO_2 levels higher than 15 mg/l will slow or inhibit bacteria growth.

Pediococcus cerevisiae (Fig. 3–11) bacteria produce histamine in some wines, predominantly red table wines. Histamine is an organic compound that stimulates

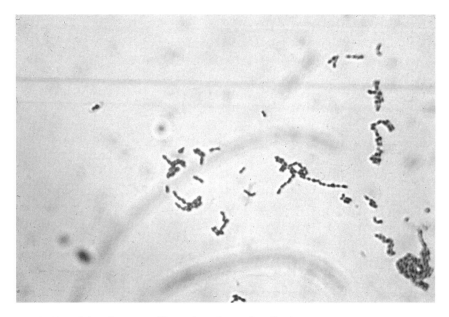

Fig. 3–10 Light microscope illustration of *Lactobacillus brevis.*

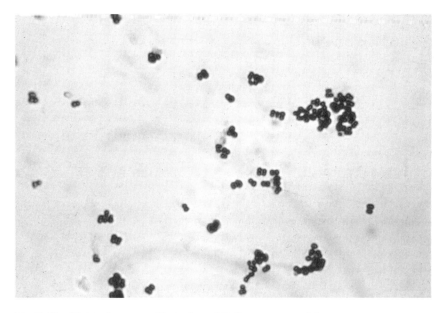

Fig. 3–11 Light microscope illustration of *Pediococcus cerevisiae.*

blood circulation and may play a role in nasal stuffiness, headache, and some allergic reactions of wine consumers.

MOLDS

Viticulturists may have to contend with such molds as black rot, downy mildew, powdery mildew, and others in the vineyard from year to year. Winemakers contend with a different set of molds—all of which affect wine indirectly. Alcohol content does not allow molds to grow directly in wine. Poor sanitation of equipment and storage vessels, however, can lead to the growth of molds. Wine exposed to these contaminated surfaces will absorb moldy odors and flavors. Unchecked winery mold growth can damage the wine—seriously deteriorating its quality.

The famous "noble mold" or, more properly, *Botrytis cinerea,* is an exception. When desired and when grown in optimal conditions, this fuzzy gray mold permeates grape skins and allows internal water to evaporate. The resulting berries are shriveled with concentrated tannins, acids and sugars. Such grapes make the renowned French *pourriture noble* (noble mold) wines of Sauternes in Bordeaux and *Edelfaule* wines in Germany. Many similar "late harvest" wines of both types are also grown in the United States.

Connoisseurs consider noble mold wines some of the most exquisite white wines made. The loss of natural grape water and the delicate techniques of making such wines successfully renders them an expensive delicacy.

With unexpected rains or improper handling in the vineyard, botrytized berries may also become infected with *Penicillium expansum*—not to be confused with the *Penicillium notatum* of medicinal fame. It emerges as a white powder-like dust on grape berries and matures to an ominous bluish green mold. Wines can be ruined with the compound trichoroanisole (TCA), which is formed in corks infected with this microorganism. (See Fig. 3–12 for a micrograph of *Penicillium.*)

Various species of *Rhizopus* molds can be found in many winegrowing regions and are generally associated with summer bunch rot in vineyards. These molds can act much like spoilage yeasts, converting sugars to ethanol and carbon dioxide gas. Under the microscope, however, these are seen as spherical sacs at the end of thread-like sporangiophores and are sometimes referred to as "pin molds."

THE MICROBIOLOGICAL WORKPLACE

Commercial winemaking requires vintners to become involved in handling microbes. Yeasts and malolactic bacteria need to be properly stored, cultured, and

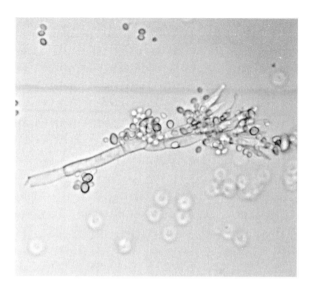

Fig. 3–12 Light microscope illustration of *Penicillium*. Courtesy: A. Dumont, Lallemand, Inc.

grown for successful fermentations. Problem microbes need to be detected and overcome. The investment need not be large. Fully adequate microbiological facilities can be achieved in a winery for less than $2,000, including a good used microscope.

The ideal location for a microbiology laboratory is a small well-lighted room, at least 50 square feet, which has washable walls, ceiling, and floor. It should have a countertop, sink, and hot and cold running water, as well as several electrical outlets.

Alternatively, microbiological tasks can be performed in a multipurpose area. A "hood," or small self-contained workplace, will protect the work area from dust and air currents and will isolate organisms from the winery. Avoid buying used chemical hoods, as they are generally designed for rapid air circulation in the wrong direction for microbiological work and are virtually impossible to sanitize.

Building a small effective hood made from plywood, plexiglass, and hinges, can be very cost-effective. The plywood back and adjoining sides are cut on a bevel and hinged together to collapse for storage. (See Fig. 3–13.) The unit is covered with primer and several coats of white high-gloss washable enamel paint. Plexiglass panels are mounted between railing strips.

A small styrofoam incubator (Fig. 3–14) will provide a continuous temperature to grow yeasts and bacteria rapidly. The temperature is controlled by the wattage of the light bulb used in a socket assembly inserted through the side of the styrofoam ice chest. This unit is inexpensive to construct and can maintain a constant temperature of 75°F with a 5 watt bulb.

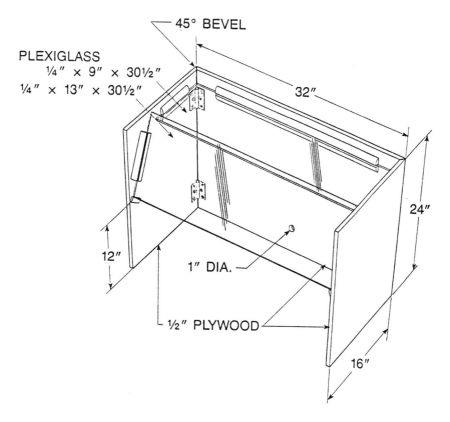

Fig. 3–13 Microbiology hood illustration.

Equipment

The following list contains equipment needed to set up a small effective microbiological laboratory:

Alcohol Burner—Glass Lamp. The primary purposes of this is to create a flame to disinfect equipment, dry microscope slides, and sterilize loops.

Aspirator Filter Pump. This device is powered by the velocity of water passing through a cold-water faucet, producing the vacuum needed to pull wine samples through a filter assembly.

Side-Arm Vacuum Flask—1,000 ml. A side-arm vacuum flask supports the filtration unit and collects filtered wine drawn through by vacuum. Size 8 rubber stoppers with a single hole are needed to connect the filtration unit to the vacuum flask. Heavy-duty black rubber tubing connects the aspirator to the vacuum flask.

Yeast and Mold Swab Test Kits. These are employed for detecting microbial contamination of winery equipment, tanks, corks, and so forth.

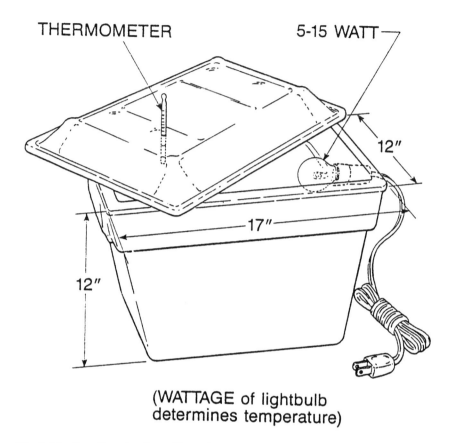

Fig. 3–14 Styrofoam incubator illustration.

Paper Chromatography Kits. Detection of malolactic fermentation is made using these kits. Alternatively, a paper chromatogram can be made by following the methodology outlined in Appendix B.

Microscope. The cost of a microscope capable of 1000 × magnification is not within every vintner's budget. Often, laboratories in universities, high schools, hospitals, and other sources have surplus used microscopes for sale at prices far below new models. From time to time, some of the suppliers listed in Appendix A offer preowned as well as new microscopes. A good used microscope can cost several hundred dollars. Depending on features, a new microscope can cost several thousands of dollars.

Microscope Accessory Kits. This is an assortment of glass slides, cover slips, lens paper, inoculating loop, and other ancillary material. These packs generally include Methylene Blue powder which is dissolved at a rate of 1 gram in 10 milliliters of ethanol and 90 milliliters of water. The resulting mixture is used to stain yeast and bacteria on slides for microscopic observation.

FACTORS AFFECTING MICROBIAL GROWTH

As with all living things, the ability of microbes to thrive is directly dependent on their environment.

- **Moisture** is essential to cell growth and activity. In extremely dry environs, actively growing cells will eventually die or go into a dormant state if they have that capability. As the moisture content of the environment increases, cellular activity also increases. In solutions where there is a very high concentration of dissolved solids, and thus a commensurately lower concentration of moisture, living cells cannot utilize low moisture levels and are thus unable to multiply. For this reason, jellies, concentrates, and brine solutions are resistant to microbial spoilage due to their "low water activity."
- **Temperature** is a controlling factor for microbial activity. Optimum growth rates for wine-related organisms are in the range 60–85°F. However, some yeasts are capable of fermenting slowly at temperatures below 30°F. Most microbial growth ceases at temperatures above 100°F, and viable cells (actively growing cells) are killed if exposed a few minutes at temperatures above 160°F.
- **Nutrients** are required by all cells, but each organism genus and species has specific requirements. Wine provides carbohydrates in the form of sugars, ethanol, and organic acids. Amino acids and ammonia are available in grapes as sources of nitrogen. Juices and musts contain vitamins and other complex factors required for the growth of microorganisms.
- **pH,** a measure of acid strength, is a major factor controlling microbiological activity. Low pH is one of the main reasons why so few microbes will grow in wine. Yeasts and some lactic acid bacteria will tolerate pH as low as 2.8. Wine organisms are increasingly active as pH increases to 4.0 Because wines experience serious nonmicrobial problems at pH higher than 4.0, a high pH is not a factor in inhibiting microbial growth in juice, must, or wine.

COMMERCIAL STARTER CULTURES

Almost every yeast starter culture marketed commercially has been indexed to ensure consistent results in fermentation. Commercial yeasts are generally capable of fermenting normal levels of grape sugar to dryness and are active in SO_2 levels as high as 50 milligrams per liter. They should perform in juice and must with a pH as low as 2.8 and withstand alcohol levels as high as 14% by volume.

As mentioned previously, commercial yeast starters are available which can produce a protein "killer" factor which inhibits undesirable wild yeasts. Some yeast strains reduce malic acid and others help to release fruity terpene flavors. There are specialized yeasts which have proven superior for white wine, others for reds, and still others for the secondary fermentation of sparkling wines.

Commercial lactic acid bacteria cultures should tolerate pH levels as low as 3.2 and efficiently convert malic acid to lactic acid. Resulting wines should have less fruit aroma and more complexity in diacetyl and acetoin flavors, but free of sauerkraut and musty odors.

It is recommended that fresh starter cultures be purchased before each vintage season from a reliable supplier. Store all yeast cultures in a refrigerator; lactic acid bacteria cultures must be stored in a freezer. Buy enough to inoculate all anticipated juice and must lots for the entire season. Taking from one fermentation to inoculate another is a poor practice, as contamination can be both introduced and spread. Many vintners select one yeast strain for cool-temperature white fermentations and another for the warmer red fermentations.

Follow yeast culture supplier instructions faithfully for the best results. Ensure that the temperature of the juice or must is within 10°F of the starter culture at the time of inoculation. The use of yeast nutrient can help avoid stuck (unfinished) fermentations and can reduce the generation of hydrogen sulfide by nitrogen-stressed yeasts.

MICROBIAL SPOILAGE IN JUICE AND MUST

The primary problem facing fresh juice processors, in both bulk storage and packaged final products, is yeast fermentation. According to U.S. Food and Drug Administration (FDA) regulations, a level of 0.5% (by volume) or greater of ethyl alcohol exceeds permitted concentrations for juice labels. A secondary problem is bacterial spoilage producing increased levels of acetic acid.

The presence of gas in juice is evidence of yeast spoilage. As this spoilage progresses, sour odors and flavors develop. Bulk juice should be monitored for ethyl alcohol concentration. Juice samples should be cultured on appropriate media to determine the numbers of viable yeast cells, along with an observation under a microscope, in order to confirm the presence of budding yeast cells.

Wine Preservation

Sterile Filtration

This is the most modern and widely used method for preserving wine in larger wineries, although smaller units have recently become available to smaller wineries. Sterile filtration is effective only if all filters and equipment have been properly sanitized and if the environment is free of contaminating microbes. These criteria are difficult to meet even with the most modern stainless steel equipment. Consequently, sterile filtration should be supplemented with chemical or heat treatment to ensure adequate shelf life. To achieve sterile filtration, a membrane filter of less than 0.50 microns porosity is required.

Pasteurization

This archaic method of wine preservation is rarely found in small wineries. It requires specialized heat-exchange equipment and a constant supply of live steam—both of which are generally out of reach in small estate winery budgets. Pasteurization is performed by rapid heating of the wine to 190–200°F, holding in this temperature range for about 3 minutes, and then rapidly cooling. Pasteurization is effective in killing both yeasts and bacteria, but usually at the expense of distorting delicate flavors and eventual browning of color.

Chemical Agents

Chemical preservation of wine is the most widely used method employed by estate vintners. The use of sulfur dioxide and sorbic acid for wine preservation is discussed in Chapter 8.

MICROBIAL SPOILAGE OF JUICE CONCENTRATES

When juice has been concentrated to at least 68° Brix, sufficient water has been eliminated to achieve resistance to spoilage due to the low level of water activity discussed earlier. One problem which can develop when concentrates are moved in and out of refrigerated storage is the condensation of moisture on the inner surfaces of containers. This condensate can serve to dilute the concentrate to provide higher levels of water activity—and thus create a fermentable condition for yeasts or mold growth.

When concentration is less than 68° Brix, slow yeast fermentation can occur which is easily detected by the presence of carbon dioxide gas and a reduction in Brix level. Mold growth can be detected by small fuzzy islands on the surface of the concentrate, along with a musty odor. Concentrates over 60° Brix should be stored at 30–32°F, whereas lower concentrations of juice should be stored at 23–25°F. The condition of concentrates should be inspected at least once per week; storage for more than several months is not recommended.

MICROBIAL SPOILAGE OF BULK WINE

Although dry wines fermented up to at least 11% alcohol are relatively resistant to microbial growth, they are still susceptible to spoilage by the growth of undesired organisms. In all cases, the higher the pH in the wine, the more likely microbial growth will occur, the faster it will occur, and the greater intensity of spoilage will result.

Detection

Brettanomyces yeast growth usually takes place during barrel aging. It is often slow to develop and goes undetected until the wine is severely damaged with unpleasant "barnyard" odors and flavors. Even residual sugar levels as low as 0.02 percent can support growth of this spoilage yeast. This organism may, or may not, produce noticeable gas or a surface film. Under the microscope, *Brettanomyces* is seen as small lemon-shaped cells which can be difficult to distinguish from cultured *Saccharomyces cerevisiae* cellars. (See Fig. 3–15) Yeast mold agar with 100 ppm cycloheximide will allow the growth of creamy white colonies of *Brettanomyces* within 4–7 days after inoculation while inhibiting all other wine yeasts.

Surface yeast growth is observed as a powdery, sometimes blistered, film on the surface of wine. *Hansenula* will appear under the microscope as medium-sized oval- or oblong-shaped yeast cells. *Pichia* is usually a short rod-shaped yeast, often very tolerant of sulfur dioxide. *Candida* produces long rod-shaped cells with buds taking form at the ends of the mother cells.

Acetic acid spoilage will be indicated by a vinegar odor and a slimy, rubbery film on the surface of the wine. Significant increases in the volatile acidity will be found and stained bacteria cells will be gram negative. (See Appendix B for volatile acidity and staining procedures.) Culture of these organisms can be difficult and usually unnecessary due to the above indicators.

Fig. 3–15 Petri dish culture of *Saccharomyces cerevisiae*.

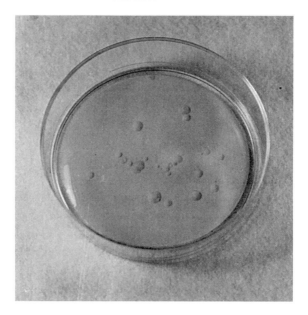

Malolactic fermentation is accompanied by small amounts of gas, decreased total acidity, and increased pH—and sometimes a significant increase in volatile acidity. Paper chromatography will indicate increased lactic acid and decreased malic acid. Large numbers of bacteria will be distributed throughout the wine during fermentation, with these cells settling to the bottom after fermentation is completed. Modified Rogosa agar with 100 ppm of cycloheximide is the most common culturing medium—cultures taking about 7 to 10 days to produce visible colonies.

Control

The single most important factor in maintaining quality in bulk wine is the ability to maintain storage containers absolutely full. Flushing headspaces with inert gases such as carbon dioxide or nitrogen helps for short-term storage but is risky for long-term storage without control by sophisticated gas metering equipment.

Secondary is temperature and chemical preservation. The colder the storage temperature, notwithstanding freezing, the longer it will take for microbial damage to occur. Free sulfur dioxide levels should be made appropriate for the pH of the particular wine. (See Appendix B for SO_2 analysis and molecular intensity.)

When dealing with surface yeast and bacterial spoilage, filter the wine through at least 0.5-micron nominal filters into sanitized containers, adjust SO_2, and bottle as soon as possible. Surface yeast spoilage is particularly difficult to control and often reappears even after filtration.

MICROBIAL SPOILAGE OF BOTTLED WINE

In a properly filled and sealed bottle, acetic acid bacteria are rapidly inhibited. The most common microbial problem in bottled wine is yeast refermentation of wines which have residual sugar. Haziness and profuse gas in the wine after the cork is pulled are characteristic of yeast spoilage. Some yeast can grow slowly in the bottle, producing sticky clumps which settle rapidly when disturbed. *Brettanomyces* yeasts are capable of growing slowly after bottling, rarely producing more than a small fine sediment but causing significant deterioration of wine aroma and flavor. Lactic acid bacteria can grow rapidly in corked bottles, producing hazy wine and small amounts of gas, usually seen as slowly rising fine bubbles coming steadily to the surface after the cork is pulled. Table 3–1 presents a summary of the causes of microbial spoilage.

Detection

In all cases, microscopic examination of the bottle sediment should reveal the organisms responsible for the spoilage. *Brettanomyces* is positively identified by its ability to grow upon agar containing cycloheximide.

Table 3–1 Determination of causes of microbial spoilage.

	Yeast	Bacteria: Acetic acid bacteria	Bacteria: Bacilli	Bacteria: Lactic acid bacteria
Visual appearance	Fine haze or precipitate or film; wine may be gassy	Gray film on surface	Fine haze or precipitate or silky, streaming cloud when shaken; wine may be dull	Fine haze or precipitate or silky, streaming cloud when shaken; wine may be dull
Odor	Not characteristic	Vinegary (acetic acid and ethyl acetate)	—	Wine may be slightly gassy. Sauerkraut or diacetyl character.
Microscopic appearance	Greater than 4–5 mμ	Ellipsoidal or rods (involutionary forms may be present)	Less than 4–5 mμ; Rods.	Less than 4–5 mμ. Rods or cocci (like tangled mass of hair = *Lactobacillus trichodes*).
Growth on				
Basic medium[a]	+	+	+	+
Basic medium + cycloheximide[b]	0 or +[c]	0 or +	0 or +	(Ferments malate to lactate = malolactic bacteria) +
Catalase test[d]	—	+	+	—
Spore formation	0 or +	—	+	—
Calcium carbonate plates[e]	—	Clearing.	No clearing	No clearing.

[a]Basic medium: The basic culture medium contains 2.0 g/100 ml of Bacto tryptone, 0.5 of Bacto peptone, 0.5 of Bacto yeast extract, 0.3 of glucose, 0.2 of lactose, 0.1 of liver extract (Wilson), 0.1 ml of 5% aqueous Tween 80, 100 ml of diluted and filtered tomato juice. To prepare tomato juice, dilute 4-fold with distilled water, filter through Whatman No. 1 paper using a Büchner funnel and Super Cel filter aid. To prepare the basic medium dissolve solid ingredients in diluted tomato juice by heating (avoid scorching by frequent agitation of the flask). When cool, adjust pH to 5.5 with concentrated hydrochloric acid, add 2 g agar/100 ml, and autoclave 15 min at 15 psig.

[b]"Acti-dione": 1 ml containing 10 mg of cycloheximide is added to each 100 ml of medium.

[c]Growth is presumptive evidence of *Brettanomyces* or *Dekkera* yeast.

[d]Basic medium with cycloheximide: Add a drop of 3% of fresh hydrogen peroxide to colonies. If gas is evolved the organism is catalase positive.

[e]Basic medium with cycloheximide minus the glucose and lactose and containing 2% calcium carbonate and 3% ethanol (the ethanol is added after the autoclaving): If acetic bacteria grow they will cause clearing of this cloudy medium around the colonies.

Source of data: Tanner and Vetsch (1956) and personal observations of authors.

Source: From Amerine et al. 1980. Technology of Winemaking, AVI Publishing Company, Westport, Conn.

Control

Free SO_2 levels at bottling should be slightly above the amount required to provide 0.8 molecular sulfur dioxide at the wine's pH.

When residual sugar levels in the wine are higher than 0.2 percent, a level of at least 180–200 parts per million of sorbic acid (1 gram per gallon) is recommended, along with the appropriate level of sulfur dioxide, to inhibit any further growth of yeast. Wines with less than 10% alcohol should be treated with 250 ppm sorbic acid.

As mentioned previously, pasteurization may be used to preserve wine, but it is rarely used because delicate wine flavors are easily damaged during heating.

Sterile filtration without any use of preservatives at bottling is risky but can be effective if the filter and all tubing, containers and closures, as well as the bottling equipment and environment are properly sanitized. When using this method, it is essential that bottled wines are evaluated for sterility.

WINERY SANITATION

Evaluating the effects of winery sanitation procedures requires frequent careful visual and sensory examinations of the entire area. Grape stem and seed particles in the crusher–stemmer and drains, mold residue on walls, tanks, pumps, filters, and hoses, along with particles of cork and paper on the floors are visual signs of an ineffective sanitizing program. Foul odors from drains and from the inside of tanks and hoses is another sure indication that better control is needed.

Careful cleaning and sanitizing of all surfaces exposed to fresh juice, must, or wine is essential. Appendix B provides analytical procedures for control of winery sanitation. Figure 3–16 illustrates a sanitation detection kit.

To sanitize surfaces with heat, temperatures greater than 161°F are necessary for at least 15 min duration. If possible, attempt to reach temperatures up to 200°F. Critical areas are tubing connections and rubber seals or gaskets; these may have crevices which are more difficult to clean and which may be damaged with excessive heat. Chemical sanitizing is effective for surfaces, but less effective than heat.

Sodium hypochlorite, which is simply household bleach, is an extremely toxic agent and will kill all yeasts and bacteria on contact if the proper concentration of active material is available. Hypochlorite solutions are not as effective when soap or excessive dirt residues are present. Caution is advised as hypochlorites are destructive to wood surfaces and generate unpleasant fumes in the atmosphere. Chlorine residues are extremely damaging to wine sensory quality.

NOTE: Hypochlorite solutions must never come in contact with basic (alkaline) compounds, as poisonous chlorine gas may be generated.

Fig. 3–16 Sampler and Swab Test Kit.
Courtesy: Millipore® Corporation, Bedford, MA.

For cellar floors and other production area floors, walls, and work surfaces, as well as in tubing or on other juice contact surfaces, ¼ cup of sodium hypochlorite solution per gallon of water is recommended. Surfaces should be thoroughly rinsed with scalding hot water and then left to dry. Another rinsing with scalding hot water should precede further use in contact with juice or must.

Sulfite and Citric Acid Sanitizing

Sulfurous acid, not to be confused with the highly toxic and corrosive sulfuric acid, is simply sulfur dioxide dissolved in water. In most wineries, this is accomplished by adding potassium meta-bisulfite to water—generally about 2 ounces in 10 gallons. When 4 ounces of citric acid are also added, the resulting mixture is a more compatible sanitizing agent than hypochlorite and can be used more safely for barrels, corks, and filter pads, as well as plastic and stainless steel surfaces. **NOTE: Prolonged exposure of stainless steel surfaces to sulfur dioxide solutions can result in pitting and rusting.**

Dissolving 3.0 grams of potassium meta-bisulfite and 12 grams of citric acid in 1 gallon of water will result in a 500 ppm concentration of sulfurous acid at a pH of about 3.0, which is very effective as a winery sanitizing solution.

CHAPTER 4

ENOLOGY (WINEMAKING)

Sooner or later, vintners will encounter the term *enology*, which is sometimes spelled in the British version as *oenology*. In either case, it is best simplified by its Greek origin of *oinos* (wine) and *logos* (logic). Wine logic is based on the application of analysis, equipment, materials, and technology in designed methodologies to achieve predetermined results. Enology is constructed upon a gastronomic art form—heavily supported by the disciplines of biochemistry, microbiology, and mechanical engineering. A modern-day winemaker is often referred to as an *enologist*.

The U.S. federal regulatory body for wine is the Bureau of Alcohol, Tobacco, and Firearms (ATF) which defines wine as "the product of the juice of sound, ripe grapes." There are also many famous wines made from unripened grapes, and even more made from overripened grapes. More importantly, the ATF definition overlooks any stipulation for the all-important fermentation process necessary to generate "the product."

Wine, as the word used in the wine industry is defined, is "the product of fermenting and processing grape juice or must." Crushed grapes, generally with all or some portion of the stems removed, are known as *must*.

The ATF requires that commercial wines made from any other fruit than grapes must be qualified by fruit type on the label. Wine made from peaches, for example, must be labeled as "peach wine"; wine made from blackberries must be labeled as "blackberry wine"; wine made from plums is labeled as "plum wine"; and so forth. Wine made from any species or variety of grape can be commercially labeled with the word "wine" standing alone.

Wine is the result of fermenting and processing grapes or some other fruit—sometimes even vegetables. For virtually all white wines, the juice is pressed out prior to fermentation. Most reds are made by fermenting crushed grapes (must) before pressing, the skins extracting red pigments into the new wine. Pink wines, often called "blush wines" nowadays (they are also known as "rose" wines), are generally made by very limited skin contact during fermentation, then quickly pressed, so that only a small amount of color is extracted from the grape skins.

There are countless wine "recipes" and grandfathered methods which are treasured by some winemakers. These are encouraged but are not always easily justified by wine logic. The most consistent high-quality wines generally emerge from winemakers who use analysis and logic for processing decision making.

Grapes and other fruits vary in sugar, acidity, and flavor constituency each year, requiring adjustments which have a sound reasoning. Wines from previous vintages may have properties which need fine-tuned adjustments in later processing. The enological approach makes good sense for effective record keeping, too.

GRAPE AND WINE COMPONENTS

With sugar and water having no flavor, note in Table 4–1 that only about 3% of the total wine composition accounts for flavor and color. In other words, the difference between the very finest and poorest of grapes and wines is determined by a small percentage of its constituency.

Table 4–1 indicates a rather straightforward conversion of sugar to alcohol during fermentation. The increase of water percentage during fermentation is due to several biochemical pathways. The Embden–Meyerhof reaction diverts some of the sugars to the production of lactic acid, creating water as a by-product. Yeasts produce enzymes which convert pyruvic acid to carbon dioxide gas and water.

PECTIC ENZYMES

Pectins are generally classified as a group of polyuronides consisting of D-galacturonic acid units in long-chain polymers. The most important form of pectin in winemaking is pectinic acid, which is the operative component which reacts with sugar and fruit acids when heated and which forms the gel consistency as temperature is lowered. It is pectin which is added in fruit processing to make

Table 4–1 Composition of grapes and natural table wine.

Component compound	Approx. % in grapes	Approx. % in wine
Water	75.0	86.0
Sugars (fructose, glucose, with minor levels of sucrose)	22.0	0.3
Alcohols (ethanol, with trace levels of terpenes, glycerol, and higher alcohols)	0.1	11.2
Organic Acids (tartaric, malic, with minor levels of lactic, succinic, oxalic, etc.)	0.9	0.6
Minerals (potassium, calcium, with minor levels of sodium, magnesium, iron, etc.)	0.5	0.5
Phenols (flavonoids such as color pigments, along with nonflavonoids, such as cinnamic acid and vanillin)	0.3	0.3
Nitrogenous compounds (protein, amino acids, humin, amides, ammonia, etc.)	0.2	0.1
Flavor compounds (esters such as ethyl caproate, ethyl butyrate, etc.)	Trace	Trace
TOTAL	99.0	99.0

Source: Adapted from Amerine et al., 1987, The Technology of Wine Making, fourth edition, *Westport CT: AVI,* pp. 111–112.

jelly. Arabans, dextrans, galactans, and other gum-like compounds formed during fermentation are also sometimes classified as pectin materials.

Pectinic acid can be a hurdle in the pathway of optimal clarity and yield. Some natural depectinization occurs during fermentation—generally more in reds than whites due to greater proportions of methyl esterase found in red grapes.

For many white wines, an appropriate treatment with pectic enzyme (pectinase) will serve to break down pectin, increasing free-run juice yield and ease in filtration while reducing the potential for troublesome fining processes. In the case of blush wines, pectinase treatment at the crusher is essential as it permits a more balanced extraction of tannin and pigment phenolics during the brief fermentation period on the skins. In most red grape varieties and cultivars, this provides a better ratio of phenolics, resulting in a more attractive pink hue as compared to dull tawny (brownish orange) color values.

SUGARS AND SWEETNESS

Wine publications often relate to reducing sugar, residual sweetness, invert sugar, dextrose, glucose, fructose, fermentable sugar, nonfermentable sugar, and several other terms in reference to this group of carbohydrates. This terminology and its proper context can sometimes be confusing and, more importantly, lead to mistakes.

Sugars are categorized in biochemistry as *saccharides*. A *disaccharide* such as *sucrose* yields two (hence, the prefix *di*) molecules of simple sugars when hydrolyzed. The broad definition of sugar *hydrolysis* is the decomposition of crystals in water. In the case of enology, sucrose hydrolysis results from decomposing one 12-carbon molecule ($C_{12}H_{22}O_{11}$) into two 6-carbon molecules. These are often referred to as hexoses—expressed in chemical formulary as $C_6H_{12}O_6$. Note that the recombination of two hexoses results in the addition of two atoms of hydrogen and one atom of oxygen, or H_2O, which is, of course, the water taken from hydrolysis. In that this cleavage of sucrose into two hexoses is made by invertase enzymes, the transformation is referred to as *inversion*. Sometimes, we find liquid sugar being sold as "liquid invert," which refers to the product being comprised of water in which 6-carbon sugars are dissolved—usually in a very heavy syrup-like concentration.

Grapevine roots store sugars produced from photosynthesis in the sucrose form. As maturation of the fruit progresses, the sucrose is typically inverted to a 1 : 1 ratio of glucose and fructose as it is translocated to the grape berries following *veraison*. Fully ripened grapes may contain up to 1% of uninverted sucrose, which is inverted later by yeast-synthesized invertase during fermentation.

Glucose and fructose are *monosaccharides;* the "mono" indicating that the 6-carbon sugar molecules cannot be hydrolyzed to a simpler form. These are the sugars of greatest concentration in grapes and are essential substrates (source materials) for the fermentation process in winemaking. The literature often expresses these sugars as D-glucose and D-fructose, which are prefixes referring to the rotation angle variation with respect to the wavelength of light transmitted—dextro (D) the right and levo (L) to the left. Dextrose and D-glucose are identical compounds and are often referred to as corn sugar, grape sugar, honey sugar, and so on; whereas D-fructose is also known as levulose, or "fruit sugar." Both glucose and fructose are chemically symbolized by the same $C_6H_{12}O_6$ structure and, thus, both have a molecular weight of 180.09. The difference is that glucose has a melting point of 146°F and for fructose it is 105°F.

The definition of *reducing sugars* is less straightforward. Wines fermented to "absolute bone dryness" still contain between 0.1% and 0.2% of unfermented sugars. The most common reducing sugars are glucose and fructose, but the precise reasoning why they do not completely convert during fermentation remains unclear. About one-third of the reducing sugars can be attributed to various *poly-*

saccharides, which are complex forms of sugar unfermentable by yeasts but may be substrates in malolactic fermentation.

Many wines are designed, to contain some specific amount of sugar in order for the wine to finish with some particular degree of sweetness. Traditionally, this *residual sweetness* results from natural grape hexoses. In modern times, added cane or beet sugar, both of which are sucrose, are permitted. ATF regulations (27 CFR, Part 4, Section 24.179) allow natural grape wines not more than 17% total solids (21% for natural fruit wines) by weight, provided the alcohol content does not exceed 14% by volume. There are additional provisions for specially sweetened natural wines.

The language of sugar and dissolved solids in wine production is expressed in terms of Brix, Balling, and extract. Ripe grapes should arrive at the winery at about 22° Brix, which is roughly equivalent to percent total solids by weight. Native American *Vitis labrusca* grapes may achieve considerably less than 22° Brix. Late-harvest *Vitis vinifera* and French-American hybrid grapes may often reach more than 35° Brix. Although nearly all dissolved solids in grape juice or must are sugar, there are also some solids attributed to acids, minerals, and phenols.

Balling is an indicator of wine density or weight on the human palate—often referred to in sensory terms as "mouth feel." Increasing dissolved solids have the property of increasing the weight of wine, as they are heavier than water. As fermentation progresses, the increasing alcohol content decreases the weight of wine, as ethyl alcohol is lighter than water. The combination of solids and alcohol taken together is expressed in terms of Balling. A dry wine with virtually no dissolved solids would have a Balling less than 0, which is the Balling of water.

Brix and Balling are terms often misunderstood and thus often misused—perhaps because they can both be measured using a simple hydrometer in analysis (see Appendix B). This is best exemplified by considering that it would be impossible to have a negative Brix.

Extract is virtually the same as Brix except it refers to total dissolved solids in wine—or Balling with the effect of the alcohol removed.

LATE-HARVEST GRAPES

Despite the lack of wide market appeal, vintners respond to the connoisseur niche by offering an increasing number of late-harvest wines in U.S. markets. Some of these wines are made from grapes naturally raisinized by excessive sun exposure in the vineyard. Sluggish fermentation rates in higher Brix musts resulting from late-harvest grapes can be attributed to an increase in water activity, which is an osmotic force taking up water which places a dehydration strain called *zytorrhysis* on yeast cells. An increase from 20° to 50° Brix can reduce yeast cell volume by about 50%. Minimizing Brix levels, and therefore reducing the impact of this phenomenon, has brought closer vintner scrutiny on identifying a rather precise amount of ethanol, residual sweetness, and acid balance desired to meet perceived consumer demand.

Among moderate temperatures and higher humidity, *Botrytis cinerea,* the "noble mold," can appear as fuzzy gray spots on grape skins during the harvest season. Botrytis spores bore through the skins, creating a porosity from which berry waters can evaporate. This serves to concentrate retained sugars, acidity, glycerol, and flavors. The infection process actually reduces some sugar and acidity (increases glycerol and mucic acid), but the net effect is a concentration. Botrytized fruit flavors undergo change due to the formation of ethyl esters of hydroxy-, keto-, and dicarboxylic acids. Yeast enzyme synthesis and activity rates determine how these acids are metabolized into the ultimate ester flavor profile. Some natural fruit flavors, such as the aromatic monoterpenes, geraniol, and linalool, common to the Johannisberg Riesling, Gewürztraminer, and Muscat varieties, may be destroyed by botrytis mold cells.

The mold can also form laccase, an enzyme capable of oxidizing important phenols, including anthocyanins. This accounts for the golden brown colors in most botrytized wines. Botrytis infection can be quantified using a laccase assay available in laboratory kits. Samples are initially treated with polyvinylpolypyrrolidone (PVPP) to reduce polyphenol interference and then mixed in a spectrophotometer cell with a syringaldazine–ethanol solution, along with a buffer. The change in absorbance is recorded over several minutes and calculated by formula to provide laccase activity measured in laccase units per milliliter

The botrytis fungus can reduce up to half of the protein, amino acid, and free ammonium nitrogen (FAN) constituency of grapes. Constrained FAN can induce deamination activity by yeasts on protein and amino acid constituents, causing the development of unpleasant hydrocarbon (kerosene-like) flavors and other maladies of flavor. FAN is a more essential element in active yeast growth, and deficiencies can be supplemented with diammonium phosphate and other food-grade sources.

A more serious problem is a reduction of thiamine and other B-complexes necessary for decarboxylation in the pyruvic acid cycle and the synthesis of various keto acids mentioned earlier. These vitamins can also be supplemented, often as components in concert with FAN in commercially available products. Yet another concern is the growing evidence which indicates that botrytis mold can produce trace amounts of antibiotics that may be toxic to yeasts.

Botrytis infection produces polygalacturonase which hydrolyzes pectins into polysaccharides. Mucic acid development can react with calcium to form slow-developing precipitates. Both are hurdles to effective fining—evidenced by the formation of haziness in the wines after bottling. Proper use of glucose oxidase enzymes followed by appropriate applications of kieselsol fining mentioned below can provide additional help in overcoming problem clarifications.

Moderation of these perils and pitfalls remains best controlled by close monitoring of botrytis development in the vineyard and processing techniques in the cellar.

ACIDS, ACIDITY, AND PH

The principal acids involved in winemaking are tartaric, malic, lactic, and acetic. The sum total of these and other minor acids determine the amount of tartness a wine will deliver on the palate. Although some people can detect hints of apple-like flavor in malic acid and cheese-like flavor in lactic acid, the aggregate flavor contribution by acids in wine is minuscule.

Total acidity, often abbreviated as T.A., is a measurement routinely made by analysis in winery laboratories in order to quantify tartness in juice, must, or wine. Levels less than 0.500 g/100 mL are generally considered bland. T.A. levels exceeding 0.800 g/100 mL are usually ranked as sharp. Sweetness levels tend to mask T.A. and vice versa. It is the relativity of sweetness and T.A. in wines which is the basis for determining gustatory balance in sensory evaluation.

The measurement of the active, or effective, acid "strength" in juice, must, or wine is expressed in terms of pH. The determination of pH is made by measuring the hydrogen ion concentration in a given solution. On a scale of 0 to 14, a pH of 7 would be neutral, as is pure water. From 7 to 0, the hydrogen ion concentration increases and the solution becomes increasingly acidic. From 7 to 14, the hydroxyl ion concentration increases and the solution become increasingly basic. These progressions are not linear, but logarithmic. As a consequence, each gradient from 7 toward either 0 or 14 becomes more intense. For example, a change in pH from 3.40 to 3.30 is a major increase in effective acidity. This, however, is not nearly as large an increase logarithmically as from 3.30 to 3.20.

The determination of pH is particularly important during the harvest season, as it can be used as a standard for grape ripeness. Tartaric acid is found in the highest concentration in grapes and is a relatively "strong" acid—resulting in lower pH ranges. As grapes ripen, the tartaric acid commences to diminish and the pH increases. Sufficient pH data collected from vineyards over several vintage seasons may justify setting standards from which to predetermine optimal harvest time.

Lactic acid is a product of malolactic fermentation—a bacterial fermentation in which malic acid is transformed to lactic acid, diacetyl, and carbon dioxide gas. This is usually considered beneficial to red wines and some whites, as the "buttery" flavor of the diacetyl adds tone and complexity to the overall flavor profile. Inoculations of cultured malolactic bacteria are commonplace in the wine industry.

Acetic acid is the vinegar flavor result of oxidized alcohol by and undesired bacterial fermentation. It almost always results from poor winemaking procedures.

Citric acid is found in grapes only in trace amounts but is an important additive used by winemakers. It is not added to unfermented juice or must, as it can slow fermentation by inhibiting enzymatic functions in the glycolysis pathway. Citric acid is active in metal chelation, aiding in reducing the probability of iron and/or copper haziness. Being a tricarboxylic acid in the Krebs cycle, it also has an overall stabilizing effect. Consequently, it is usually added to wines which are nearing the end of processing—just before bottling.

Analytical procedures for total acidity, volatile acidity, and pH are found in Appendix B.

OXYGEN AND OXIDATION

The literature indicates that, although musts which have not been treated with sulfur dioxide contain high levels of polyphenoloxidase and resultant pigmentation oxidation, the delay of sulfiting ultimately results in wines containing reduced total phenolics and improved sensory quality. Appropriate additions of ascorbic acid can aid in minimizing persistent browning due to oxidation.

Free oxygen is rapidly utilized by cultured species of *Saccharomyces* wine yeasts (facultative anaerobes) at the outset of logarithmic growth—with full depletion of oxygen resulting in anaerobic fermentation. This condition results in the redistribution of certain lipid and sterol compounds essential for yeast membrane construction during cell division and membrane mechanics during the glycolytic functions of fermentation. Some winemakers actually supplement oxygen by agitating white wine fermentations when activity commences to diminish—more often in combination with FAN supplements. Unfortunately, this is a treatment frequently overdone, which results in an excessive level of oxygen.

Commercial strains of wine yeasts fully adapt to continuing fermentation without any available oxygen in the must. This anaerobic condition results in the redistribution of certain lipid and sterol compounds essential to yeast membrane construction during cell division, as well as membrane mechanics during the glycolysis functions taking place during fermentation. Under normal conditions with minimal oxygen present, these redistribution dynamics should cause no alarm or unusual difficulty.

With excessive oxygen present, however, the new wine is in jeopardy of color browning and flavor degradation. The greater a wine's buffering capacity, or its content of oxidizable compounds, the greater its aging potential. Wine aging is, thus, a controlled process of oxidation. For example, a young wine racked (decanted from less sediment) promptly after fermentation to cool stainless steel tanks, then clarified, and coarse filtered, would be much less susceptible to oxidation than a wine allowed to assimilate yeast and fruit pulp compounds by extended lees contact. Wood storage containers contribute nonflavonoid phenols, which are oxidizable. Higher storage temperatures (more than 55°F) and higher pH levels are conducive to higher oxidation rates.

Consequently, wines containing higher levels of yeast autolysis compounds (wine fermented *sur lies,* or treated with significant additions of yeast hulls) and/or containing an appropriate level of free SO_2 are chemically equipped to bind with greater quantities of oxygen. The result is often measured in wines having less browning, with more complexity and structure in the flavor profile. Higher levels of phenols, as found in long-term macerated musts, also influence a higher buffering capacity and account for the major reason why red wines, particularly at lower pH levels, generally take far longer to age to maturity.

Minimizing oxygen exposure includes storage vessels being regularly "topped-off" to replace ullage. Cold-temperature short-term aging in stainless steel tankage, nitrogen sparging, and adequate maintenance of free SO_2 levels are essential for lower oxygen risk. Attention and care in racking, fining, filtering, and other transfer operations can reduce splashing and other mechanical sources of oxygen. Tanks should be sparged with carbon dioxide or nitrogen gas. Close monitoring of temperatures and sulfur dioxide levels are also part of the quality assurance routine.

SULFUR DIOXIDE

The use of sulfur in wine production has several centuries of history; yet it remains one of the most controversial of wine treatments, one of the most misunderstood, and its capabilities oftentimes miscalculated.

The greatest variables exist in definition and communication. The term *sulfur* is the rhombic yellow sulfur solid employed in vineyard sprays, sulfur sticks, and so forth and is the elemental form from which hydrogen sulfide and sulfur dioxide is derived.

Hydrogen sulfide, H_2S, has the odor of "rotten eggs," which often results in wines made from grapes having been dusted with sulfur in the vineyard. There is evidence that excessively high inoculations with an overabundance of yeast cells can also contribute to H_2S generation. Hydrogen sulfide can also arise from sluggish, nitrogen-deficient and/or extended low-temperature fermentations, or any other condition which may inhibit healthy yeast growth. Yet another source of H_2S is from lees substrates, which is a common occurrence in poorly managed *sur lies* regimens.

Sulfur dioxide is a gas having a very prickly, sharp pungency in the nose. Heavy doses of sulfur dioxide gas leave an unpleasant powdery residue on the palate. This gaseous form of SO_2 is generally used by larger wineries, where greater quantities are required. In most smaller wineries, SO_2 is sourced from meta-bisulfite anions. This exists in two general forms: The bisulfite form, or "free SO_2," has a rather "burnt match" odor and is the form which has the desired antimicrobial and antioxidant properties. The "bound SO_2" form includes bisulfites bound to aldehydes, pectic compounds, various phenol complexes, certain proteins, and other molecular sites. Bound sulfur dioxide has comparatively little distinguishable flavor. The free and bound forms combine to make "total SO_2," which is currently regulated by the ATF to a limit of 350 mg/L, or 350 parts per million as found in 27 CFR, Part 4, Section 4.22 (b) (1).

It remains rather commonplace for vintners to add 30–90 mg/L of free sulfur dioxide to grapes dumped in the crusher, or early on in the resulting must. The traditional rationale for sulfite additions at the crush is that this dosage inhibits or kills wild yeasts and bacteria attached to the *bloom* (waxy coating) of grape skins and provides some measure of protection from oxidation. There is growing evidence which indicates that SO_2 additions at the crusher may actually increase browning in resulting white wines.

Although "killer" yeasts are not new to enology, their role in reducing or eliminating SO_2 at the crusher is an application that is relatively recent. Killer yeasts are species that kill sensitive members of their own species and frequently those of other species, as well. They function by secreting a plasmid-coded protein toxin that binds with 1,6-beta-D-glucan receptor components in the cell walls of sensitive strains. This toxin interacts directly with protein components of the cell membrane and, in turn, disrupts the normal state of cell activity. Recent research suggests that killer yeasts are immune due to a precursor protein that functions as an inhibitor of toxin in its cell membrane metabolic processes.

Killer yeasts are particularly effective in reducing infection from *Brettanomyces* yeasts. This spoilage microorganism is most often identified with red wine spoilage in the form of acetic, isobutyric, and isovaleric acids, which emerge as "barnyard" flavors. *Brettanomyces* can resist mid-range dosages of free SO_2, is insensitive to sorbic acid, and may go unnoticed until growth has become widespread. Consequently, musts undosed with SO_2 require immediate inoculation with killer yeast cultures in order to achieve maximal protection. Some enologists inoculate at the crusher hopper—taking advantage of heavy oxygen demand by the yeasts to lower the oxidation potential.

It has been reported that the killer system occurs in some natural yeast strains. Musts inoculated with a sensitive cultured yeast strain can be dominated by wild killer strain populations, causing stuck fermentations. The resulting wine can suffer from reduced ethanol yield, high volatile acidity, formation of H_2S, and contaminant flavors caused by acetaldehyde, fusel oils, and lactic acid.

Vintners continue to evaluate a growing availability of killer yeast strains. Popular strains of killer yeasts are *Champagne 111* and *Montrachet 1107,* the latter more popularly known as Prisse de Mousse. These and other strains of cultured killer yeasts are frequently used by progressive winemakers.

Conversely, there are some vintners who prefer to allow natural yeasts to conduct the fermentation in order to achieve enhanced flavor complexity. This practice has some perils and pitfalls which should be avoided, and particularly so by new vintners.

Another point of support against SO_2 treatment at the crusher is borne upon increased astringent phenols being extracted from the must. On the other hand, some traditional red wine vintners actually prefer this added dimension of phenolic complexity. Further, there is evidence that an absence of sulfur dioxide at the crusher permits yeast-inhibiting bacteria to flourish, which, in turn, contributes to stuck fermentations.

In the case of clean, freshly harvested grapes, it is becoming increasingly common to find the initial dose of SO_2 being made during the first racking. At this time, CO_2 has significantly diminished and SO_2 is needed to achieve an appropriate degree of bacterial and oxidation protection. Little further complexity is usually desired in the case of most white and blush wines. Reds may require considerable malolactic bacterial activity and taming of harsh tannins via oxidation. Consequently, minimal SO_2 treatment is essential.

The proper calculation of SO_2 treatments depends heavily on the pH of the wine being treated. The active portion of free SO_2 is called molecular SO_2 and is much more active at lower pH levels. Wines with a pH of 3.50, for example, require about twice the SO_2 addition to achieve the same preservative effect as wines with a pH of 3.20. For dry table wines at a pH of 3.25, it is typical to add 20–30 mg/L of free SO_2 at the first racking and another addition just before bottling—raising the level up to 30–40 mg/L. On the other hand, some winemakers may have just cause to feel a need for upward of 80 mg/L of free SO_2 at the time of bottling high-pH wines, particularly those having a history embattled with spoilage organisms or other suspect problems. The question always begged for distressed wines is whether or not they should be bottled at all.

Table wines having residual fermentable sugar should be dosed to levels of free SO_2 about 50% greater than for dry wines, in tandem with a sufficient treatment of sorbic acid, with a ATF maximum of 300 mg/L, as provided in 27 CFR, Part 4, Section 24.246. There are plenty of indications in the literature, as well as empirical evidence, that lower free SO_2 levels result in wines which have greater fruit character and less harshness on the palate.

The manner of how potassium meta-sulfite (KMS) additions are made is also not without controversy. One rather archaic method is to dissolve KMS in a small amount of water first. Adding KMS directly to the wine allows the lower pH in the wine to disassociate SO_2 gas, which conserves the amount of bisulfite ions, or free SO_2, lost to evaporation.

PHENOLS, PHENOLICS, AND POLYPHENOLS

As mentioned earlier, table wines are comprised of about 85% water and 12% ethyl alcohol—leaving only 3% or so remaining to account for color, flavor, and body. A major group of compounds found in this small but essential portion of wine are called *phenols*—sometimes referred to as *phenolic* substances and/or *tannins*. Technically, each is a different term. The chemistry of these structures is very complex and diverse, each having important relationships in winemaking.

Phenols refer to the fundamental form of phenic acid, phenylic acid, and oxybenzene. A phenol very common to wine is hydroxybenzene. Phenolics are more complex structures formed from primary phenols, such as in ripening grapes and maturing wines. Tannins are condensed giant phenolic polymers which can be precipitated by proteins such as the mucoproteins on the human palate—accounting for the "puckery" astringent effect when high-tannin wines are tasted.

Yet another family of phenolics are extracted from aging wines in barrels—varied due to the species of oak selected, by the intensity of "toasting" given during construction, and by the ratio of time/temperature employed in the cellar.

The entire family of phenols, phenolics, and tannins can be referred to as *polyphenols*. On average, about 65% of grape polyphenols are found in the seeds; another 22% are in the stems, 12% in the skins, and only 1% in the pulp. The quality and quantity of polyphenols in grapes is derived from species and variety, vineyard locale, and cellar management.

Polyphenols are categorized into two major groups: *flavonoids* and *nonflavonoids*. Flavonoids are large polymer molecules involved with wine color and tannins, and the highly publicized *resveratrol*, which is thought to be effective in reducing low-density cholesterol in the human body. Nonflavonoids are generally smaller molecules which are primarily associated with oak flavor extracts such as cinnamic acid and vanillic acid.

Color

The pigmentation in grape skins, which eventually determines wine color, are phenolic flavonoids called *anthocyanins*. These are a group of five closely related anthocyanidin compounds constructed in combination with sugar residues called *glucosides*. There are monoglucosides having a single sugar residue, and diglucosides with two. Anthocyanidins decompose rather rapidly without a glucoside component. Consequently, the stability of wine color is largely dependent on anthocyanin glucoside structure.

It is principally the monoglucoside form which is associated with wines made from *Vitis vinifera* grapes—generally more stable than the diglucoside form identified with the red muscadine varieties belonging to the *Vitis rotundifolia* species. Varying density of purple hues are typical to the predominance of the monoglucoside anthocyanin *malvidin* in Cabernet Sauvignon, Merlot, and other viniferas. Some *Vitis labrusca* types, such as Concord, also possess a significant level of monoglucoside malvidin, but in concert with the other four monoglucoside anthocyanins, *cyanidin, delphinidin, peonidin*, and *petunidin*. Interspecific hybrid grapes such as the French-American cultivars each have a unique phenolic color profile inherited from their breeding lines.

Higher total acidity and lower pH values found in slightly immature grapes are generally associated with superior color hue and stability. These factors, however, may combine to create a specific color hue and the intensity of that hue is primarily a function of viticultural response. Vines stressed in cooler climates and poorer soils can be expected to yield grape berries of smaller diameter. This leads to the famous "struggling vine" theory which postulates that comparatively smaller berries resulting from challenging vineyard environs provide a greater ratio of skin surface area per unit of resulting wine and, thus, greater intensity of color.

Virtually all of the anthocyanin pigments are extracted from grape skins by the time a properly conducted on-skin fermentation has reached 5.0° Balling. No enhancement of color is gained by prolonged skin contact during maceration.

Color can be enhanced, however, by employing the *saignee* system of must concentration. This involves some portion of the free-run juice being separated from the freshly crushed red grapes and processed as white or blush wine. The remaining must is then either fermented by itself or added to other crushed grapes of the same type to increase the availability for increased pigment extraction. Exacting control over the extent of extraction is necessary in order to avoid generating distorted flavor profiles and saturated concentrations of phenolics, as well as excessive astringency, bitterness, and eventual color precipitates.

A more simple method to achieve red wine color enhancement is by simple blending of a "teinturier," a dense, inky wine made from Salvador, Colobel, and other heavily pigmented grape varieties. The wine extracted from the highest pressure can also be a ready source of supplemental pigmentation. Vintners differ in their approach to this, as some feel that blending strays from the ideals of varietal purity; others point out that many classic reds, such as Bordeaux and Rhones, are blends among different varieties anyway.

New products made possible by membrane separation processing techniques have resulted in the isolation of concentrated pigments in retentate form. Although this may carry similar concerns of varietal purity, it does so on a far smaller scale, as comparatively little pigment is required to achieve favorable results.

NITROGENOUS COMPOUNDS—PROTEINS

Amino acids, peptides, and proteins are all nitrogenous compounds derived from elemental nitrogen—as are amides, amines, and peptones. Although the aggregate of these comprise less than 1% of grape composition, they are, nevertheless, essential in making wine.

Yeasts have no known direct requirement for amino acids. However, fermentation requires amino acids as a catalyst in synthesizing nitrogen into the free ammonium state which is required by yeasts.

There are more than 25 different amino acids in most wines. The precise level of each depends on grape variety, vineyard locale, and climate, among other factors. Amino acids can be condensed into peptides and proteins which creates the common problem of protein instability in white table wines.

Proteins are complex nitrogenous compounds consisting of up to 20 different amino acids bound together by peptide linkages, forming polymer chains which can achieve molecular weights which exceed 150,000. Heavier, less soluble proteins are formed in higher-temperature (over 75°F) fermentation and/or storage. Higher alcohol levels also aggravate protein solubility. The greater levels of phenolic compounds extracted in red table wines also precipitate protein complexes. To the extent that phenolics in reds may be excessive, proteins in the form of egg white albumin, casein, and gelatin are commonly used in finings.

Higher temperatures, higher alcohol levels, and greater concentrations of phenolics are not generally acceptable in delicate white table wine flavor profiles. Consequently, white wine fining regimens must be very carefully determined and carried out in order to achieve clarity at minimal expense to flavor loss. Typically, gelatin/kieselsol, bentonite, or Sparkolloid® is employed for white wine clarification.

MALOLACTIC FERMENTATION

The principal effect of malolactic fermentation is a reduction in total acidity, along with the development of a buttery-like flavor known as diacetyl. As a rule, malolactic fermentation is desired in more complex table wines, but is undesired in lighter types, which express greater fruit flavor profiles. Most cultures of malolactic bacteria are found in the *Leuconostoc* and *Lactobacillus* species, with generally higher levels of diacetyl resulting from *Pediococcus* strains.

Along with pH levels higher than 3.25 and virtually no free sulfur dioxide content, cultured malolactic bacteria typically require supplemental B-complex vitamins for adequate conduct fermentation. Yeast cytoplasm extracts refined from autolysates (dead yeast cells) are a natural source of these substrates and are the rationale on which the prolonged exposure to dead yeast cells following barrel fermentation of white wines (sur lies) is based. Similar substrates are extracted by extended skin contact time (maceration) for reds. Alternatively, the ATF regulations found in 27 CFR, Part 4, Section 24.246 provide for additions of yeast cytoplasm "to facilitate fermentation of juice/wine" in amounts not exceeding 3 lbs. per 1000 gal.

GRAPE FLAVORS

It is imperative that a winemaker be familiar with the nature of wine flavors. The literature often refers to chemical structures of flavor and an increasing amount of casual discussion is borne upon flavor chemistry.

One of the most advanced wine flavor sources is Jean Lenoir who manufactures the *Le Nez du Vin* (The Nose of Wine) aroma identification kits which are highly recommended for the neophyte and even for more advanced enophiles. (See Fig. 4–1.)

In many white grape varieties we find what Lenoir calls *primary floral aromas*. Among these are acacia, grapefruit, honeysuckle, linden, peach, pear, rose, peony, and violets. These flavor values are most often found in grapes which have been harvest slightly before maturity—prior to when primary fruit aromas have formed during full maturation.

Many of the primary floral and fruit aromas exist in the form of higher terpene alcohols such as citronellol, linalol, and geraniol. These are commonly found in Johannisberg Riesling, Gewürztraminer, Vidal Blanc, Vignoles, and most of the Muscat varieties. Other aromas are identified in the form of esters. Among these are the isoamyl acetate associated with bananas, ethyl propionate associate with apples, and the celebrated methyl anthranilate found in the native American *Vitis labrusca* species exemplified by the cultivars Niagara and Concord.

Primary fruit aromas in red grapes are often identified as cherry, black currant or cassis, strawberry, raspberry, and plum, which are often used as flavor descriptives in the evaluation of wines made from Chambourcin, Gamay, and Pinot Noir.

Fig. 4–1 Le Nez du Vin.
Courtesy: David Ferguson.

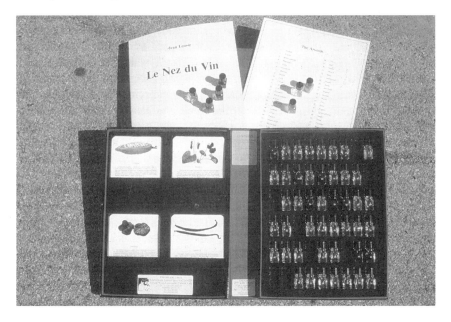

Many of these are found in ester forms. Strawberry flavor is attributed principally to several butyrate compounds; raspberry is identified with ethyl caproate.

Heavy floral and fruit aromas extracted from mature or slightly overripe grapes require particularly close attention in order to avoid oxidation and microbial infection. This involves culling out bad clusters and berries (what the French call *triage*), 50–55°F fermenting in sanitized stainless steel, rackings sparged with nitrogen, absolutely full tankage for storage, minimal finings, cold storage stabilization, and prompt bottling.

Another flavor group is called *primary vegetal aromas*. Good examples of these are anise, green pepper, tobacco leaf, and mint. This group is often structured in the form of carbonyl compounds, such as methoxy–isobutyl–pyrazine which is the principal aroma component associated with bell peppers. These compounds can be very complex, such as the aroma associated with Cabernet Sauvignon, 2-methoxy-3-isobutylpyrazine. Vegetal aromas can also exist as phenols, such as cinnamic acid found in grass and tobacco. This character is most often desired in very heavy red wines such as those traditionally made from Cabernet Sauvignon, Chancellor, and Merlot.

Yet another group, *primary wood aromas,* is exemplified by briar, cedar, hazelnut, resin, oak, and eucalyptus. These generally exist in the form of phenolic compounds such as vanillin which is the aroma component associated with oak. Wood aromas can also be carbonyl compounds such benzaldehyde, the aroma associated with bitter almond.

As mentioned several times earlier, a vinegar flavor in wine is generally associated with the decomposition of ethyl alcohol to acetic acid by various species of spoilage bacteria. Ethyl acetate, described as a "paint thinner" aroma, is often associated with acetic acid formation. Ethyl acetate can, however, metabolize independently from action by certain species of yeast. Wine bacteria and yeasts are discussed in Chapter 1.

Also mentioned several times previously is an earthy character, sometimes described as "barnyard," attributed to fermentation products resulting from the action of *Brettanomyces* yeast. Some wine judges call this condition simply "Brett." The occurrence of Brett in white wines is almost always considered a flaw, although judges often differ in opinion as to whether a little of this flavor contributes positively to the complexity in some reds.

Another set of spoilage flavors found in wine are due to one of the forms of sulfur degradation. These conditions usually arise from elemental sulfur dusted on vines, essential in controlling various types of mildew and molds in vineyards. Sulfur is also an important natural constituent in the synthesis of essential proteins in grapevines. Whether dusted or natural, elemental sulfur on or in grapes can be transformed by both cultured and wild yeasts into hydrogen sulfide, or a "rotten eggs" character, in wine.

Sulfide compounds can be further degraded by bacterial action into one of the mercaptan compounds which emerge in various types of complex compounds. At best, this is manifested in an "asparagus," "cabbage," or "green bean" flavor. As may be expected, this can sometimes be confused with primary vegetal flavors. At worst, mercaptan spoilage is expressed as a "rubbery" aroma, or a foul "wet dog" smell, or, worst of all, a "skunky" character.

Moldy–musty flavors arise from either moldy grapes, moldy barrels, or some other exposure of the wine to a source of mold. Molds will not grow directly in wine due to its alcohol content.

Flavors that seem "cooked" or "brown apple" in character are often due to oxidation of wine resulting from extended exposure to air and/or higher temperatures (over 65°F) in processing and storage. These flavors can also be categorized with "prune" flavors that come from grapes that are overripened and/or have decayed. Some wines, such as Madeira, are purposely heavily oxidized to produce high levels of aldehyde compounds that result in a pronounced "nutty–caramel" flavor.

WINE BLENDING

History indicates that the blending of wines has gone on since ancient times. At first, it may have been more wine "mixing" than blending—wine mixed with sea water, honey, tree resins, and other additions to dilute spoiled wine so it did not taste quite so bad. The salt in the water was also an aid in preservation.

In one way or another, most of the great wines in the world are the product of blending. Even small vineyards in Burgundy are blended from barrique to barrique each vintage. Large Bordeaux estates blend vineyard plot by vineyard plot and free-run with press wine in order to achieve the perfect chateau style each year. In Champagne, blending is made on a large scale with wines from various vineyards from different villages, and sometimes across several vintages, in *assemblage* of the "house" cuvée.

There is no magic to be found in blending. Successful blending, in its purest form, is synergistic—the result is greater than the simple sum of its individual components. Consider the following blending postulates:

1. Blending reduces the character of individual components.
2. Blending increases the complexity of the resulting product.
3. Blending a faulted wine with a sound wine results in a faulted wine.
4. Blending two stable wines can result in an unstable wine.
5. **Always** make a lab blend first.

Blend Expression Rationale

It is the creative ability of each individual winemaker which ultimately determines blending success. The form and character of each wine expresses the manner by which that product has been conceived to parallel market demand. There are, of course, blends which are "monuments" to various egoist expressions. Often, these motives are not shared by consumers and, consequently, such wines may find difficulty in the marketplace.

There are some fundamental blending parameters which seem to find general acceptance by wine enthusiasts:

Dry Whites	Greenish straw hues
	Lighter color densities
	Lighter body
	Crisp acidity
	Moderate fruit flavor
	Simplicity
Sweet Whites	Golden amber hues
	Heavier color densities
	Moderate–heavy body
	Sweetness balanced with acidity
	Complexity
Blush	Pink (not orange) hues
	Lighter color densities
	Lighter body
	Delicate acidity and sweetness
	Light fruit flavor
	Simplicity
Dry Reds	Crimson–tawny hues
	Color density correlated with body
	Tannins correlated with aging design
	Moderate acidity
	Broad complexity
Sweet Reds	Scarlet–purple hues
	Dense color intensity
	Heavy body
	Sweetness balanced with acidity
	Heavy fruit flavor
	Simplicity

Key ATF Blending Regulations

Of major concern is that adequate supplemental records exist in the winery such as to permit ATF inspectors to readily account for the varieties, geographic ori-

Enology (Winemaking) **113**

FORM APPROVED: OMB NO. 1512-0059 (2/29/91)

DEPARTMENT OF THE TREASURY – BUREAU OF ALCOHOL, TOBACCO AND FIREARMS

FORMULA AND PROCESS FOR WINE

(See Instructions and Conditions on Reverse)

1. NAME, ADDRESS AND WINERY NUMBER OF PROPRIETOR

WONDERFUL WORLD WINERY
1234 PARADISE WAY
PLEASANTVILLE, CA
99999
U.S. BONDED WINERY NO. 9999

2. CLASS AND TYPE OF PRODUCT

OTHER THAN STANDARD CRANBERRY FLAVORED WINE

3. ALCOHOL CONTENT OF FINISHED PRODUCT (range may be shown)

11 – 13 PERCENT BY VOLUME

4. FORMULAS SUPERSEDED:

WINERY NUMBER	FORMULA NUMBER

5. FORMULA (List ingredients quantitatively. If more space is needed, use reverse or separate sheet)

DRY WHITE WINE UNDER 14 PERCENT ALCOHOL BY VOLUME	80-90 GALS
DRY RED WINE UNDER 14 PERCENT ALCOHOL BY VOLUME	10-20 GALS
CANE OR BEET SUGAR	20-30 LBS
XYZ CORP. CRANBERRY NATURAL FRUIT ESSENCE NO. AB-12	1-2 GALS
OR	
ABC CORP. CRANBERRY NATURAL FRUIT ESSENCE NO. YZ-22	1-2 GALS
POTASSIUM META-BISULFITE	1-3 OZ

6. PROCESS OF PRODUCTION (If more space is needed, use reverse or separate sheet)

THE FOLLOWING FORMULA IS SUBMITTED IN ACCORDANCE WITH 27 CFR, SUBPART J, SECTION 24.218:

TO EACH 100-GALLON BLEND OF BASE WINE BLENDED IN MIXING TANK NO. 1 OR 2, THE SUGAR AND ESSENCES WILL BE ADDED AND THOROUGHLY MIXED.

AFTER ANALYSIS, POTASSIUM META-BISULFITE WILL BE ADDED AS A PRESERVATIVE AND THE RESULTING WINE WILL BE ANALYZED AGAIN, THE WINE WILL THEN BE FILTERED AND BOTTLED.

7. DATE	8. PROPRIETOR	8a. BY (Signature and Title)
JAN 24, 1997	BETTY B. GOOD AND DAVID B. GOOD	*David B. Good*
		DAVID B. GOOD, WINEMAKER

FOR ATF USE ONLY (Items 9, 10 and 11)

9.

☐ Approved subject to the provisions of 27 CFR _____ .

☐ Natural and Artificial flavor, predominently natural for ATF labeling (Contains not more than 0.1 percent top note material.)

☐ Contains more than 0.392 gram CO_2 per 100 ml of wine.

LABELING (finished product only)

☐ The designation of the product must include a truthful and adequate statement of composition such as:

_____

☐ The finished product label must indicate the use of:

☐ caramel ☐ certified color (other than FD&C Yellow #5)

☐ artificial flavor ☐ FD&C Yellow #5

☐ _____

10. DATE	11. APPROVED – ATF SPECIALIST FOR THE DIRECTOR, BUREAU OF ALCOHOL, TOBACCO AND FIREARMS	12. FORMULA NUMBER

ATF F 5120.29 (6-90) REPLACES 698 - Supplemental which is obsolete.

Fig. 4–2 Illustration of completed ATF F 5120.29.

Enology (Winemaking)

5. FORMULA (Continued)

6. PROCESS OF PRODUCTION (Continued)

INSTRUCTIONS

General Instructions

The proprietor shall submit a separate ATF Form 5120.29 for each formula covering the production of special natural wine, agricultural wine, effervescent wine in which more than 10 percent of the volume is finishing dosage, essences and extracts produced on bonded wine premises, and other (than standard) wine except for distilling material and vinegar stock. The proprietor shall forward the completed Form 5120.29, in triplicate (Quintuplicate for applicable wine products produced in Puerto Rico), to the Product Compliance Branch, Bureau of Alcohol, Tobacco and Firearms, Washington, D.C., 20226. Production may commence upon the receipt by the proprietor of an approved formula on Form 5120.29.

Specific Instructions

Item 1. Enter the name, complete address and winery number of the applicant.

Item 2. The class and type shall conform to one of the class and type designations in the regulations issued under the Federal Alcohol Administration Act. Wine products containing less than 7 percent alcohol by volume shall be designated in accordance with the labeling regulations of the Federal Food, Drug and Cosmetic Act. Do not show a brand or fanciful name in item 2. Nonbeverage wine will be designated as such, since products are labeled under the Federal Food, Drug and Cosmetic Act.

Item 3. If the product is to be bottled at more than one alcohol content within the same tax rate, state the alcoholic content as a range to include all alcoholic contents at which the product will be bottled.

Item 4. Enter the winery number and formula number for each formula which is superseded.

Item 5.

(a) Specify the kind and quantity of each and every material or ingredient to be used in the formulation of a batch of the product, e.g., 100 gallons, 1000 gallons, etc.

(b) Identify flavoring or blending materials by the name of the flavor or blender, name of the flavor or blender manufacturer, the manufacturer's product number (if none, so indicate), drawback formula number (if none, so indicate), date of approval of the nonbeverage formula, alcohol content of the flavor or blender (if nonalcoholic, so indicate), a description of any coloring material contained in the flavor or blender.

Item 6. Show in sequence each step employed in producing the product including the step at which the specified materials will be added and the approximate period of time to complete production.

Item 12. Number the formulas in sequence, starting with number "1." Nonbeverage formulas shall be prefaced with the symbol "NB."

Conditions for Granting Approval

This approval is granted under 27 CFR Part 24 and does not in any way provide an exemption from or a waiver of the provisions of the U.S. Food and Drug Administration regulations relating to the use of food and color additives in food products.

PAPERWORK REDUCTION ACT NOTICE

This request is in accordance with the Paperwork Reduction Act of 1980. The information collection is used by ATF to determine that wine is processed in an authorized manner and meets one of the standards of identify for wine. The information is required to obtain a benefit.

The estimated average burden associated with this collection of information is 2 hours per respondent or recordkeeper, depending on individual circumstances. Comments concerning the accuracy of this burden estimate and suggestions for reducing this burden should be directed to Reports Management Officer, Information Programs Branch, Room 7011, Bureau of Alcohol, Tobacco and Firearms, 1200 Pennsylvania Avenue, N.W., Washington D.C., 20226, and the Office of Management and Budget, Paperwork Reduction Project (1512-0059), Washington D.C., 20503.

ATF F 5120.29 (6-90)

Fig. 4–2 (*continued*)

gins, vintages, quantities, and other appropriate information relating to each blend made.

Blending grape wine with wines made from any other fruit or vegetable, and/or flavoring components, requires the application of Form ATF F 5120.29, FORMULA AND PROCESS FOR WINE, with ATF for approval in order to make such *special natural wines.* (See Fig. 4–2.) Rules are found in 27 CFR, Part 4, Sections 24.195–198.

Key ATF Blending Regulations for Labeling

27 CFR, Part 4, Section 4.23a *Grape type designations.*

(a) *General.* The names of one or more grape varieties may be used as the type designation of a grape wine only if the wine is also labeled with an appellation of origin, as defined in Section 4.25a.

(b) *One variety.* The name of a single grape variety may be used as the type designation if not less than 75 percent of the wine is derived from grapes of that variety, the entire 75 percent of which was grown in the labeled appellation of origin area.

(c) *Exceptions.*
  (1) Wine made from any *Vitis labrusca* [native American] variety (exclusive of hybrids with Labrusca parentage) may be labeled with the variety name if:
     (i) Not less than 51 percent of the wine is derived from grapes of that variety;
     (ii) the statement "contains not less than 51 percent (name of variety)" is shown on the brand label, back label or a separate strip label (except that this statement need not appear if 75 percent or more of the wine is derived from grapes of the named variety);
     (iii) the entire qualifying percentage of the named variety was grown in the labeled appellation of origin area.

  Exceptions.
  (2) Wine made from any variety of any species found by the Director upon appropriate application to be too strongly flavored at 75 percent minimum varietal content may be labeled with the varietal name if:
     (i) Not less than 51 percent of the wine is derived from grapes of that variety;
     (ii) the statement "contains not less than 51 percent (name of variety)" is shown on the brand label, back label or a separate strip label (except that this statement need not appear if 75 percent or more of the wine is derived from grapes of the named variety);
     (iii) the entire qualifying percentage of the named variety was grown in the labeled appellation of origin area.

(d) *Two or three varieties.* The names of two or three grape varieties may be used as the type designation if:
  (1) All of the grapes used to make the wine are of the labeled varieties;
  (2) the percentage of the wine derived from each variety is shown on the label (with a tolerance of plus or minus 2 percent);

(3) if labeled with a multicounty appellation, the percentage of the wine derived from each variety from each county is shown on the label; and
(4) if labeled with a multistate appellation, the percentage of the wine derived from each variety from each state is shown on the label.

27 CFR, Part 4, Section 4.25a *Appellations of origin.*

(a) *Definition.*
   (1) *American Wine.* An American appellation of origin is:
      (i) The United States;
      (ii) a State;
      (iii) two or no more than three States which are all contiguous;
      (iv) a county (which must be identified with the word "county", in the same size of type, and in letters as conspicuous as the name of the country);
      (v) two or no more than three counties in the same State; of
      (vi) a viticultural area (as defined in paragraph (e) of this section).

(b) *Qualification.*
   (1) *American wine.* An American wine is entitled to an appellation of origin other than a multicounty or multistate appellation or a viticultural area, if:
      (i) At least 75 percent of the wine is derived from fruit or agricultural products grown in the appellation area indicated;
      (ii) it has been fully finished (except for cellar treatment pursuant to Section 4.22(c), and blending which does not result in an alternation of class or type under Section 4.22(b)) in the United States, if labeled "American"; or, if labeled with a state appellation, within the labeled State or an adjacent State; or if labeled with a county appellation, within the state in which the labeled county is located; and
      (iii) it conforms to the laws and regulations of the named appellation area governing the composition, method of manufacture, and designation of wines made in such place.

(c) *Multicounty appellations.* An appellation of origin comprising two or no more than three counties in the same State may be used if all of the grapes were grown in the counties indicated, and the percentage of the wine derived from grapes grown in each county is shown on the label, with a tolerance of plus or minus 2 percent.

(d) *Multistate appellation.* An appellation of origin comprising two or no more than three States which are all contiguous may be used, if:
   (1) All of the grapes were grown in the States indicated, and the percentage of the wine derived from grapes grown in each State is shown on the label, with a tolerance of plus or minus 2 percent;
   (2) it has been fully finished (except for cellar treatment pursuant to Section 4.22(c), and blending which does not result in an alteration of class or type under Section 4.22 (b)) in one of the labeled appellation States;

(3) it conforms to the laws and regulations governing the composition, method of manufacture, and designation of wines in all of the States listed in the appellation.

(e) *Viticultural area*
   (1) *Definition.*
      (i) *American wine.* A delimited grape growing region distinguishable by geographical features, the boundaries of which have been recognized and defined in Part 9 [in this chapter of the ATF wine regulations]

27 CFR, Part 4, Section 4.26   *Estate bottled.*

(a) *Conditions for use.* The term "Estate bottled" may be used by a bottling winery on a wine label only if the wine is labeled with a viticultural area appellation of origin and the bottling winery:
   (1) Is located in the labeled viticultural area;
   (2) grew all of the grapes used to make the wine on land owned or controlled by the winery within the boundaries of the labeled viticultural area;
   (3) crushed the grapes, fermented the resulting must, and finished, aged, and bottled wine in a continuous process (the wine at no time having left the premises of the bottling winery).

(b) *Special rule for cooperatives.* Grapes grown by members of a cooperative bottling winery are considered grown by the bottling winery.

(c) *Definition of "Controlled".* For purposes of the section, "Controlled by" refers to property on which the bottling winery; has the legal right to perform, and does perform, all of the acts common to viticulture under the terms of a lease or similar agreement of at least 3 years duration.

27 CFR, Part 4, Section 4.27   *Vintage wine.*

(a) *General.* Vintage wine is wine labeled with the year of harvest of the grapes and made in accordance with the standards prescribed in classes 1, 2, or 3 of Section 4.21. At least 95 percent of the wine must have been derived from grapes harvested in the labeled calendar year, and the wine must be labeled with an appellation of origin other than a country (which does not qualify for vintage labeling). The appellation shall be shown in direct conjunction with the designation required by Section 4.32 (a)(2), in lettering substantially as conspicuous as that designation. In no event may the quantity of wine removed from the producing winery, under labels bearing a vintage date, exceed the volume of vintage wine produced in that winery during the year indicated by the vintage date.

Meritage Blending

An association of California vintners, following a nationwide contest in search for an appropriate term for rather Bordeaux-like whites and reds, established Meritage as an approved ATF label designation in 1989 (see Fig. 4–3).

MERRYVALE

MERITAGE
NAPA VALLEY
72% SAUVIGNON BLANC 28% SEMILLON

PRODUCED AND BOTTLED BY MERRYVALE VINEYARDS,
ST. HELENA, CALIFORNIA, USA, ALCOHOL 13.5% BY VOLUME

Fig. 4–3 Example of commercial Meritage label.
Courtesy: David Ferguson.

The adoption of the category came with a set of rules which, loosely summarized, are as follows:

1. Membership in the Meritage Association is required in order to use the word Meritage on wine labels.
2. Meritage wines must be either white or red (no blush wines) and only grape varieties approved for classified Bordeaux commercial winegrowing may be used.
3. Wines labeled as Meritage must be the vintner's finest in its type—often measured as the most expensive in the line.
4. Meritage wines cannot also be varietal labeled, although appropriate variety percentages can be listed subordinately.
5. A limit of 25,000 cases of production is permitted for each type from each vintage.

The track record of Meritage wines indicates that a market niche has been created and is being expanded.

Proprietary Blending

Whereas the Meritage rules above disallow the use of any grape varieties other than of classic Bordeaux origin, vintners can and are encouraged to pursue the feasibility of producing and marketing *proprietary* wines. These are products which are identified by a name which, after ATF approval, becomes the exclusive property of the vintner.

Blending Cost Rationale

The typical need for blending is to provide an element missing in a base wine. Even more typically, the unit cost of a blend component may be much more than the unit cost of the base wine. Consider the following simple calculation:

Color deficient Chambourcin	950 gal @	$5/gal =	$4750
Dark red blending wine	50 gal @	$15/gal =	750
Total:	1000 gal		$5500

The first notion of paying $15 or more per gallon for blending wine can bring trepidation to the foundations of good judgment. A closer look reveals that the blend has increased the unit cost from $5.00 to only $5.50 per gallon—or just 10 cents per bottle. Of great virtue in this exercise is that the resulting wine may be much more attractive to consumers—perhaps a dollar, or two, or more per bottle. In this context, the 10 cent per bottle blending investment seems very shrewd.

This type of rationale can also be exemplified with lackluster white and blush wines which can be greatly enhanced with a comparatively small percentage blend addition of fruity–spicy Muscat Canelli or Muscat de Frontignan.

FINING

Young white wines are typically clouded with suspended proteins which require clarification. The extraction of phenolic compounds from grape skins usually neutralizes the proteins in red wines—leaving them clear but harshly astringent.

Consequently, the clarification and softening processes in commercial winemaking are known as *fining*. It is chemically complex, involving molecular interaction, bond formation, and both absorption and adsorption.

The materials and techniques for fining are discussed in Chapter 8.

DETARTRATION

Most vintners achieve tartrate stability in new wines by the proper use of fining agents together with the immediate precipitation of potassium bitartrate crystals in cold storage.

The precipitation of unstable potassium bitartrate (KHT), or "cream of tartar," is performed best in controlled atmosphere of constant temperature—usually for at least 4 weeks in a temperature range of 25–28°F. This time can be cut in half by "seeding" KHT unstable wines with specially prepared tartar crystals which act as nuclei focal points on which new crystals are more readily formed.

Wines which have high calcium content (over 50 ppm) can be frustratingly difficult to stabilize. Even with months in a cold-room storage, precipitates can form later in bottled wines. It is for this reason that calcium carbonate treatments for reducing juice or wine acidity are avoided.

An alternative to this procedure is an *ion-exchange* process in which unstable potassium ions are exchanged for stable sodium ions on tartrate crystals—much the same as what takes place in common water softeners. Major hurdles with ion exchanging wine include a comparatively high cost for the equipment and resins required, quality control of metal content through expensive atomic absorption analyses, and the creation of a product which does not align well with low sodium diets.

BARREL AGING*

Winemakers can find an abundance of literature relating to the composition of oak extractives and their effect on flavor of oak-aged wines. More than 200 volatile components of oak wood have been identified. Components of an oak barrel are illustrated in Figure 4.4.

Probably the most thoroughly studied group of volatile oak extractives is derived from oak lignin. These are complex organic compounds comprised mainly of higher alcohols and make up a significant share of wood weight. After the oak timber is harvested, lignin can be broken down by hydrolytic and microbiological reactions. Another major group of volatile oak wood components exist as a result of thermal degradation of the oak polysaccharides. These sugar-based compounds, such as maltol, which can impart a sweet toasty flavor, comprise more than 50% of dry wood weight. The third major group of oak wood components are lactones which impart coconut-like flavors but, when exposed to organic solvents found in wine, express the caramel–vanilla character which is desirable in wine—to one degree or another.

There are other flavors which are also attributed to oak barrel aging:

Bitter almond	Leather
Caramel	Nutty
Cedar	Pencil shavings
Cigar box	Sap
Clove	Smoke
Coffee	Tar
Dusty	Toast

*Adapted from *Barrels and Oak Aging of Wine,* by Lincoln Henderson, Bluegrass Cooperage Company, Louisville, KY.

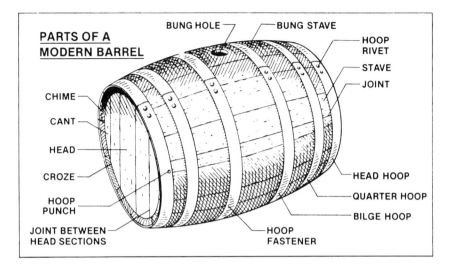

Fig. 4-4 Components of a wine barrel.
Courtesy: Sebastiani Vineyards

Barrel fermentation has become a popular technique employed by winemakers over the past decade. The general notion is that bitter phenolic compounds existing in the oak react with fermenting juice proteins, resulting in more oak flavor extraction and less astringent tannin extraction. The result is wines with slightly less fruit intensity and increased levels of higher alcohols.

Lees contact, or *sur lies,* also has a significant effect on the complexity of barrel-fermented white wine. Greater frequency and duration of the *batonnage,* or lees stirring, exposes the wine to greater influence of enzymes created by yeast autolysis and the flavors generated from yeast protoplasm. This is a treatment which requires daily surveillance to guard against the early development of acetic acid (vinegar), dirty-mustiness (wild yeasts), ethyl acetate (paint thinner), mercaptans (garlic–skunky), sulfides (rotten eggs, manure), and oxidative color changes. Lees provide a wealth of substrates for the growth of spoilage microorganisms.

Cost Implications

French Oak barrels costing upwards of $600 each removes any question about the serious cost involved with barrel aging.

There are essential elements in aging/storage equations, such as container cost or lease rate, maintenance costs, depreciation schedules, salvage value, capacity, and the measure of space/time requirements needed to achieve desired results in improved wine quality.

Let's consider 2000 gal of wine capacity in two different manners:

- a single 2000-gal stainless steel tank coded as SSN
- forty 50-gal French Oak barrels coded as FOB

The SSN with single wall (no jacket), a manway door, and three valves (outlet, racking, and vent), on short legs, should cost about $4000. (See Fig. 4–5 for the com-

Fig. 4–5 Components of a wine tank.

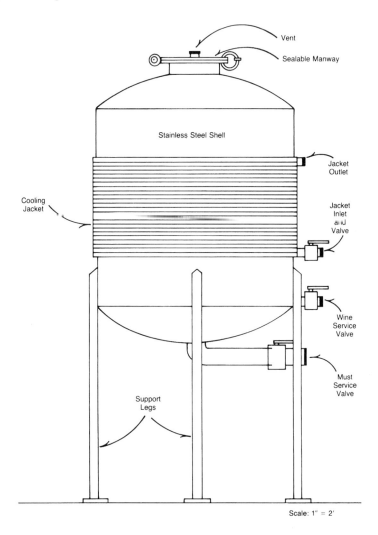

Scale: 1" = 2'

ponents of a wine tank.) This is a ballpark price—stainless steel is rather commodity-like and prices fluctuate. It is typical to find vintners depreciating such tanks over 5 years, although accountants may advise otherwise for given operational strategies.

It must be remembered that this investment of $4000 over 5 years also has a cost, called "opportunity cost" by agricultural economists. If the $4000 was borrowed, the opportunity cost is in the form of paid interest. If the $4000 was taken from owned resources, then the opportunity cost is in the form of lost interest. Opportunity cost calculated at 10% compounded over 5 years adds $1274 to the price, or a total outlay of $5274 for the tank. At 5 years of age, a salvage value of about $2000 could be expected if the tank remains in good condition. Subtracting this from the $5274 total outlay results in an actual cost of $3274. Dividing this figure by 2000 gal results in an actual cost of $1.64/gal.

Use the same rationale to calculate the cost of French Oak barrels. Typically these vary from 50 to 60 gal in capacity, but 50 gal will be used for this exercise.

Forty new 50-gal FOB will cost about $600 each, or a total of $24,000 for the 2000 gal of total capacity. Opportunity cost at the same rates over the same 5-year period amounts to $7644, or a total outlay of $31,644. Retiring barrels after 5 years leaves a common salvage value of about $150 apiece, or $6000 for the entire lot. Subtracting $6000 from the $31,644 results in an actual cost of $25,644, or $12.82/gal.

The overall cost/value table can now be started:

	$/gal capacity
SSN	1.64
FOB	12.82

Tanks and barrels take up cellar space which must be assigned a "rent" cost. Rent should be calculated as the total cost per square foot or per cubic foot of cellar space per year. Figure total fixed costs and variable costs appropriate to the cellar in which tanks and barrels are to be housed, and then divide by the total footage. For this example, a rent of $6.00/ft^2 is assigned. The SSN will occupy about 60 ft^2, which will cost $360 per year, and $1800 for 5 years. Divided out, this becomes a per-gallon cost of $0.90. The 40 FOB, stacked in tiers three barrels high, will occupy about 120 ft^2 and thus cost $720 per year—totaling $3600 for 5 years. Divided out, this becomes $1.80/gal.

	$/gal capacity	$/gal rent
SSN	1.64	0.90
FOB	12.82	1.80

It takes a lot more time to clean and sanitize FOB than SSN. On the other hand, most vintners age their wine in barrels for a longer period of time. Estimate a need to clean and sanitize a SSN about four times per year. Calculating 1 man-hour at $8/h per treatment results in a labor cost of $32 per year, or $160 for 5 years—$0.08/gal.

By the time barrels are received, cleaned, treated, and placed in service, cleaning and sanitizing treatments require about 1 man-hour per barrel per year. At the same $8/h wage rate, this results in an annual barrel cleaning cost of $320, or $1600 for 5 years—$0.80/gal.

The table is now expanded with these maintenance costs:

	$/gal capacity	$/gal rent	$/gal maintenance	$/gal total
SSN	1.64	0.90	0.08	2.62
FOB	12.82	1.80	0.80	15.42

The common reaction to French Oak barrels costing more than fivefold that of stainless steel tanks is negative—that barrel aging is just too expensive. Obviously, the next step is to look at the ratio of cost/benefit achieved with this treatment.

Wines made and marketed in the "squeeky clean" manner often associated with stainless steel aging can satisfy a market demand which is limited. Most blush wines are made with no barrel aging and many fruitier whites and reds, particularly *nouveau* types and all wines made from native American *Vitis labrusca* varieties, are also the better without wood complexity.

A common usage of new FOB is for white wines such as Chardonnay or Seyval Blanc for the first year or two—and then reds such as Cabernet Sauvignon or Chambourcin for the next 3 years. The benefit of FOB is needed in a measure of cost so it can be compared with the treatment cost determined above.

Most of the popular "fighting varietal" (Chardonnay and Cabernet Sauvignon) wines are priced in the $4 to $6 range. As a rule, these are good everyday wines made with little, if any, benefit of barrel aging. This is a fiercely competitive market. With grapes valued at $1000 per ton, juice price alone is about $1.15 per 750-mL bottle. Common cellar processes add another $0.60 to $0.80 per bottle, and packaging increases the cost by another $1 or so. It is difficult to find less than $3 per bottle in total cost before marketing. Excise taxes, shipping, and wholesale markup result in yet another $1 per bottle. Retail markup can easily bring the shelf price to more than $5.

Consider the $2.62 total 5-year per gallon cost figure for SSN calculated above. A typical fighting varietal red wine process/aging regime would be 2 years, or 40% of this $2.62, or $1.50/gal. Dividing that by 5, as there are approximately five bottles in a gallon, we find $0.21 of cost attributed to SSN containment.

Using the same formula the $15.42 total 5-year per gallon cost figure for FOB over 2 years is $6.16. Dividing that by 5 results in a FOB aging cost of $1.23 per bottle. Subtracting $0.21 for the SSN cost reveals that 2 years of FOB aging costs about $1.02 per bottle more than the same 2 years in SSN.

Can this added cost burden be justified in terms of the final product being more attractive in the marketplace? An initial response to this is that the barrel-aged product will need a packaging treatment which attracts attention to its barrel-aging hierarchy—perhaps punted bottles, embossed labels, longer corks, tin alloy capsules, and so forth. Combined, this can easily add another $0.50 cost per bottle and, added to the 2 years of FOB aging cost of $1.02, the total added production cost becomes $1.52 for each bottle of barrel-aged wine.

The fighting varietal production cost estimated at $4 per bottle above included SSN tankage cost. For the same wine to be aged in FOB, the cost would advance by $1.52—for a total cost of $5.52 per bottle.

A secondary response to market acceptance addresses profitability. It is common to assign a 30% margin over cost for the vintner's profit. Thus,

SSN @ $4.00/bottle × 1.30 profit = $5.20 winery price

FOB @ $5.52/bottle × 1.30 profit = $7.18 winery price

The addition of a wholesale channel cost at 25%, which includes Federal Excise Tax, freight, and wholesale margin, is as follows:

SSN @ $5.20/bottle × 1.25 wholesale = $6.50 to retailers

FOB @ $7.18/bottle × 1.25 wholesale = $8.98 to retailers

Finally, the addition of a retail markup of 35% results in the following consumer shelf price comparison:

SSN @ $6.50/bottle × 1.35 retail = $8.78

FOB @ $8.98/bottle × 1.35 retail = $12.12

The question begged by the $8.78 price is how the vintners can sell such wines at $4 to $6 and still make a profit? Virtually all of the answer is found in the size and scale of larger operations—where hundreds of thousands of production gallons and rapid shelf movement can dilute costs to just fractions of what smaller vintners must pay.

Do consumers feel that French Oak barrel aging makes a wine worth $3.34 more per bottle than a fighting varietal wine made in stainless steel? The many thousands of French Oak barrels which exist in the cellars of successful vintners around the United States and the entire world of commercial winemaking attest to a positive response to that question. It is impossible for most small-scale vintners to compete with the large fighting varietal vintners, even when their wines are marketed exclusively at the winery. Consequently, the role of modest-sized vintners has become largely a "boutique" operation—high-quality, barrel-aged wines which target classified European wines as direct competitors.

American Oak Versus French Oak

There are numerous studies comparing various species of oak, such as the French and American oak barrels illustrated in Figure 4.6, and many more controversial debates on the topic. Prior to the 1970s, the "toasting" of oak barrels, which is a charring treatment, was employed throughout the Bourbon whiskey industry, but was virtually unknown in wine production. Barrel aging for wines was made on bare oak surfaces treated only with one or another chemical soakings and rinsings.

Today, with toasting regimes ranging for very light to very dark, there is a growing school of thought that the level of toast given an oak barrel has more flavor influence than the species of oak itself. This has, as might be expected, brought even more research and debate to the topic.

Wood Species and Generally Accepted Utility

Allier	Dry whites	Light–medium toast
	Dry reds	Medium–heavy toast
American	Dry heavy-bodied reds	Medium–heavy toast
Limousin	Dry heavy-bodied reds	Medium–heavy toast
Nevers	Dry whites	Light–medium toast
	Dry reds	Medium–heavy toast
Troncais	Dry whites	Light–medium toast
	Dry light-bodied reds	Medium–heavy toast
Vosges	Dry delicate whites	Light–medium toast

Winemakers are encouraged to study and gather a database on oak aging in order to best fit the wine style desired. With new American Oak barrels sold at about one-third the cost of new French Oak barrels, the cost implications are obviously profound.

Conditioning*

Water quality is perhaps the single most important factor in maintaining sound cooperage. If using heavily chlorinated water, a charcoal filter is strongly recom-

---

*Adapted from *Barrel Care and Maintenance Manual,* Barrel Builders Inc. St. Helena, CA.

Fig. 4–6 French Oak barrels racked on left row; American Oak barrels stacked on right row.
Courtesy: Dry Creek Vineyard.

mended. If using well water or water from other sources which is high in iron and elemental sulfur, a potassium permanganate filter or other ion-exchange method which will remove these minerals is required. Residual iron can cause iron haze "casse" in wine—and residual sulfur can be a substrate for the formation of sulfides.

Wooden wine vessels are porous and can never be absolutely sterile. Conditioning treatments are employed to clean, detartrate, and discourage microbial growth.

When adding any treatment compound to a barrel, ensure that it is completely dissolved in water first. When adding both citric acid and potassium metabisulfite (KMS) to a barrel, ensure that these solutions are dissolved separately with the citric acid first. Citric acid lowers the pH of the water, which serves to make the sulfur dioxide emanating from the KMS more effective.

New Barrels

For obvious reasons, it is recommended that a barrel be watertight before filling it with wine. Before following the procedures below, be sure the hoops are snug. As a barrel swells, the hoops may be tightened a little at a time. If the barrel is excessively dry, a water mist sprayed inside and outside the barrel will conserve water and hasten the tightening. Boiling hot water should not be used to swell a barrel, as it may cause the inside of the stave to swell faster than the outside—possibly causing staves to crack.

PROCEDURE:

1. Rinse the barrel to be sure it is free of dust and debris.
2. Fill the barrel halfway with cold water.
3. Add 90 g of citric acid to the water and stir until fully dissolved. Mixing may be performed by rocking barrels back and forth.
4. Add 45 g of KMS solution to the water and stir until fully dissolved.
5. Fill the barrel completely with water.

It usually takes 3–5 days to completely soak up a barrel. Each barrel should be topped off daily with clean water. When barrels are completely soaked up, empty and rinse well several times before filling with wine.

Although a barrel may be tight when filled with water, it may seep when filled with wine. Wine has a lower specific gravity—a property which can allow it to leak through smaller pathways. It is best to introduce a fermenting must or a non-filtered wine to the newly soaked barrel, as the higher solids content will help seal off these pathways. If the barrel does seep a little, keep it topped with wine and wash off the outside in order to avoid microbe growth.

Maintenance of Empty Barrels—Short Term

Any chemical introduced to wooden barrels will penetrate the wood. Using hot water as a solution for these chemicals will increase the rate of this penetration. Citric acid additions function as neutralizers of residual chemicals in the wood.

Plain water can stagnate if left in a barrel for more than several days. If a barrel must be filled with water for more than 2 days, either empty the barrel, rinse, and refill with fresh water every 2 days thereafter, or add 180 g of citric acid and 90 g of KMS to the soaking water to help preserve it. This treatment is recommended only for the short term, several weeks maximum, as a holding solution will leach some of the oak flavor from the wood.

Maintenance of Empty Barrels—Long Term

Barrels left empty after wine or water storage will soon be invaded with bacteria and/or mold growth—which will eventually ruin them for any further wine usage. Consequently, it is important to create an atmosphere of sulfur dioxide gas inside empty barrels in order to discourage microbial growth.

PROCEDURE:

1. Clean barrel thoroughly and rinse thoroughly several times with cold water. Drain dry because residual water standing in the barrel will combine with SO_2 gas to form sulfurous acid which will soak into the wood.
2. Hang sulfur stick ($1/2$ stick for each 50-gal barrel) from bottom of sulfur bung. Light, ensure stick is burning properly, and insert into barrel, with bung left loosely in bung hole. An alternative to sulfur stick burning is injection with SO_2 gas from a pressurized canister for 3–5 s.
3. Allow to burn at least 3–5 min. When completely burned, remove sulfur bung and replace with regular bung driven tightly into bung hole.
4. Store barrel in a cool, dark, dry place for storage. Repeat treatment every 3–4 weeks to keep sulfur fresh.
5. If barrel is to remain empty for a long period of time, it should be soaked overnight with fresh water prior to burning a sulfur stick to help keep it tight. If a barrel begins to dry and the staves shrink, air will enter the barrel, displacing the sulfur and creating an environment for destructive microbial activity.
6. When preparing the barrel for wine use, be sure it is tight and proceed with the new barrel treatment outlined above.

Soda Ash Treatment

Soda ash is a caustic chemical and a harsh treatment for distressed barrels. Its main function is to dissolve tartrate crystal deposits and attempts (usually not totally successful) to bleach out red wine color. Another major purpose of soda ash is to clean and leach out undesirable odors and flavors caused by microbial growth. Soda ash, in conjunction with chlorine treatment, is also used to sanitize barrels.

Treatments with soda ash will also leach out the oak flavor of a barrel and is therefore not recommended for use on a new barrel. Soda ash should be dissolved in hot water, 140–180°F. Protective clothing and handling with care is strongly recommended. Soda ash used in strengths exceeding $1/4$ oz./gal of water has burned oak staves.

PROCEDURE:

1. Thoroughly clean and rinse barrel, ensuring it is tight with no leakage.
2. Fill barrel halfway with hot water.
3. Add soda ash up to $1/4$ oz./gal of water and mix by rocking barrel back and forth.
4. Fill barrel completely with hot water. Bung may be placed loosely in bung hole.
5. Allow to stand for 24 h, empty, and rinse several times with cold water.
6. For severe cases of microbial spoilage, barrel may then require supplemental chlorine treatment outlined below. Otherwise, it should be neutralized with citric acid treatment also outlined below.
7. Maintain barrel with short-term treatment outlined above and inspect carefully to ensure undesirable odors and flavors have been removed. If not, repeat soda ash treatment. If so, thoroughly rinse several times with cold water and immediately fill with wine.

Chlorine Treatment

This treatment should be employed only as a last resort to save microbe-infected barrels. Weak treatments of chlorine can form trichloroanisole (TCA) which can produce disagreeable "corky–musty" odors and flavors.

Chlorine treatments should be used in a strength of 200 ppm available chlorine.

PROCEDURE:

1. Thoroughly clean and rinse barrel, ensuring it is tight with no leakage.
2. Fill barrel half-way with 200 ppm chlorine cold water solution and place bung loosely in bung hole.
3. Allow to stand 48 h, empty, and rinse thoroughly several times with cold water.

Citric Acid Treatment

A citric acid treatment is used to neutralize any residual chemicals in barrels and is essential to the maintenance of sound cooperage.

PROCEDURE:

1. Rinse barrel thoroughly several times with cold water.
2. Fill barrel halfway with cold water.
3. Add 180 g of citric acid and mix by rocking barrel back and forth.
4. Fill barrel completely with cold water.
5. Allow to stand 48 h, empty, and rinse thoroughly several times with cold water.
6. Maintain barrel with short-term treatment outlined above and inspect carefully to ensure undesirable odors and flavors have been removed. If not, repeat soda ash treatment. If so, thoroughly rinse several times with cold water and immediately fill with wine.

FILTRATION

Filtration has two important factors to consider. First, of course, is the degree of clarity desired. Second, and arguably more important, is the amount of color and flavor lost.

All filtering operations remove some portion of color and flavor—justifying some red wine labels which prominently state "Unfiltered" as a proud indicator of higher quality. This is also used sometimes in combination with a disclaimer statement to excuse any haziness or precipitates which may have occurred in the bottle.

The advancement of food science and technology has given us many new processes and techniques readily applicable to producing higher-quality wines with greater efficiency. One of the most meaningful developments in recent years is the ability to filter liquids through submicron porosity media. In fine wine production, however, the application of this must focus on adequate or optimal filtration in order to conserve precious color and flavor constituency.

Pressure and flow rate are important parameters of filtration performance. High pressure generally indicates improperly clarified wines causing difficult filtration with slow flow rates, resulting in excessive removal of color and flavor. Ideal wine filtration is a near-constant low pressure and moderate flow rates. Pressure surges can easily breech filter media—particularly delicate membranes.

Filtration efficiency depends on three major elements of filter media: its properties, surface area, and porosity.

Primary filtration of very cloudy wines is typically performed with diatomaceous earth (DE) powders ranging in porosity of just several microns to very coarse. Used to excess, this media can take a considerable toll on color and flavor loss. DE is most efficient when employed in a depth filter in which suspended wine particles are entrapped as powder slurries are "dosed" at specific intervals and collect on screens inside the filter tank chamber. Some plate filters have frames available for DE filtrations—most often used for the filtration of lees. (See Fig. 4–7.) Wines which are obviously brilliant from good fermentation and clarification technique do not generally require a primary filtration.

Secondary, or "final," filtrations in small wineries are usually performed with pads in frame filters or cartridges in canister chambers—both of which are screen-type filtration devices. Pads and cartridges are nothing new, although the continued development of new plate and canister design, along with increasingly efficient pad and cartridge media, results in better economies of operation. State-of-the-art filters have the capability to conduct both primary and secondary filtration, either individually or with a divider plate, simultaneously.

Commercial wine production on a smaller scale, with good quality control, can get along quite well with cartridge media—usually constructed from paper, glass fibers, cotton, hemp, and cellulose. Cartridges range from rather coarse 40-μ porosity to less than 1-μ. Although the unit gallon cost of filtration with cartridges is quite high, it can be justified when compared to the much higher cost of investing in a frame filter apparatus.

Fig. 4–7 Two-stage stainless steel plate and frame filter with divider plate. *Courtesy: Prospero Equipment Company.*

The most common difficulty with pad and cartridge filtration is a surface area which is too small for the task at hand. Vintners often seek using pads or cartridges along with more filter time and higher pressure. This is a false economy, as the added time and energy are usually more costly than the media, plus the resulting wine quality will suffer undue color and flavor loss. A good rule of thumb is that the differential filter pressure should not exceed half of the recommended rate published by the media manufacturer.

Microfiltration and Ultrafiltration

Microfiltration is the type of final filtration most often employed in the wine industry. It requires very low pressure, generally from 5 to 15 psi, and employs a semipermeable membrane for the removal of suspended wine particles at less than 0.5-μ porosity—well within nominal "sterile" filtration limits. Ultrafiltration utilizes a higher pressure range, from 10 to 150 psi, and employs less porous membranes which can separate high-weight molecules out of solution—actually a molecular "sieve."

Microfilters are designed for a comparatively slow product flow across the membrane until an appropriate pore size allows the wine to pass through. This "cross-flow" technique reduces particle buildup on the membrane media, increasing throughput while reducing cleaning time. Continued improvement of both the filters and membrane media, in tandem with a growing demand for microfiltration by the wine industry, has resulted in unit wine cost having reduced considerably.

Unfiltered Wines

Experienced wine consumers actually expect older reds to have a sediment and ritualize the decanting procedure. It is probably in this market segment that most of the rationale and justification for producing unfiltered wines is found. Neophytes and occasional wine consumers may find some virtue in unfiltered wines as being somewhat ecologically friendly—and there may be an association to such wines being more "organic" or "natural." There can, however, be no question as to unfiltered red wines tasting bigger and richer. Considerable color and flavor may be conserved with a significantly lesser share of finings given over a commensurately longer time to perform. More and more vintners are returning to the time-honored method of egg white treatments for red wines. A light addition of tannic acid beforehand will help prevent the formation of phenolic degradation compounds. Despite the conventional wisdom of bypassing red wine cold stabilization, a few weeks under refrigeration could possibly enhance clarity and eliminate the nasty, ugly tartrate crystalline formation on the cork. Acid balance could be regained with a small addition of citric acid, which, as a tricarboxylic acid, provides superb stability and gives the resulting wine a little lift to its fruit component.

PACKAGING

Having potential consumers visit a winery tasting room is, of course, a big advantage over having to depend solely on the attention-getting of packaging on the shelves in an independent retail store. Vintners often make the mistake of paying too little attention to the importance of attractive packaging in order to maximize consumer appeal. Dull, drab, unprofessional packaging takes the excitement out of even the finest of wines. Worse yet, some vintners get caught up in egoistic label schemes which consumers do not understand and/or with which they cannot identify.

Bottles

There is not much consumer acceptance for nontraditional bottle shapes. There is plenty of evidence to show that packaging Cabernet Sauvignon in a Burgundy bottle, or Pinot Noir in a Bordeaux bottle just does not work.

One pathway to attention-getting bottles is via attractive colors. The traditional Burgundy shape in "dead-leaf" green continues to be very popular for U.S.-grown Chardonnay and Pinot Noir during the 1980s. Since then, this same bottle has been widely employed for fine-quality Rhone varietals, Fumé Blanc, Gamay, Seyval Blanc, Vignoles, Chambourcin, and Marechal Foch.

The traditional Bordeaux shape in colorless "flint" was widely adopted for Sauvignon Blanc, white Meritage, Vidal Blanc, and pink "blush" wines during the 1980s. The same shape in "Champagne" green has been used for Cabernet Sauvignon, Merlot, and, of course, the Bordeaux chateau wines for many years. A new "antique green" serves well, too, for red Meritage, Sangiovese, and other heavier reds. There is a slighted tapered version of this which is also accepted for top-of-the-line wines.

Punted Burgundy and Bordeaux bottles, those with "push-up" indentations in the bottom, are another attribute of fine bottle quality and consumer appeal.

The tall, narrow, "hock" bottle shape is required to be green in the Mosel–Saar–Ruwer region of Germany, but German regulations require it to be amber in the Rheinland. In the Nahe region, it is often blue in color. The extremely tall and narrow versions are not practical as they are easy to tip over, do not stack on their sides, and are difficult to stand up on a shelf or in a refrigerator.

Most important is bottle quality. Some of the imported glass does not meet the standards set by U.S. bottle manufacturers—either in its composition or in dimension tolerances. This is particularly important in the neck of the bottle where a given cork diameter may be either too loose, causing leakers, or too tight, causing breakage. It is also important in the body of the bottle where varied mold sizes can serve to significantly effect fill levels.

Closures

Many consider screw-cap types to be the perfect closure for wine bottles. Unfortunately, these are not acceptable to most fine wine buyers and screw-caps are therefore generally relegated to lower priced wines.

Natural cork (Figure 4.8) remains the closure of choice throughout most of the wine world, but the advancing technology of synthetic corks made of various

Fig. 4–8 Illustration of cork bark harvest and cork production.
Courtesy: Sebastiani Vineyards.

ethylene–vinyl acetate compounds is encouraging many vintners to take a closer look. Manufacturers say that most problems of difficult driving, and even more difficult pulling, are now ironed out. They advertise new technology, such as virtual immortality of their ethylene–vinyl acetate copolymer (EVA) formulation, improved lubricants, variable resiliencies, and even the possibility of scavenging headspace oxygen and SO_2. Considerable testing of new types continues both with vintner trials and university research. Although there are only several vintners which have totally committed all production to synthetic corks, there are many poised to move forward with the conversion in varying degrees. In 1994, sales of more than several million synthetic corks were reported—hardly significant—but that level was surpassed in just the first month of 1995. For the venturesome vintner, synthetic corks may be worth a try.

The key to cork selection is size—the proper diameter to ensure a good balance between ease in pulling and optimal seal. Perhaps the only thing more denigrating to a good packaging concept than an oversized cork virtually impossible to pull is a sticky bottle caused by an undersized leaking cork. Wines designed for long periods of bottle age, reds in particular, are usually bottled under longer corks.

Capsules

Traditionally, capsules have been a very important part of packaging—primarily for consumer appeal. Manufacturers have made much about the ability of capsules to "seal" bottles, but, in fact, most capsules can be removed and replaced rather easily. In the case of most hard poly types, it is simply a matter of pulling them off and putting them right back on again.

Until the late 1980s, lead and lead-alloy capsules were the epitome of fine wine capsules—durable, efficient, rich in appearance, and expensive. With several controversial studies indicating the possibility of trace amounts of lead migrating from capsule into poured wine, lead was banned by the ATF.

Inexpensive capsules generally portray themselves in much the same way—as cheap wine. Great advances have been made in the appearance of heat-shrink PVC film capsules, and they offer great resistance to tampering. Even more improvement has been achieved in replacing lead with tin–aluminum and aluminum polylaminate roll-on capsules.

As with corks, capsule diameter must be precisely fitted to the bottle neck in order to avoid folding, wrinkling, or tearing when applied.

The notion that it is better to have no capsule at all than a bad one is gaining momentum in the United States. The revolutionary B-Cap® closure system features a custom-designed, drip-resistant, flange-lip bottle mouth. Corks are driven as usual, but instead of a capsule, a small decorative paper circle is applied to cover the top of the cork. The full length of the cork is perfectly visible; some vintners preferring to partially mask the cork with a narrow garnish strip applied around the bottle neck. The B-Cap marketers advertise it as "functional, economical, attractive, user friendly, and environmentally correct."

Label Rationale

Many vintners, particularly those who market chiefly to winery visitors and tourists, operate on the notion that they have a captive audience. This often takes the form of cheap, in-house labeling borne upon the "letting my wine speak for itself" syndrome. Although the wine may, indeed, be very good, the question is whether or not it is salable. Is the label something which can exude pride in the consumer's cellar? Is the packaging something serious wine imbibers want to present at their dinner table with guests? Is the product something which qualifies as a nice gift? If the answer to any of these questions is in doubt, then results are likely to be a bottle or two bottles of "courtesy" sales rather than full-case sales and repeat orders. Figure 4.9 illustrates some well-designed labels which have proven to be successful.

Fig. 4–9 A selection of well-designed commercial wine labels. *Courtesy: David Ferguson.*

Vintners should consider that it is not what **they** think about **their** packaging which is important. It is only the **consumer**'s impression of whether or not a package sufficiently attracts their money.

Another important part of label rationale is accoutrements, such as tags, stickers, and other supplemental attention-getters. There is no reason not to flaunt a prestigious award or medal or critical accolade. It is common to find vintners taking advantage of medals by significantly increasing the price for medal-winning wines. This could slow movement of such wines, but, more importantly, it brings attention to vintner quality across the entire product line. Most importantly, the price increase generates additional profits, which may be needed to lower prices to make nonmedal winners more attractive.

Label Shapes

Although there are many labels shapes in use which radically depart from the traditional square or rectangle, prospective vintners should understand that odd shapes are a common cause of labeling problems and increased labeling cost.

Design and printing costs are generally higher in that odd shapes require custom dies for cutting and, one way or another, the cost of these dies will be passed on to the vintner. Nontraditional shapes cause greater paper waste, too—yet another cost borne by the vintner. Labeling machines usually require expensive custom parts fabricated in order to handle custom shapes.

Label Colors

There can be little question that color and its integration in wine packaging should be placed in the hands of professionals. This is expensive but absolutely essential in order to properly address each wine to its target market.

Color psychology seems to differs with each psychologist who feels qualified to write about it. Several references have been distilled down to this general set of consumer truisms:

White	Clean—winter—delicate
Black	Formal—death—night
Brown	Heavy—religion—obscure
Red	Hot—danger—stop
Blue	Cool—thin—distant
Green	Life—summer—proceed
Yellow	Floral—spring—fresh
Orange	Citrus—autumn—ripe
Violet	Royal—soft—expensive

This begs the notion of color combinations. Red and green are, of course, traditional decor for the Holiday Season, but that is hardly a combination of heat and summer. Adding white to the red and green brings to mind things Italian. The combinations are virtually endless and, as mentioned, are best dealt with by professionals.

A chief concern for color, and the entire realm of label design, is that of vintner ego expressed to a fault. Obviously, wine labels should be totally representative of the product and the producer, but there have been many wine failures because of overdone labels.

Perhaps even more important is that of underdone labels, with such a lack of character that there is insufficient consumer attraction to make a sale. This condition exists widely throughout the wine industry—principally from vintners who are ignorant of the importance of packaging in consumer appeal and/or vintners who feel they are saving money by minimizing label cost. This is a very poor economy. Vintners must understand that no matter what the quality of their wine, it is the package which creates the desire for a prospective consumer to try it. There are gold-medal-winning wines which go begging because of inept packaging. Conversely, there are mediocre wines which sell actively because of clever packaging.

Label Humor

There are some wine labels which are funny and not meant to be, which is truly unfortunate. On the other hand, some vintners set out to generate label smile appeal. Although there are many wine marketing successes which can be attributed to snappy label design, there is a danger that vintners can become typecast and the winery regarded as flippant and insincere. Consumers may not take any wines in the line seriously, and a solid reputation may erode to novelty.

Label Experiments

A case can be made for every wine label being on a pathway of eventually becoming dated. Some of these, of course, have a shorter life span than others. Conversely, there are also labels in the marketplace which have not markedly changed for decades and remain successful. Among these are classic European wines, such as some chateau labels which have not changed since before the turn of the century.

Few U.S. vintners can claim classic heritage. Thus, wine packaging, like any other consumer product, undergoes trends which are dynamic. It makes good sense to closely observe consumer reaction, both in the winery tasting room and in wine shops and restaurants. Obviously, positive appeal coupled with positive sales is not of immediate concern. When the proverbial "red flags" of consumer disinterest appear, it behooves the vintner to seriously consider new label design.

Again, it makes sense to consult with packaging professionals to ascertain what possibilities for change may be recommended. A good packaging company or label-printing firm should have a design team available for the vintner at little or no cost. Their incentive is, of course, continued business with that vintner.

It is impractical for most vintners to market-test new packaging concepts in stores and restaurants. The cost for small print runs and the complexity of obtaining ATF approval for each new label discourages this concept. On the other hand, mock-ups can be shown to interested visitors and tourists in the winery tasting room. Comments can be gathered over a sufficient period of time from which rational decisions can be made for the change of choice.

Key ATF Labeling Regulations

27 CFR, Part 4, Section 4.32 *Mandatory label information.*

(a) There shall be stated on the brand label:
 (1) Brand name, in accordance with Section 4.33.
 (2) Class, type, or other designation, in accordance with Section 4.34.
 (3) Alcohol content, in accordance with Section 4.36.

(b) There shall be stated on any label affixed to the container:
 (1) Name and address, in accordance with Section 4.35.
 (2) Net contents, in accordance with Section 4.37.

27 CFR, Part 4, Section 4.33 *Brand names.*

(a) *General.* The product shall bear a brand name, except that if not sold under a brand name, then the name of the person required to appear on the brand label shall be deemed a brand name for the purpose of this part.

(b) *Misleading brand names.* No label shall contain any brand name, which, standing alone, or in association with other printed or graphic matter creates any impression or inference as to the age, origin, identity or other characteristics of the product unless the Director finds that such brand name, either when qualified by the word "brand" or when not so qualified, conveys no erroneous impressions as the age, origin, identity, or other characteristics of the product.

27 CFR, Part 4, Section 4.34 *Class and Type.*

(a) The class of the wine shall be stated in conformity with Subpart C of this part if the wine is defined therein, except that "table" ("light") and "dessert" wines need not be designated as such. In the case of still grape wine there may appear, in lieu of the class designation, any grape-type designation, semigeneric geographic type designation, or geographic distinctive designation, to which the wine may be entitled. In the case of champagne, or crackling wines, the type designation "champagne" or "crackling wine" ("petillant wine", "Frizzante wine") may appear in lieu of the class designation "sparkling wine". In the case of wine which has a total solids content of more than 17 grams per 100 cubic centimeters the words "extra sweet", "specially sweetened", "specially sweet" or "sweetened with excess sugar" shall be stated as a part of the class and type designation. The last of these quoted phrases shall appear where required by Part 240 of this chapter, on wines sweetened with sugar in excess of the maximum quantities specified in such regulations. If the class of the wine is not defined in Subpart C, a truthful and adequate statement of composition shall appear upon the brand label of the product in lieu of a class designation. In addition to the mandatory designation for the wine, there may be stated a distinctive or fanciful name, or a designation in accordance with trade understanding. All parts of the designation of the wine, whether mandatory or optional, shall be in direct conjunction and in lettering substantially of the same size and kind.

(b) An appellation of origin such as "American," "New York," "Napa Valley," or "Chilean," disclosing the true place of origin of the wine, shall appear in direct conjunction with and in lettering substantially as conspicuous as the class and type designation if:
  (1) A grape type (varietal) designation is used under the provisions of Section 4.32a;
  (2) A semi-generic type designation is employed as the class and type designation of the wine pursuant to Section 4.24b;
  (3) A product name is qualified with the word "Brand" under the requirements of (j); or
  (4) The wine is labeled with the year of harvest of the grapes, and otherwise conforms with the provisions of Section 4.27. The appellation of origin for vintage wine shall be other than a country.

27 CFR, Part 4, Section 4.35   *Name and address.*

(a) *American wine.* On labels of containers of American wine, there shall be stated the name of the bottler or packer and the place where bottled or packed (or until January 1, 1985, in lieu of such place, the principal place of business of the bottler or packer if in the same State where the wine bottled or packed, and, if bottled or packed on bonded premises, the ATF registry number of the premises) immediately preceded by the words "bottled by" or "packed by" except that:
  (1) If the bottler or packer is also the person who made not less than 75 percent of such wine by fermenting the must and clarifying the resulting wine, or if such person treated the wine in such manner as to change the class thereof, there may be stated, in lieu of the words "bottled by" or "packed by," the words "produced and bottled by," or "produced and packed by."
  (2) If the bottler or packer has also either made or treated the wine, otherwise than as described in paragraph (a)(1) of this section, there may be stated, in lieu of the words "Bottled by" or "Packed by," the phrases "Blended and bottled (packed) by," "Rectified and bottled (packed) by," "Prepared and bottled (packed) by," "Made and bottled (packed) by" as the case may be, or, in the case of imitation wine only, "Manufactured and bottled (packed) by."
  (3) In addition to the name of the bottler or packer and the place where bottled or packed (but not in lieu thereof) there may be stated the name and address of any other person for whom such wine is bottled or packed, immediately preceded by the words "Bottled for" or "Packed for" or "Distributed by" or other similar statement; or the name and principal place of business of the rectifier, blender, or

maker, immediately preceded by the words "Rectified by," "Blended by" or "Made by," respectively, or, in the case of imitation wine only, "Manufactured by."

(c) *Form of address.* The "place" stated shall be the post office address (after December 31, 1984 the post office address shall be the address shown on the basic permit or other qualifying document of the premises at which the operations took place; and there shall be shown the address for each operation which is designated on the label. An example of such use would be "Produced at Gilroy, California, and bottled at San Mateo, California, by XYZ Winery"), except that the street address may be omitted. No additional places or addresses shall be stated for the same person unless (1) such person is actively engaged in the conduct of an additional bona fide and actual alcoholic beverage business at such additional place or address, and (2) the label also contains in direct conjunction therewith, appropriate descriptive material indicating the function occurring at such additional place or address in connection with the particular product.

(d) *Trade or operating names.* The trade or operating name of any person appearing upon any label shall be identical with the name appearing on the basic permit or notice.

27 CFR, Part 4, Section 4.36   *Alcoholic content.*

(a) Alcoholic content shall be stated in the case of wines containing more than 14 percent of alcohol by volume, and, in the case of wine containing 14 percent or less of alcohol by volume, either the type designation "table" wine ("light" wine) or the alcoholic content shall be stated. Any statement of alcoholic content shall be made as prescribed in paragraph (b) of this section.

27 CFR, Part 4, Section 4.37   *Net contents.*

(a) *Statement of net contents.* The net contents of wine for which a standard of fill is prescribed in Section 4.73 shall be stated in the same manner and form as set forth in the standard of fill.

27 CFR, Part 4, Section 4.38   *General Requirements.*

(a) *Legibility.* All labels shall be so designated that all the statements thereon required by Section 4.30 through 4.39 are readily legible under ordinary conditions, and all such statements shall be on a contrasting background.

(b) *Size of type.* (1) Containers of more than 187 millimeters. All mandatory information required on labels by this part, except the alcoholic content statement, shall be in script, type, or printing not smaller than 2 millimeters; except that if contained among other descriptive or explanatory information, the script, type, or printing of the mandatory information shall be of a size substantially more conspicuous than that of the descriptive or explanatory information.

(f) *Additional information on labels.* Labels may contain information other than the mandatory label information required by Sections 4.30 through 4.39, if such information complies with the requirements of such sections and does not conflict with, nor in any manner qualify statements required by this part. In addition, information which is truthful, accurate, and specific, and which is neither disparaging nor misleading may appear on wine labels.

27 CFR, Part 4, Section 4.39 *Prohibited practices.*

(a) *Statements on labels.* Containers of wine, or any label on such containers, or any individual covering, carton, or other wrapper of such container, or any written, printed, graphic, or other matter accompanying such container to the consumer shall not contain:

(1) Any statement that is false or untrue in any particular, or that, irrespective of falsity, directly, or by ambiguity, omission, or inference, or by the addition of irrelevant, scientific or technical matter, tends to create a misleading impression.

(2) Any statement that is disparaging of a competitor's products.

(3) Any statement, design, device, or representation which is obscene or indecent.

(4) Any statement, design, device, or representation of or relating to analyses, standards, or tests, irrespective of falsity, which the Director finds to be likely to mislead the consumer.

(5) Any statement, design, device or representation of or relating to any guarantee, irrespective of falsity, which the Director finds to be likely to mislead the consumer. Money-back guarantees are not prohibited.

(6) A trade or brand name that is the name of any living individual of public prominence, or existing private or public organization, or is a name that is in simulation or is an abbreviation thereof, or any graphic, pictorial, or emblematic representation of any such individual or organization, if the use of such name or representation is likely falsely to lead the consumer to believe that the product has been endorsed, made, or used by, or produced for, or under the supervision of, or in accordance with the specifications of, such individual or organization; *Provided,* That this paragraph shall not apply to the use of the name of any person engaged in business as a producer, blender, rectifier, importer, wholesaler, retailer, bottler, or warehouseman of wine, nor to the use of any person of a trade or brand name that is the name of any living individual of public prominence or existing private or public organization, providing such trade or brand name was used by him or his predecessors in interest prior to August 29, 1935.

(7) Any statement, design, device, or representation (other than a statement of alcohol content in conformity with Section 4.36), which tends to create the impression that a wine:

(i) Contains distilled spirits;
(ii) Is comparable to a distilled spirit; or
(iii) Has intoxicating qualities.

However, if a statement of composition is required to appear as the designation of a product not defined in these regulations, such statement of composition may include a reference to the type of distilled spirits contained therein.

(8) Any coined word or name in the brand name or class and type designation which simulates, imitates, or which tends to create the impression that the wine so labeled is entitled to bear, any class, type, or permitted designation recog-

nized by the regulations in this part unless such wine conforms to the requirements prescribed with respect to such designation and is in fact so designated on its labels.

(b) *Statement of age.* No statement of age or representation relative to age (including words or devices in any brand name or mark) shall be made except (1) for vintage wine, in accordance with the provisions of Section 4.27; (2) references relating to methods of wine production involving storage or aging in accordance with Section 4.38(f); or (3) use of the word "old" as part of a brand name.

(Authors' Note: Providing no individual State regulation prohibits the practice, 27 CFR, Subpart C, 4.50, permits the use of vintage labels, without any appellation of origin, upon wines made from out-of-state fruit—provided that wine is sold exclusively within the state boundaries of the bottler.)

27 CFR, Part 4, Section 4.50   *Certificate of label approval.*

(b) Any bottler or packer of wine shall be exempt from the requirements of this section if upon application the bottler or packer shows to the satisfaction of the Director that the wine to be bottled or packed is not to be sold, offered for sale, or shipped or delivered for shipment, or otherwise introduced in interstate or foreign commerce. A "Certificate of Exemption from Label Approval under the Federal Alcohol Administration Act" (Form 1650) shall be issued by the Director upon application upon the form designation "Application for Certificate of Exemption from Label Approval under the Federal Alcohol Administration Act" (Form 1648) properly filled out and certified to by the applicant.

(c) *Statement of bottling dates.* The statement of any bottling date shall not be deemed to be a representation relative to age, if such statement appears in lettering not greater than 8-point Gothic caps and in the following form: "Bottled in ———" (inserting the year in which the wine was bottled).

(e) *Simulation of Government stamps.* (1) No labels shall be of such design as to resemble or simulate a stamp of the United States Government or any State or foreign government. No label, other than stamps authorized or required by the United States Government or any State or foreign government, shall state or indicate that the wine contained in the labeled container is produced, blended, bottled, packed, or sold under, or in accordance with, any municipal, State or Federal Government authorization, law, or regulation, unless such statement is required or specifically authorized by the laws or regulations of a foreign country. If the municipal, State, or Federal Government permit number is stated upon a label, it shall not be accompanied by any additional statement relating thereto.

And, any wine bottle on or after November 18, 1989, must include a statement on the label precisely as follows:

Government warning: (1) According to the Surgeon General, women should not drink alcoholic beverages during pregnancy because of the risk of birth defects.

(2) Consumption of alcoholic beverages impairs your ability to drive a car or operate machinery, and may cause health problems.

WINEMAKING PERILS AND PITFALLS

Every winemaker will encounter wines which may fall short of expectations, or express faults which are disappointing. Some defects can be averted if they are detected in time. The best way to avoid such problems is, of course, to ensure that equipment, materials, and methods are in proper order beforehand. Wines with serious problems are generally the products of grapes and/or procedures with equally serious problems.

Table 4–2 is provided to help identify some of the more common difficulties in finished wines. A comprehensive wine laboratory service should be employed to examine and advise upon all wines felt to be out of order.

Table 4–2   Common difficulties in finished wines.

Effect/symptom	Cause/comments
Visual Mode	
Browning	Exposure to air
	Low free sulfur dioxide level
Cloudy	Yeast fermentation
	Bacterial fermentation
	Insufficient pectic enzyme treatment
	Insufficient fining
	High iron level
Color deficiency	Overcropping of grapes
	Unripe grapes
	High free sulfur dioxide levels
	Amelioration dilution
Color instability	Grape rot
	High pH level
	Oxidative enzyme reaction
	High free sulfur dioxide level
	Bacterial fermentation
Effervescence	Yeast fermentation
	Bacterial fermentation
Floating solids	Cork dust
	Cork paraffin
	Equipment lubricants
	Insufficient sanitation
Haziness	High copper level
	Protein instability

Table 4–2 (*continued*)

Effect/symptom	Cause/comments
Precipitate	Autolyzed yeast
	Breached filter media
	Potassium bitartrate crystals
	Calcium tartrate crystals
	High iron level
	Insufficient sanitation
Suspended solids	Filter fibers
	Mold from bottling equipment
	Case dust
	Insufficient sanitation

Olfactory (Flavor) Mode

Effect/symptom	*Cause/comments*
Aroma deficiency	Overcropping of grapes
	Unripe grapes
	Amelioration dilution
Burn nasal passages	High alcohol
Burnt match	High sulfur dioxide level
Buttery	Malolactic fermentation
Cooked	High fermentation temperature
(burnt caramel)	Exposure to air
Hydrogen sulfide	Sulfur residues from vineyard
(rotten eggs)	Wild yeast fermentation
	Insufficient yeast nutrients
	Excessive lees contact
	Insufficient sanitation
Mercaptans	Extended hydrogen sulfide condition
(skunky)	Bacterial fermentation
	Insufficient sanitation
Moldy/musty	Moldy grapes
	Moldy barrel
	Decayed barrel
	Bacterial fermentation
	Decayed cork
	Insufficient sanitation
Oxidization	Wild yeast fermentation
(stale bread)	Exposure to air
	Low sulfur dioxide level
Paint thinner	Bacterial fermentation
	Low sulfur dioxide level
Prune-like	Dehydrated grapes
	High pH
	High fermentation temperature
Rubbery	Sulfide compounds

Table 4-2 (*continued*)

Effect/symptom	Cause/comments
Vinegary	Bacterial fermentation
	Low sulfur dioxide level
Wet wool	Sulfide compounds
	Wild yeast fermentation
	Bacterial fermentation
	Low sulfur dioxide level
	Insufficient sanitation
Woody	Insufficient new barrel treatment
	Excessive barrel storage

Gustatory (Taste) Mode

Effect/symptom	Cause/comments
Astringency	Unripe grapes
	High phenol level in grapes
	Insufficient destemming
	High sulfur dioxide level at crusher
Bitter	Unripe grapes
	High phenol level in grapes
	Cracked grape seeds
	High sulfur dioxide level at crusher
Coffee-like	Sulfide compounds
	Insufficient new barrel treatment
	Excessive barrel storage
	Insufficient sanitation
Flat—insipid	Overripe grapes
	Overcropping
	High pH
	Amelioration dilution
	High sweetness level
Metallic	High copper level
	High iron level
Nutty—toasty	Oxidation
	Wild yeast fermentation
	Exposure to air
	Low sulfur dioxide level
Thin body	Overcropping
	Amelioration dilution
Vegetal	Sulfide compounds
	Bacterial fermentation
	Low sulfur dioxide level
Vinegary	Bacterial fermentation
	Low sulfur dioxide level
Woody	Insufficient new barrel treatment
	Excessive barrel storage

CHAPTER 5

WINE CLASSIFICATION

Wine experts generally classify wine into five categories, with the distinctions among the classes based primarily on major differences in their manner of vinification.

TABLE WINES

The overwhelming majority of the wine produced in the world falls into the table wine category. These range from the obscure and ordinary to the most expensive and celebrated classics. As the name suggests, table wines are designed for use at the table as a complement to good food. For this same reason, table wines are sometimes referred to as "dinner wines."

There are white, blush (pink), and red table wines. These are the base wine needed in order to make every other wine type. Whereas neophytes may prefer wines with a little sweetness, the overwhelming preference among veteran wine consumers is for dry wines. Sugar blunts crisp acidity and has a tendency to mask delicate wine flavors.

Two of the most common terms used in connection with table wine identification are *varietal* and *generic*. Varietal wines are labeled with the variety of grape or fruit used in making the wine, such as Chardonnay or Johannisberg Riesling. Generic wines are named for a specific geography of origin, such as Chablis, part of Burgundy, one of only several locales in France where Chardonnay is permitted by regulation to be grown commercially. In the United States, we have the freedom to plant Chardonnay, or any other vine species, anywhere. In similar fashion, we often see wines labeled as Rheingau or Rheinpfalz from the German Rheinland, where Johannisberg Riesling is widely cultivated. The Riesling is also planted in many states across the United States. Bureau of Alcohol, Tobacco and Firearms (ATF) regulations permit the commercial use of a select list of European generic names by U.S. vintners, but, as may be expected, this is very controversial.

Varietal wines are, consequently, more commonly used on U.S. wine labels, whereas European wines are typically generic. The following is a list of a few examples of generic wines—along with the varietals which are cultivated in those regions:

GENERIC	VARIETAL
White Table Wines	
Alsace (France)	Gewürztraminer and Pinot Blanc
Chablis (Burgundy in France)	Chardonnay
Gavi (Piemonte in Italy)	Cortese
Graves (Bordeaux in France)	Sauvignon Blanc and Semillon
Rheingau (Germany)	Johannisberg Riesling
Orvieto (Umbria in Italy)	Trebbiano and Malvasia
Red Table Wines	
Barolo (Piemonte in Italy)	Nebbiolo
Beaujolais (Burgundy in France)	Gamay
Medoc (Bordeaux in France)	Cabernet Sauvignon and Merlot
Chianti (Tuscany in Italy)	Sangiovese
Cote d'Or (Burgundy in France)	Pinot Noir
Rhone (France)	Grenache and Syrah

Among the most widely consumed American table wines are the "jug" wines that can be found in most markets. Unfortunately, these often have less than complimentary reputations—owing to the old myth that "only expensive wines can be good." Jug wines are defined as any wine commercially marketed in a container of 1.5 L capacity or larger. The bag-in-a-box packages can also loosely be considered as jug wines. Most jug offerings are California generic and varietal table wines, with a few from eastern American and European vintners.

To some extent, jug wines epitomize the many advancements in wine production technology. Research has met consumer demands for freshness, flavor, and overall appeal with new techniques which allow for wines to be made more quickly at less cost. To these specific ends, jug wines are primarily designed for immediate consumption and usually do not improve with further aging. One frequently hears that jug wines are "America's answer to the *ordinary* wines of Europe."

ATF rules require table wines to have an alcohol content not exceeding 14% by volume. Commercial table wines must remain "in bond" (federal excise tax unpaid) at the winery until the FET is paid to the ATF. Vintners producing less than 100,000 gal per calendar year pay a rate of $0.17/gal FET. Vintners producing more than 100,000 gal annually are subjected to a sliding FET rate scaled up to a maximum of $1.07/gal, which is paid by all vintners producing 250,000 gal or more each year.

SPARKLING WINES

These are wines which "effervesce," or bubble. The first sparkling wines may have been made in the mid-1600s by a French monk named Dom Perignon. He is also thought to have been the first to use plugs made from cork tree bark for bottle stoppers. Apparently, the seal was so good that wine bottled in the winter, before it was completely finished fermenting, resumed fermenting in the bottle during springtime temperatures. This resulted in some buildup of CO_2 gas pressure inside the bottle during storage, which was later released as "sparkling" bubbles when the cork was removed.

Nowadays, there are people who still use the terms "Champagne" and "sparkling wine" interchangeably, but this is incorrect. Champagne is a sparkling-wine-producing region in France and is, thus, a generic term. All Champagnes are sparkling wines, but not all sparkling wines are Champagne.

Needless to say, fermentation pressure in a table wine bottle is a very dangerous situation. Modern methodology for the traditional Champagne process, or *methode champenoise,* is by a highly controlled secondary fermentation of a low-alcohol table wine. Pressure may build up to 100 lbs. psi in specially designed bottles.

In the Champagne region, the French are permitted by regulation to use only Chardonnay, Pinot Noir, and Pinot Meunier grapes—the latter two having blue-black skins but white juice inside the berry. *Blanc de Blancs* on the label indicates that the white sparkling wine is made exclusively from white grapes, and *Blanc de Noirs* is a white made exclusively from black grapes.

Sparkling wine fermentation in tanks is a process called *Charmat,* or the "bulk process." Due to European Economic Community (EEC) regulation, sparkling wine made in Germany must be labeled as *Sekt,* whereas in Italy, it is referred to as *Spumante,* and in Spain, as *Cava.* In the United States, it can legally be referred to as "California Champagne," or "Napa Valley Champagne," or with some other ATF-approved preface for identification.

The driest of sparkling wines is generally referred to as *naturel;* dry is *brut* or *sec.* These three are the most popular. *Demi-sec* is noticeably sweet and *doux* very sweet. One also finds *dry* and *extra dry* on sparkling wine labels, too, and these are often somewhere between *sec* and *demi-sec* in sweetness level.

The ATF federal excise tax (FET) rate on naturally fermented sparkling wines is $3.40/gal for all producers, regardless of annual output. Artificially carbonated wines are taxed at a rate of $2.40/gal.

Vintners are advised to make sparkling wines only after a high level of proficiency with table winemaking has been achieved. Sparkling wine fermentations require precise calculations and remain very dangerous even in the hands of the most experienced masters. A good place to start is by reading advanced texts on the subject. Seek out seminars and workshops presented by academic and industry groups. Visit other small wineries operated by people who are willing to illustrate the essentials in their sparkling wine operations. Best of all, endeavor to enlist the help of an experienced winemaker through the first batch or two.

DESSERT WINES

As the name implies, these are wines generally consumed with, or instead of, dessert courses. Credit is given to the medieval Moors occupying Spain and Portugal, despite their Islamic prohibition, for inventing the distillation process from which "wine spirits," or brandy, is made. But it was the British, in the 17th and 18th centuries, who perfected the addition of brandy to wines grown in Portugal which are now exemplified by sweet, rich Ports. Wines having undergone a brandy addition are sometimes referred to as "fortified." Sherry from Spain, Madeira from the isle of Madeira, and Marsala from Sicily are also good examples of fine dessert wines.

Dessert wines are usually made by the addition of grape brandy to a fermenting juice or must, less often to a completely fermented table wine. The brandy addition usually increases alcohol content up to 19–20% by volume—not to exceed 24% by ATF regulations found in 27 CFR, Part 4, Section 24.233. Dessert wine FET is $1.57/gal for all commercial vintners.

The brandy used in wineries is "high-proof" wine spirits refined to more than 95% ethanol and, following approval by the ATF, is added in bond so as to avoid paying the comparatively very expensive brandy FET.

Vintners are cautioned to investigate state winemaking regulations carefully as some states prohibit high-proof brandy storage and additions.

APERITIF WINES

These are wines which are designed to serve as appetizers to prime the palate before a special meal. Dubonnet®, Lillet®, and Pernod® are well-known proprietary names of light and dark aperitifs in Europe. The consumption of aperitif wines in the United States is largely as cocktail "mixers," such as dry vermouth in martinis and sweet vermouth in manhattans.

Aperitif wines are fortified with brandy, generally up to a level of 17% or 18% alcohol by volume—carrying the same $1.57 FET as for dessert wines. ATF approval of winery applications is required prior to making these "special-natural" wines, as they are sometimes referred. Essences from various barks, herbs, peels, roots, and/or spices combined in a special closely guarded, sometimes patented, recipe are added to create a consistently distinctive wine.

The history of this category may date to the ancient Greeks who added, among other things, honey, sea water, and wood resins to their wines. Coniferous resins are still added to Greek wines called *retsina*.

POP WINES

This is a wine type thought by some to be named after its *pop*ularity among young adults and ethnic groups. Others will testify that pop wines have a similarity to *soda pop*. Perhaps both apply. In any event, it is a category which has emerged just during the past several decades.

Commercially produced pop wines may be made at alcohol levels under 14% by volume, in which case they are subject to the table wine FET rate, or from 14% to 24% by volume, in which case they are subject to the dessert wine FET rate.

In short, pop wines closely resemble aperitif wines, except that the added essences are more exotic, typically from pronounced fruit and/or berry flavors.

CHAPTER 6

WINERY DESIGN

Many people drawn toward commercial wine production have more rewards in mind than making a profit. A flair for the romantic life or some set of Bacchanalian notions lead them to the idea of starting their own winery. The prospect of realizing such a dream is almost always driven by eagerness—the excitement of awaiting the first set of drawings and the ensuing anxiety to get the structure operational.

This type of enthusiasm is as it should be, as long as such passions do not dilute prudent judgment in working with winery design concepts. The point here is to resist the "add-ons syndrome," which can burden a workable plan with untimely cost.

An essential key in winery start-up is to remain as financially "liquid" as possible, with a tenacious attitude toward conserving capital. One old adage is that "in order to make a small fortune in the wine business, one must start with a large fortune."

Most aspiring commercial vintners do not have large fortunes, and finding alternative ways of getting things properly accomplished with significant reductions or replacement of dollar outlays is absolutely imperative.

MANAGEMENT

The first order of new vintner management is the pursuit of exhaustive research and planning. This embraces the market audit, business strategy, site selection, production design, financial support, and all the clever synergies which will eventually separate winners from losers. A careless approach is an invitation to disaster. New wineries can be overdone but never overplanned.

It is not the way of modern business to forgive poor management. Whether it is blamed on bad information, poor production quality, insufficient capital, superior competition, or whatever excuses may be considered, once the final analysis of a failed business is made, it will *always* reflect upon the inability of management to overcome adversity.

Even with ideal plans in place, winery managers can expect changes to be made in order to deal with unexpected events. This is called "contingency management" and is a particularly important factor in steering new wineries from harm's way. There are vintners who will attest to their business remaining constantly in the contingency mode.

The gathering and processing of salient production, marketing, and administrative information is tantamount to effective decision making. Utilizing the full benefit of one or another of the integrated computer software programs for wineries is the first step recommended in assuming this responsibility.

These programs start with accounting for grapes received at the winery and then a quick and easy recording of origin, weight, units, Brix, acidity, pH, lot number assignments, grower settlements, must and juice adjustments, and most any other particular needs a vintner may fashion as necessary. The system will then track fermentation, racking, materials and treatments, blending, fining, aging, filtration, and bottling, all with attending analyses. Each lot of wine should be singled out to show its share of both variable and fixed costs. Tracking will continue for each wine through warehousing, sales and depletion, customer history, order processing, and printing of bills of lading and invoices, all while keeping track of complex discount rates, promotional costs, sales incentives, and performance rates. Integration permits the creation of all required ATF-reporting documents, as well as up-to-the-minute reports for cost accounting, accounts receivable, accounts payable, and ledgers. Computer software is available for tracking all aspects of production and for cost analysis and record keeping. (See Figs. 6–1 and 6–2.)

Good management will see to it that such systems are provided all the inputs necessary to generate outputs which can go a long way toward effective decision making for any part of the winery business.

The standard expectations for managers of any business are tireless pursuits of tough choices, protection of assets, achievement of goals, exceptional foresightedness, and inspiring leadership. This is classic entrepreneurship. Those who are interested in 8-to-5, 5 days per week, should consider doing something else.

WORK ORDER

Press

From: z (not specified)
Via: UPS
For/By: (not specified)

29-Sep-95
39

Description Product Code	Container	Lot Number	Location	Add	Remove		Unit Wt Quality	Count/Unit Count/UnitWt		OK?
Grapes, Pinot Gris *P. Gris.Grapes....*			Press	0	5.19	Tons	0	0	0	
Juice, Pinot Gris *P. Gris.Juice....*	Tank..136	tg1……	Main Warehouse	802	0	gallons	0	0	0	

Press lot tg1 to tank 136. No SO2 addition. chill to 45 - 50 deg, gas head space. hold 24 to 48 hrs. - before yeast - raise ta before yeast

WORK ORDER

Rack

From: z (not specified)
Via: .
For/By: (not specified)

02-Oct-95
41

Description Product Code	Container	Lot Number	Location	Add	Remove		Unit Wt Quality	Count/Unit Count/UnitWt		OK?
Juice, Pinot Gris *P. Gris.Juice....*	Tank..136	tg1……	Main Warehouse	0	802	gallons	0 OK	0	0	
Juice, Pinot Gris *P. Gris.Juice....*	Tank..127	tg1……	Main Warehouse	822	0	gallons	0	0	0	
Yeast *Yeast.P.Dm....*		chem3…….		0	2.2	lbs	0	0	0	
Tartaric Acid *Tartaric Acid……*		chem2…….	Main Warehouse	0	4.9	lbs	0	0	0	

tank 136 - rack to tank 127 - add 2.2 lbs. p.dm yeast, 4.9 lbs tartaric acid

Fig. 6–1 Printout of computerized grape receiving and press record. *Courtesy: Breckenridge Software, Littleton, CO.*

155

Inventory Activities In Progress

15-Jul-96

Reference Product	Activity Lot Name Container	Organization Location	Quantity	Supplier Ttl Costs	Carrier Unit Cost	Wt/Unit	Weight	Volume	Quality Saleable
				z (not specified)					
				UPS					

Friday, September 29, 1995

39	Press	(not specified)							
Juice, Pinot Gris P. Gris.Juice....	tgl........ Tank..136	Main Warehouse	+ 802 gallon − 0	$2,595.00 $0.00	$3.24 /gallon $0.00	0	6416 0	802 0	
		Total 39	+ 802	$2,595.00			6416	802	
Press lot tg1 to tank 136. No SO2 addition. chill to 45 - 50 deg. gas head space. hold 24 to 48 hrs. - before yeast - raise ta before yeast			− 0 /ar	$0.00 $2,595.00			0 6416	0 802	
29-Sep-95			+ 802 − 0 /ar	$2,595.00 $0.00 $2,595.00			6416 0 6416	802 0 802	

Fig. 6–2 Printout of computerized inventory cost analysis. *Courtesy: Breckenridge Software, Littleton, CO.*

FEASIBILITY AND FINANCE

The structure of the winery business is another essential element facing anyone considering operating a commercial winery. Once the decision is made to move forward with a new winery project, it is a good idea to employ the services of both an attorney and an accountant. They should be able to provide answers to the many questions which surround whether or not such a venture is feasible or realistic. Further, they should explain all the hurdles, perils, and pitfalls which are involved with starting a business. This can be challenging, even frustrating, and is often where the whole idea of being a commercial vintner ends. If not, they can suggest the best manner in which the business may be structured to best serve an individual, a family, a partnership, or a group of investors.

A business plan should be written which explains the rationale for projecting how the winery shall succeed. This includes a market audit in which perceived product demand is identified, tested, and measured. Product design parameters and production levels should create a supply which parallels the perceived demand. Marketing strategies should be outlined in detail. Pro forma income statements, cash-flow analyses, and balance sheets should be prepared, along with detailed personal financial statements and vita. Although several key sets of supporting statistics and other specific testimony can be part of a good business plan, this should be kept at a minimum. Bankers already have ready access to industry figures and performance.

Table 6–1 is a typical format for a pro forma income statement for a new small winery in its second year of operation. The income statements in Table 6–1 project a poor performance as related to most new businesses. Nevertheless, breakeven is generally considered favorable as compared to the actual experience of most winery operations in the second year.

Construction costs and operating expenses are the major needs for capital in a new winery. Unfortunately, they are also the most difficult for which to obtain credit. The manner of how loans may be arranged are heavily dependent on the financial condition of those who will be principals in the business. The analysis of this situation is typically the crux upon which banker–vintner financial negotiations commence.

A traditional problem in obtaining loans for commercial wine operations are prospective vintners who have had limited winery management experience. Despite being perhaps an award-winning winemaker or an accomplished wine marketer, bankers often consider the entrepreneurship of winery management yet another matter.

Bankers are placing an increasing emphasis on cash-flow financing, as compared to collateral lending. The focus of this is placed on the soundness of the business to repay instead of the value of the assets. Consequently, business plans which show dynamic revenues and retained cash liquidity are more apt to be fi-

Table 6–1 Pro forma income statement—Second Fiscal Year

XYZ Winery

Gross sales:	
1,000 cases (12 × 750 mL bottles ea) @ retail ($100/case)	$100,000
1,000 cases @ wholesale ($50/case)	50,000
Gift shop sales @ retail	25,000
Total gross sales	$175,000
Less: Transportation out (wholesale @ $5/case)	5,000
State excise taxes @ $0.05/gal	240
Federal excise taxes	809
Returns and allowances @ 3% of gross sales	5,250
	11,299
Net sales:	$163,701
Cost of goods sold:	
Grapes—32 tons @ $800 (150 gal/ton net yield)	$25,600
Cellar materials (KMS, enzymes, fining agents, etc.)	1,800
Packaging materials (bottles, labels, corks, etc.)	3,000
Gift shop merchandise	17,500
Transportation in	4,000
Real estate taxes	3,000
Insurance	2,400
Utilities	4,400
Depreciation (building, equipment, fixtures, etc.)	20,000
Repairs and maintenance	3,000
Direct labor (2,200 hours @ $7/hour)	15,400
Fringe benefits @ 20% of labor	3,080
Miscellaneous manufacturing expense	3,000
Cost of goods sold	106,180
Gross profit	$57,521
Marketing expense:	
Samples (50 cases @ $40/case cost)	$2,000
Pamphlets (10,000 @ $0.35/each)	3,500
Travel	3,500
Public relations (tasting room materials)	3,500
Sales promotion (retailer activities)	2,500
Advertising	2,000
Direct labor (2,200 hours @ $7/hour)	15,400

Table 6-1 (*continued*)

Fringe benefits @ 20% of labor	3,080
Miscellaneous marketing expense	2,000
Marketing expense	$37,480
Marketing profit	$20,041
General and administrative expense:	
Office supplies and materials	$2,000
Computer supplies and materials	1,500
Telephone	2,400
Postage	600
Bonds, licenses and permits	2,000
Professional services	6,000
Travel	3,500
Miscellaneous general and administrative expense	2,000
General and administrative expense	$20,000
Operating Income	$41

nanced. Another field of preparation should be in the development of contingency planning. Bankers will devise "worst-case" scenarios and expect alternative action plans to be in place.

One of the most successful pathways for small winery loans is via the Small Business Administration (SBA). This federal bureau does not loan money directly but does guarantee a major percentage of each loan made by cooperating banks. Typically, the SBA will guarantee about 75%. This, of course, relieves banks of undue risk, especially with start-up winery operations. At this writing, the SBA charges a sliding-scale interest rate (on top of the bank's interest rate) starting with 3.0% for the first $250,000 and increasing to 3.5% for the next $250,000, and so forth. Repayment schedules for working capital loans are maximized at 7 years; for equipment loans, 10 years; and for real estate loans, 25 years. Here again, the advice and counsel of an attorney and accountant can help a great deal in procuring such loans.

INSURANCE

Prospective vintners must understand that winery operations embrace more than general liability. Beyond this are various forms of product liability which have increasingly become litigated. Negligence in the winery, such as poorly marked steps, loose railings on stairs, wet floors, uncovered drains, and so on, can, of course, be avoided with good management. It is, however, a mistake to assume that a new winery cannot be held legally responsible for events over which it has no control. Health problems, broken-bottle cuts, and wine stains are among hundreds of such suits brought to court. Personal injury and property damage caused in automobile accidents by drivers having left a winery tasting room have resulted in vintners losing their entire estate. Consequently, membership in the American Vintners Association (AVA) should be considered. This organization is located at 1301 Pennsylvania Avenue, NW, Suite 500, Washington D.C. 20004. The AVA sponsors most forms of vineyard and winery insurance.

LOCATION

The ultimate effectiveness of administrative, production, and marketing functions are essential to winery location. Some of the key administrative parameters involve regulatory matters, both federal and state, and sometimes local, too.

Because wineries produce an alcoholic beverage, prospective vintners can expect, over and above other small businesses, an extra list of state and local concerns attached to every potential winery site. There are, of course, different beverage control regulations for each state which must be carefully considered when choosing the winery location. These often include minimal distances from the highway serving a winery and distance restrictions between the winery and other buildings such as churches and schools. Local codes and statutes are often a bit more political in nature and may involve permanent zoning variances, utility services, wastewater treatment, building permits, and highway access. Figure 6–3 illustrates an ideal new estate winery facility.

Along with all the legal site parameters, it is a good idea to have the blessing of all local officials before committing to any site. A good public relations program can go a long way toward receiving a warm welcome and operating assistance in a community. Being at odds with the local gentry can result in insurmountable hurdles. Embracing all of this is the general attitude the community may have about a new winery in the area. A good locale can result in ongoing support toward ultimate success. Persistence in the establishment of a facility within hostile territory can be very aggravating, frustrating, and, most importantly, an invitation to financial disaster.

The premiums for adequate insurance coverages, particularly fire insurance, are generally heavily influenced by winery location. Attracting quality employees is also easier when it is convenient for them to get to work.

Production parameters also need careful scrutiny when choosing a winery site. A readily accessible supply of 220-V, three-phase electric power is preferred, along with plenty of quality water. There must also be an adequate method to readily dispose of effluents which satisfies both the ATF and Alcohol Beverage Commission (ABC) or local authorities. The winery should ideally be located rel-

Fig. 6–3 A newly constructed estate winery and cave.

ative to vineyards so that delivery time can be minimized during the harvest season. This reduces time in which fruit quality can diminish and allows for maximizing personnel deployment during the concentrated seasonal demands in both vineyard and winery. Ideally, winery location should have close access to freight pickup and delivery. Trade services such as electricians, plumbers, mechanics, air-conditioning and refrigeration technicians, processing engineers, and other specialists should be close at hand.

Marketing is generally the management area in which new wineries struggle or fail. Most newly established vintners must depend on immediate revenue to help stem the tide of capital outlay in building and establishing their winery. Retail sales at the winery are essential.

Locating the winery adjacent to a metropolitan area has the potential for several added market resources. Highway traffic to and from the city provides a high-volume source for commuters, business travelers, and, most importantly, tourists. Promotional efforts at the winery, such as spring celebrations, summer concerts, harvest festivals, and holiday programs, have a much greater chance for success with a large populace to draw from within a reasonable distance. An urban concentration of wine retail outlets and restaurants enables vintners to operate, if state ABC regulations permit, more efficiently as their own wholesale channel. Access to professional marketing services such as label design and printing, brochure development and production, advertising creation and placement, sales promotion programming, and other resources are necessary to optimize revenue generation.

Having stated these general ideas, there are no profound rules in determining winery location. Sometimes, prospective vintners already own property which is well known in the area, such as a restaurant or a farmer's market in which the addition of a winery may seem feasible.

On the other hand, there is jeopardy for poor judgment of what "well known" really is—often masked by personal bias. Forcing a justification for establishing the winery at a remote location is a common error in site selection. A farm situated back from a secondary road is rarely a good choice for a winery site. Conversely, a site situated within a mile or so of an intersection to a busy highway typically has a far better chance of attracting sufficient visitors from which to achieve adequate retail sales.

Another major factor in site selection is lot size. All too often prospective vintners provide inadequate room for parking, for outdoor events, and, most importantly, for long-term growth.

SCALE AND SIZE

Perhaps no other manufacturing business is as forgiving for the diseconomies of small size as is wine production.

The higher-price categories of wines comprise a very small share of the total market, but sufficient to make economic sense of many small winery projects.

Most aspiring vintners have a well-defined plan for what they are going to produce long before any pro forma business statements are drafted. If a thorough market study reveals that a new winery may successfully market 3000 cases (7200 gal) of two white wine types during the first year of operation, then the winery structure and its equipment should be geared to produce that amount, plus an appropriate capacity for processing and short-term growth. This might, for example, consist of the following 10,150 gal of tankage:

6—1000-gal closed jacketed stainless steel tanks
6—500-gal closed jacketed stainless steel tanks
2—300-gal refrigerated stainless steel mixing tanks
10—55-gal stainless steel drums

If, instead, the same amount of two red wine types is to be produced, winery space and capacity will need to be significantly greater, as red wine aging regimens generally require at least one additional year of inventory for barrel aging. Consequently, this tankage might consist of the following 15,400 gal of tankage:

12—500-gal stainless steel tanks with dome manways
2—300-gal refrigerated stainless steel mixing tanks

10—55-gal stainless steel drums
150—55-gal oak barrels

Usually both white and red table wines are included in the planned product portfolio. Producing a total of 3000 cases over a line of, say, two whites and two reds provides more efficiency and flexibility in tankage. Consequently, one might find the following 12,275 gal of capacity sufficient to fulfill combined short-term needs:

4—1000-gal closed jacketed stainless steel tanks
6—500-gal stainless steel tanks with dome manways
2—300-gal refrigerated stainless steel mixing tanks
10—55-gal stainless steel drums
75—55-gal oak barrels

The basic rationale of this synergy exemplifies the possibility of the 500-gal tanks serving double duty as both red wine fermenters and white wine storage vessels. Each time another wine type is added to the product line, there is a corresponding demand placed on space and flexibility of winery capacity—both in tankage and in cased goods storage.

Depending on design motif, a new 20,000-gal winery should require a minimum of 5000 ft^2, and with side walls 12 ft high, about 60,000 ft^3 of space. This is, of course, only a broad estimate. Tight design modes can greatly reduce this and extensive tasting–retail areas can likewise increase the space needed.

DESIGN MOTIF

It is essential that winery design fit the personality of the vintner and the environment. Most any chosen architectural flavor can be incorporated into an efficient floor plan. Style modes should all support a central theme. Visitors will be confused finding a winery with a Spanish-style motif producing wines with French names in a countryside which is predominantly German. Recommended is a mode which places less emphasis on ethnics and more attention on product quality. It is equally important that winery design be executed with taste and grace; overdone aesthetic interpretations are often gauche and create negative visitor impressions.

CASE STUDIES

Remodeled Small Barn

The first example is a remodeled barn winery having a capacity of about 8000 gal with a "no frills" approach. It has an interior space of 1971 ft^2, along with another 432 ft^2 of outdoor concrete apron (Fig. 6–4).

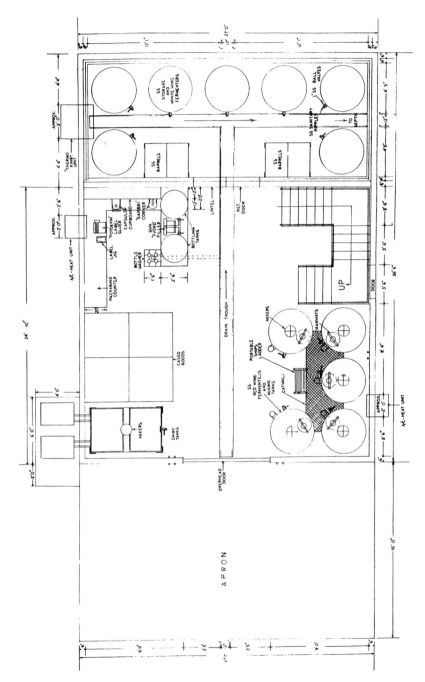

Fig. 6–4 Floor plan of remodeled barn winery.

The apron and lower floor were poured with "three-bag mix" (a superior grade) concrete reinforced with steel mesh. A floor pitch of about 1% was made toward the central drain trough in order that cleaning water could not collect and provide media for growth of microbes. Several coats of sealer were applied in order to protect the concrete from corrosive fruit acids.

The apron is utilized for crushing, stemming, and pressing; that equipment stored out of the way in an adjacent storage area during the balance of the year. Two refrigerated tanks once used for farm milk storage were overhauled and installed in this facility. One above the other, these tanks provide instant chilling for white and blush juice extracted from the press and chilling for water which can be circulated through fermenting tank jackets. The agitators allow for gentle mixing operations needed for bulk wine processing operations throughout the year.

The cold room utilizes a remodeled chilling device taken from a semitrailer refrigerated van. This can operate to maintain either 55°F for white wine fermentation during the vintage season or 26°F for stabilization. Custom-designed single-wall stainless steel tanks were chosen to maximize initial capacity.

Five 250-gal jacketed stainless steel tanks with dome manways and heavy-duty agitators serve as red wine fermenters. The jackets serve for either heating or cooling fermenting red musts. This winery is in a cool climate and often requires cold musts to be heated in order to reach the ideal 85°F temperature required by the vintner. The agitators are not essential but came as a bonus on this set of used food-processing tanks. All five tanks are mounted on legs sufficient for a gravity outlet feed into the press, which is wheeled in from the apron as needed.

The manual bottling and packaging setup is minimal in capacity and serves in saving many thousands of dollars in more sophisticated machinery. Empty bottles are brought from another barn storage area when required.

The second floor has a ground-level entry opposite the apron and contains the laboratory–office, tasting bar, retail shop, and a single rest room. This was subsequently changed when state public facilities regulations were found to require two rest rooms. Obviously, the two levels make for a rather arduous routine for tourists and transfer of cased goods for retail. See Fig. 6–5 for the layout of the interior elevations.

Simplistic Custom Design

The basic format of this concept can be used for any size of winery and is expandable to most any reasonable degree. A design sketch provided below illustrates a moderately small winery where absolute minimal cost in construction, maintenance, and operation were imperative. This division of production, with fermenting, processing, and bulk storage on one side and bottling, packaging, and cased-goods storage on the other provides for virtually limitless expansion in these opposite directions.

This building was prefabricated insulated steel construction in a simple rectangle, cleverly designed to appear complex and interesting. The floor plan (Fig. 6–6)

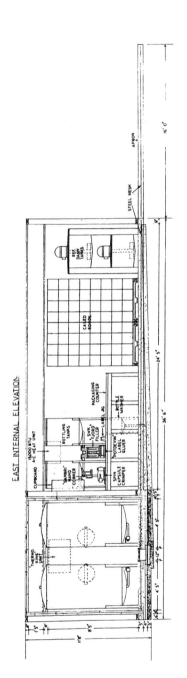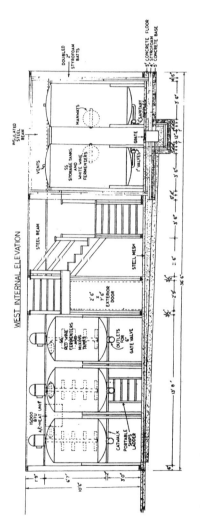

Fig. 6-5 Interior elevations of remodeled barn winery

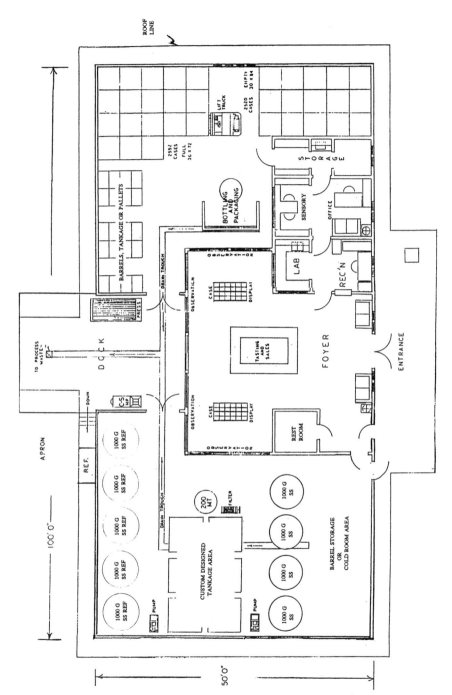

Fig. 6–6 Floor plan for simplistic design winery.

is conceived in a glass-wall-enclosed center so that observation of crushing–pressing, cellar operations, laboratory, and bottling–packaging all take place inside the hospitality–sales area—eliminating the need for guided tours. The dock area also serves as a stage for outdoor presentations for various programs and festivals held each year.

Semiautomatic bottling and packaging machines permit production of more than 300 cases per day. As the cost of such machines often places a severe financial burden on new winery ventures, it may be far less expensive to produce fewer cases daily with hand-operated machinery. This has the dual benefit of being able to display "handmade production" with part-time employees during peak visitor times, which can enhance production action and image.

This particular concept is designed for the production of several wine items and, hence, smaller-sized tanks were selected. For a smaller product portfolio, larger tanks would be needed.

The laboratory, reception, and office area are clustered, as there is often a single person fulfilling these functions in a small winery—particularly so during off-season, when this person may also serve as the tasting and sales clerk.

CHAPTER 7

REQUIREMENTS, RESTRICTIONS, AND REGULATIONS

The requirements and restrictions of wine production are regulated by an extensive set of statutes. It is highly recommended that anyone seriously considering starting a winery first obtain a complete set of both federal and state regulations. These should be read, understood, and kept readily available for future reference.

As mentioned earlier, the U.S. Bureau of Alcohol, Tobacco and Firearms (ATF) is the federal regulatory and enforcement authority for wine production in the United States.

The ATF will not issue permits for commercial wine premises in the design phase. A winery must be operational, or nearly so, before a ATF inspector will visit the facility for inspection. Consequently, ensure that winery property, buildings, and major equipment are in place prior to applying to the ATF. It often takes several inspections before the permit is issued.

With land acquisition, winery design, construction, ATF permits, Alcohol Beverages Commission (ABC) and local permits, the decision to establish a winery may need to be made more than 1 year in advance of the first grape crush.

The ATF provides a booklet entitled, *Information for New Wineries,* which can be obtained by writing to Department of the Treasury, Bureau of Alcohol, Tobacco and Firearms, Wine and Beer Branch, Ariel Rios Federal Building, 1200 Pennsylvania Avenue, NW, Washington, DC 20226. This booklet is very helpful in outlining the necessary forms and procedures required to commence the lengthy task of ATF permit application.

It is essential to understand that ATF regulations are not always straightforward. Sometimes, several individual interpretations can be drawn from a single section. Similarly, one ATF inspector may provide a different opinion on a given question than another. Although this can be a source of confusion and frustration, it is recommended that all such topics be dealt with patiently and actions fully documented in files which can be easily retrieved.

No two application processes are precisely the same. Each individual principal involved in a prospective winery may undergo a background check by ATF inspectors. Some of this investigation is addressed to personal character and funding sources for the prospective commercial winery. Obviously, everyone has a different history and thus a different length of time needed to satisfy the ATF. Application process time will vary due to the manner in which the winery will be structured—a sole proprietorship, a partnership, a closed corporation, and so forth. Another time variance exists in the potential environmental impact of the wine production facility: water consumption, disposal systems, effluent quality and quantity, distances from public waterways, and air and noise pollution, among others.

These and other ATF concerns all come together as a unique equation for every winery permit application, which is further complexed by the work load facing ATF inspectors at the time the application documents are submitted. It is poor judgment to anticipate that any given winery project is simplistic. Some prospective vintners make more work out of attempts to circumvent the application process than the actual process requires.

APPLICATION FOR BASIC PERMIT

This application is made by using ATF Form 1637. It is primarily involved with identifying the applicant, the type of business in which the applicant desires to be engaged, and its location. Instructions for completing this form are provided on its reverse side.

APPLICATION TO ESTABLISH AND OPERATE WINE PREMISES

ATF Form 5120.25 must be filled out and approved in order to establish and operate a winery. This serves to disclose the name, address, and organizational format of the applicant, such as a sole proprietorship, a partnership, or a corporation.

The work and rework necessary to create and compile all of the detailed supplemental information required by the ATF often results in frustrating hopeful vintners. Formal documentation of partnerships and corporations ownership interests, articles and charters creating the organization, and specific resolutions authorizing the application to establish and operate a winery are often required as attachments. Names and addresses of principals involved in the prospective winery must also accompany this application form. Complete instructions for completing ATF Form 5120.25 as well as detailed requirements for the necessary attachments are provided on the reverse side of this form.

ATF regulation 27 CFR, Section 24.111, requires a description of premises be attached to the application. Its language follows:

> The application will include a description of each tract of land comprising wine premises. The description will be by directions and distances, in feet and inches (or hundredths of feet), with sufficient particularity to enable ready determination of the bounds of the wine premises. When required by the regional director (compliance), a diagram of the wine premises, drawn to scale, will be furnished. The description will clearly indicate any area of the wine premises to be used as bonded wine premises, used as taxpaid wine premises, or alternated for use as bonded wine premises and taxpaid wine premises. The means employed to afford security and protect the revenue will be described. If required by the regional director (compliance) to segregate operations within the premises, the manner by which the operations are segregated will be described. Each building on wine premises will be described as to size, construction, and use. Buildings on wine premises which will not be used for wine operations will be described only as to size and use. If the wine premises consist of a part of a building, the rooms or floors will be separately described. The activities conducted in the adjoining portions of the building and the means of ingress and egress from the wine premises will be described.

As this is read, it should be clear that a diagram of the wine premises "may" be required by the regional director of the ATF. It is advised that a diagram be made beforehand in order to avoid potential delays during the application approval process. To that end, it is a good idea to contact the regional director in order to ascertain the format in which the diagram is to be rendered.

BONDS

The ATF is charged with the responsibility of protecting uncollected excise taxes on wines located on bonded premises. The application process for a commercial winery permit seeks authorization to hold wines "in-bond," or excise tax unpaid. This places a burden on the vintner to assure the ATF that all excise taxes are guaranteed payment—with a minimum bond of $1000. Unless a prospective vintner wants to make a monetary pledge covering the maximum production tax, a wine bond must be applied for with a bonding company using ATF Form 5120.36. As these are generally rather inexpensive for responsible citizens, it is a good idea to calculate the bond principal well over any perceived maximum tax exposure so that there can be no question that this is fully covered.

It is best to provide the bonding company extra copies of form 5120.36 as any erasures or alterations to the bond form require written consent to those changes. Consequently, it makes good sense to ensure that the document is executed without error. There are other important details listed in the instructions on the reverse side of the form.

ENVIRONMENTAL IMPACT

New wineries are subject to the same concerns for the environment as any other business project. The ATF administers this responsibility for compliance with two documents, ATF Forms 1740.1 and 1740.2.

The first of these, 1740.1, has two sides to complete and deals with general information relating to the composition and quantity of liquid and solid wastes, as well as noise pollution projected to be generated by the winery. It is recommended that the response to each question be calculated carefully and those figures kept on hand for ready reference when the ATF inspector visits the premises. It is also recommended that copies of local effluent, disposal, and noise codes be on file in order to avoid delays in ensuring compliance.

The second form, 1740.2, has only one side to complete and supplements the above form in regard to how liquid wastes from the winery will be handled. The primary concern is the condition and manner by which effluents may enter public waterways. It is recommended that the plat diagram discussed with the application form above have a supplemental page indicating location, size, and type of winery wastewater treatment facilities, as well as accurate distances to any nearby public waterways.

Affidavits relating to potential winery impact may be gathered, if possible, from local environmental authorities and attached as addenda to ATF Forms 1740.1 and 1740.2 when they are submitted.

Wineries planned in or near a registered historic site should have documented approval from the state historic preservation authority prior to making application.

PERSONNEL QUESTIONNAIRE

The intent of this is to help ensure that people with unsuitable backgrounds are denied responsibility in commercial wine production. Administration of this is by ATF Form

5000.9, which is a three-page format generally studied with great scrutiny by ATF officials. Approach this form with close attention to detail, including all zip codes, telephone numbers, and so forth, as inspections usually include interviews with some or all of the personal references listed. Prepare a fully detailed financial profile, including bank statements, security transactions, loan schedules, and any other pertinent documents which substantiate funding sources for the prospective winery.

Form 5000.9 must be completed by all partners and any person who has, or will have, a significant financial interest in the winery.

SIGNATURE AUTHORITY

If the winery structure is in the form of a corporation, ATF Form 5100.1 provides for a list of corporate officials who will be authorized to sign on behalf of the company. This document must be executed with the corporate seal.

For absentee owners and similar partnership situations, it may be necessary to provide authority for another person to sign official ATF documents. This can be made with a *Power of Attorney* via ATF Form 5000.8.

LABEL APPROVALS

Prospective vintners should not delay label design until after the winery application is approved. As each label must be submitted with ATF Form 5100.31 and can be subject to its own time-consuming processes, it is a good idea is to move forward with artwork and other preliminary label design activity in order to obtain label approvals as soon as possible.

Membership in the American Vintners Association (AVA) provides assistance in processing wine label approvals. For full information on this service, write the AVA at 1301 Pennsylvania Avenue, NW, Suite 500, Washington, DC 20004.

FORMULA WINES

Prospective vintners may desire to make wines in which flavorings are added or some other additive which is not naturally part of the grape or fruit. Wine jargon sometimes refers to these as "Special Natural Wines." Each of these wines require a completed ATF Form 5120.29 to be submitted for approval. Instructions on the reverse side of the form are easy to follow. As formula-approval processing can also be a source of delay in getting a new winery operation, it is advised that the research and development in devising such products be finalized and readied for executing Form(s) 5120.29 as soon as practicable. See pages 113 and 114.

STATE ABC REGULATIONS

Of course, each state has its own regulatory statues. It is beyond the scope of this book to outline them all. However, it may serve prospective vintners to peruse some of the Indiana ABC rules in order to exemplify how these are structured. These are, of course, valid only in the state of Indiana.

Key sections of the Indiana Code relating to winery start-up and operation follow:

Section 7.1-3-12-5. Scope of small winery permit.

(a) The holder of a small winery permit is entitled to manufacture table wine, to bottle wine, and to bottle table wine produced by his small winery. He is entitled to serve complimentary samples of his table wine on the licensed premises. He also is entitled to sell his table wine on the licensed premises to consumers either by the glass, or by the bottle, or both. The permittee also is entitled to sell table wine by the bottle or by the case to a person who is the holder of a permit to sell wine either wholesale or retail. The holder of a small winery permit is exempt from the provisions of IC 7.1-3-14 [Wine Retailers Permits]. The holder is entitled to advertise the name and address of any retailer or dealer who sells wine produced by his winery.

(b) With the approval of the commission, a holder of a permit under this chapter may conduct business at a second location that is separate from the winery. At the second location, the holder of a permit may conduct any business that is authorized at the first location, except for the manufacturing or bottling of wine.

Section 7.1-3-21-11. Location of premises in proximity to school or church—Prohibitions—Restrictions.

The commission shall not issue a permit for premises situated within two hundred feet (200′) from a school or church if no permit has been issued for the premises under the provisions of Acts 1933, Chapter 80.

Section 7.1-4-1-25. Small winery permit.

The annual license fee for a small winery permit is two hundred fifty dollars ($250).

Section 7.1-4-1-31.5. Supplemental retailer's permit [Sunday sales permit].

The annual license fee for a supplemental retailers permit is thirty-three and one-third percent ($33^1/_3$%) [$83.33] of the amount of the annual license fee that the applicant paid, in the aggregate, for the retailer's permit that qualifies him to hold a supplemental retailer's permit.

Section 7.1-4-4-1. Rate of tax.

An excise tax at the rate of forty-seven cents ($0.47) a gallon is imposed upon the manufacture and sale or gift, or withdrawal for sale or gift, of wine within this state.

Section 7.1-4-4-2. Beverages to which tax is applicable.

The wine excise tax shall apply to wine that contains less than twenty-one percent (21%) of absolute alcohol reckoned by volume. The wine excise tax also shall apply to an alcoholic beverage that contains fifteen percent (15%), or less, of absolute alcohol reckoned by volume, mixed with either carbonated water or other potable ingredients, or both, by either the manufacturer or the bottler, or both of them, and sold in a container filled by the manufacturer or bottler, and which is suitable for immediate consumption directly from the original container. An alcoholic beverage that is subject to the wine excise tax shall not be also subject to the liquor excise tax.

Section 7.1-5-7-13. When employment of person 18 years or older on licensed premises lawful.

The provisions of IC 7.1-5-7-12 shall not prohibit the employment of a person eighteen (18) years of age or older on or about licenses premises where alcoholic beverages are sold, furnished, or given away for consumption either on or off the licensed premises, for a purpose other than selling, furnishing, consuming or otherwise dealing in alcoholic beverages. Nor shall the provisions of IC 7.1-5-7-12, prohibit a person eighteen (18) years of age or older from ringing up a sale of alcoholic beverages in the course of his employment.

Section 905 IAC 1-15.31. Employees between eighteen (18) and twenty-one (21) years of age.

Sec. 1. (a) A minor eighteen (18) years of age or older employed at a licensed premises may ring up a sale of alcoholic beverages in the course of his employment, provided that at the time the sale is rung up, there is at least one (1) other employee on the licensed premises who is:
(1) twenty-one (21) years of age or older; and
(2) responsible for supervising said minor.

Section 7.1-5-10-1. Times when sales unlawful.

(a) It is unlawful to sell alcoholic beverages at the following times:
   (1) At a time other than that made lawful by the provisions of IC 7.1-3-1-14 {basically, Monday through Saturday 7 a.m. until 3 a.m.}
   (2) On Christmas Day and until 7:00 o'clock in the morning, prevailing local time, the following day; and
   (3) On primary election day and general election day from 3:00 o'clock in the morning, prevailing local time, until the voting polls are closed in the evening on these days.

(b) During the time when the sale of alcoholic beverages is unlawful, no alcoholic beverages shall be sold, dispensed, given away, or otherwise disposed of on the licensed premises and the licensed premises shall remain closed to the extent that the nature of the business carried on the premises, as at a hotel or restaurant, permits.

(c) It is unlawful to sell alcoholic beverages on New Years Day for off-premises consumption.

CHAPTER 8

GETTING STARTED

RECORD KEEPING

The distress of repeatedly making the same mistakes is perhaps only surpassed by the frustration of having made a good wine and forgetting how it was done. It is wise to keep a good price and expense history in the files, and it is essential to have up-to-date processing and inventory information to supplement ATF-required records. All of this obviously indicates that a functional record-keeping system is essential in operating a successful winery business. A comprehensive personal computer program as discussed in Chapter 6 is ideal. It may, however, be good education and thrift to start out with manual card files and ledgers in order to fully experience all of the wine production, marketing, and accounting operations in detail before applying computerization. A flowchart for basic record keeping is presented in Fig. 8–1.

Every lot of wine must be given adequate identity from which their history can be tracked. An easy-to-follow card file is more efficient than a ponderous journal. As mentioned in the previous paragraph, an integrated computer program, especially designed for commercial wine operations, is best of all. Information essential to the ATF can be segregated and instantly copied on a printout as a work

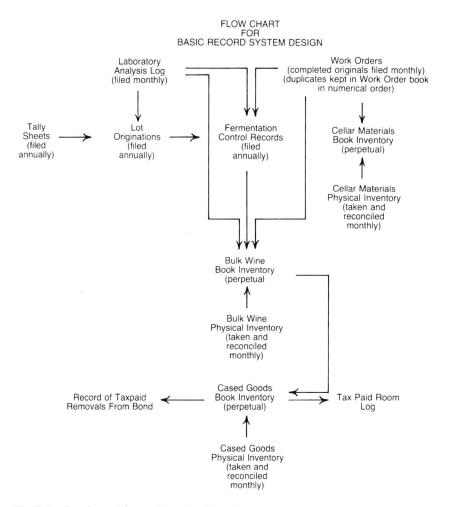

Fig. 8–1 Supplemental record-keeping flow chart.

sheet—leaving the system and the vintner free to do other things during the process of generating both routine report forms, as well as during an on-site ATF inspection. Computer systems can easily be used to generate valuable historical, trend, and other data essential to a progressive operation.

It is recommended that every vintner carefully review 27 CFR Part 4, Subpart O—Records and Reports. Following are quotations used for ATF enforcement:

> **24.300 General.** "(a) *Records and reports.* A proprietor who conducts wine operations shall maintain wine transaction records and submit reports as required by this part. . . ."

24.301 Bulk still wine record. "A proprietor who produces or receives still wine in bond, (including wine intended for use as distilling material or vinegar stock to which water has not yet been added) shall maintain records of transactions for bulk still wine. . . ."

24.302 Effervescent wine record. "A proprietor who produces or receives sparkling wine or artificially carbonated wine in bond shall maintain records showing the transaction date and details of production, receipt, storage, removal, and any loss incurred. . . ."

24.303 Formula wine record. "A proprietor who produces beverage formula wine shall maintain records showing by transaction date the details of production. . . ."

24.304 Chaptalization (Brix adjustment) and amelioration record. "(a) *General.* A proprietor who chaptalizes juice or ameliorates juice or wine, or both, shall maintain a record of the operation and transaction date. Records will be maintained for each kind of wine produced (grape, fruit, or berry). No form of record is prescribed, but the record maintained will contain the information necessary to enable ATF officers to readily determine compliance. . . ."

24.305 Sweetening record. "A proprietor who sweetens natural wine with sugar or juice (unconcentrated or concentrated) under the provisions of this part shall maintain a record of sweetening by transaction date. . . ."

27 CFR Part 4 goes on to include

24.306 Distilling material or vinegar stock record.
24.307 Nonbeverage wine record.
24.308 Bottled or packed wine record.
24.309 Transfer in bond record.
24.310 Taxpaid removals from bond record.
24.311 Taxpaid wine record.
24.312 Taxpaid wine returned to bond record.
24.313 Inventory record.
24.314 Label information record.
24.315 Materials received and used record.
24.316 Spirits record.
24.317 Sugar record.
24.318 Acid record.
24.319 Carbon dioxide record.
24.320 Chemical record.
24.321 Decolorizing material record.
24.322 Allied products record.

One of the best ways to simplify record keeping is in the use of coded lot numbers. For example, the first two digits in a lot number could indicate the vintage year, and a third digit could reveal the source of the grapes—with 1 noting estate-grown grapes, 2 being grapes purchased from XYZ Vineyards, and so forth. A fourth digit might note the processing progress of the lot, such as 1 noting fermenting juice or must, with a new digit assigned for each processing step completed, ascending to 9 which could indicate wine ready for bottling. From this code, a lot numbered say, 9618 Seyval Blanc, would instantly inform the winemaker that this is vintage 1996 Seyval Blanc, from his own estate vineyard, which may need only filtration in order to be ready for bottling. A 9725 Chambourcin would be 1997 vintage grown at XYZ Vineyards, a red wine in some middle stage of processing.

Figures 8–2 through 8–6 are forms which vintners may reprint and use in their supplemental record keeping.

A Laboratory Analysis Record (Figure 8–2) is kept as a log of all analysis and evaluations made in the winery laboratory. It is essential that the precise source, variety, and weight of grapes and fruit have documentation in the winery files. ATF inspections will require that information in order to trace production output and that label statements on the resulting wine are factual.

The second form (Fig. 8–4) which vintners should consider is that which documents the transition of grapes into fermenting must or juice. This is often called a wine's "birth Certificate." A Lot Origination (Fig. 8–3) also helps to trace information which can help to plan for more efficient time, labor, and material resource allocations for future grape receiving activities.

Once fermentation is completed, the juice or must has then, of course, become wine and should be transferred to a perpetual bulk wine record (Fig. 8–5). At that time, the Lot Origination form would be retired to a record file.

Figure 8–5 can be reprinted and used as a 5" × 7" card format for each lot of wine made. The reverse side of the card should serve to continue the record needed for each wine. As each wine lot is bottled, it should be transferred from the Bulk Wine Record to the Bottled Wine Record (Fig. 8–6).

If activity in the history of a wine goes beyond the scope of one card simply staple a new card to the front of the original and continue on in the file. When a bulk wine lot is completely bottled, the card can be terminated and retired to a "dead file." Similarly, when a bottled wine lot has been fully depleted, that card can also be retired to the "dead file."

Card records for sugar, processing chemicals, and other materials can be maintained in much the same manner.

An example of table wine histories commencing with Lot Origination, Bulk Wine Record, and Bottled Wine Record are provided at the end of Chapters 9–11 to illustrate typical white, red, and blush wine histories in a small winery.

LABORATORY ANALYSIS

DATE	TANK NO.	LOT NO.	TYPE-VARIETY	GALS.	pH	BRIX-BALL.	A_C	EXT	T.A.	V.A.	FSO2	TSO2	CU	FE	C C	N T A

From: Commercial Winemaking: processing and controls by Richard P. Vine 1981 edition, page 114
9/21/95:dlg

Fig. 8–2 Laboratory analysis record.
Artwork: Dave Gahimer.

	LOT ORIGINATION
DATE _____	CULTIVAR _____
LOT NO. _____	REMARKS: _____
SERIAL NO. _____	_____
FERMENTER NO. _____	_____
NET GALLONS _____	_____
NET TONS _____	_____
GALLONS PER TON _____	
pH _____ T.A. _____	_____
BRIX _____ ALC. _____	_____
EXT. _____	_____

AMELIORATION = _____ % @ _____ BRIX = _____ ALCOHOL

```
         GALLONS START    = _____   RESULTING TOTAL
         INVERSE OF                 GALLONS OF PRODUCT
         AMEL. PERCENTAGE

         TOTAL         @ _____ BRIX  ( _____ LBS/GAL)= _____ TOTAL
         GALLONS                                             LBS.

         START         @ _____ BRIX  ( _____ LBS/GAL)= _____ START
         GALLONS                                             LBS.

         AMELIORATION                                  _____ ADDITION
         GALLONS                                             LBS.

         SUGAR AS                     x .074 (GAL/LB)= _____ SUGAR AS
         GALLONS                                             GALLONS

         WATER
         GALLONS

         YEAST ADDITION = RATE OF _____ LBS PER _____
```

REMARKS: _____

From: Commercial winemaking by Richard P. Vine
  1981 edition, page 119
9/21/95:dlg

Fig. 8–3 Lot origination form.
Artwork: Dave Gahimer.

FERMENTATION CONTROL RECORD

TANK NO. _____ LOT NO. _____ VARIETY _____

DATE	TIME	TEMP.	BALL.	ALC.	T.A.	V.A.	C	C	N	T	

From: Commercial Winemaking: Processing and controls, by Richard P. Vine
1981 edition, page 126
9/21/95:dlg

Fig. 8–4 Fermentation control card.
Artwork: Dave Gahimer.

BULK WINE BOOK INVENTORY CARD

Lot. No.	Type-Variety			Class		Color				

REMARKS:

Date	Racked From Tank W.O. Gallons	Treatment-Disposition	Racked To Tank W.O. Gallons	PHYS INV	ALC	BALL	T.A.	V.A.	FSO2

From: Commercial winemaking by Richard P. Vine
1981 edition, page 141
9/30/95:dlg

Fig. 8–5 Bulk Wine Record card.
Artwork: Dave Gahimer.

CASED GOODS BOOK INVENTORY								
Lot No.	Vintage		Type-Variety		Size			
Class	Color		Minimum Inventory	Bottle Type	Closure		Other
Date	Cases In	Cases Out	Balance on Hand	Ref No.	Inv.	Remarks
From: Commercial Winemaking by Richard P. Vine							
1981 edition, page 207							
10/2/95:dlg							

Fig. 8–6 Bottled Wine Record card.
Artwork: Dave Gahimer.

SANITATION

The very first consideration in winemaking is for sanitation. The constant jeopardy of spoilage organisms makes it easy to turn fine grapes into poor wine. Scalding hot-water cleaning is repeatedly recommended in this book. Equally important is the drying afterwards; hot water trapped in crevices will cool and become a growth media for organisms.

QUALITY ASSURANCE

There are few vintners who look forward to the tedious demands of an effective quality assurance program. Quality control must be carefully designed and faithfully performed. More importantly, wine enthusiasts have a keen sense of value. They will not excuse low-quality wine—and they will not buy it.

An adequate quality control program does not have to be capital-intensive and it does not require extensive training. The worst-case scenario is spending the time and money to set up a workable program and *then not use it.*

Ideally, the assurance of every wine meeting prescribed standards of quality should be a task totally separated from production. This has led most larger vintners to establish a department within management framework which has an entirely autonomous quality assurance obligation and authority. Samples from every phase of production are gathered, analyzed, evaluated, approved, or disapproved. It is with the failure of a sample to pass the scrutiny of testing that the quality control system really works as protection for the vintner. Thus, a wine problem has an opportunity to be corrected before it becomes a much larger problem in the marketplace.

This divided production/quality control authority can be criticized as impractical in small wineries. This should not, however, stop any vintner from employing the spirit and philosophy of such system. Most small winery operators must assume several different roles—and the unbiased judgment of honest quality control is probably the most important of them all. In its simplest form, quality control can be a thorough sensory evaluation made outside of the winery, perhaps at home, at a time and place which creates total separation from production.

The road to achieving this as routine may take some initiative, but it does not have to be difficult. It is a matter of identifying potential sources of problems and standards of acceptance, then designing timely analytical and testing procedures to measure and record the findings. The immediate interpretation of these findings thus becomes key to any action which may be necessary to steer wine production from harm's way.

ANALYTICAL INSTRUMENTATION

Wine was made long before any testing was performed in a laboratory setting. In modern times, though, it makes sense to utilize affordable technology when it can take the guesswork out of making wine from expensive grapes.

pH Meter

Fruit acidity is one of the most important criterion in making wine. The pH test is, in simple terms, an indication of the "strength" of acidity in a juice or wine—therefore a measure of how ripeness has progressed. More precisely, pH is a measurement of hydrogen ion concentration and the relative ability to neutralize basic (alkaline) hydroxyl ions. The pH test is expressed on a scale from 0 (being the most acidic) to 14 (being the most basic), with neutrality at pH 7.

In winemaking, the pH meter is frequently used during the harvest. Whereas the taste buds of an experienced winemaster may often be the final word for when grapes are to be harvested, it is the pH test which often helps in pinpointing the optimal time. Litmus papers are inadequate for wine pH testing, as they are not precise enough to be of any value.

Most grapes are green at pH 2.70, and as ripeness progresses, pH will increase. Winemakers will often standardize pH at 2.90–3.10 for wines which are meant to have sharp, crisp acidity—particularly for *cuvée* wines earmarked for secondary fermentation into sparkling wines.

Fresh, fruity white and blush wines are made from fruit in the 3.10–3.30 range. This is usually just slightly on the green side of ripeness. Most importantly, this lower level of pH resists bacterial activity and allows for better efficiency in the use of SO_2 as a preservative.

Heavy, complex red wines are frequently harvested in the 3.30–3.50 range. Higher pH ensures complete flavor and pigmentation development in the fruit and allows for bacterial activity which has become commonplace in red wine processing.

It may take some convincing to spend $300–600 for a reliable pH meter and the several hours of time reading the instructions and practicing its proper operation. (Refer to Appendix B for the proper procedure in operating a pH meter.) Other than convenient access to pH testing from another source, there are few good alternatives, and the prevention of just one lot of good fruit from being ruined far more than justifies the expense.

Brix and Balling

These tests address the sugar content in grapes and wine and are the simplest of testing instruments essential for commercial winemaking.

A set of hydrometers, or "spindles," with the following ranges are needed:

- −5.0° to +5.0°
- 0.0° to 8.0°
- 8.0° to 16.0°
- 16.0° to 24.0°

Hydrometers float at different levels, depending on the density of the liquid in which they are placed. In pure water, the −5.0° to +5.0°-range hydrometer should float with a 0.0° reading.

The same hydrometers can be used during and after fermentation. Whereas increasing sugar levels makes the hydrometer float increasingly higher, increasing alcohol levels has the opposite effect. Consequently, the hydrometer floating in a wine sample is read as a "Balling" test and indicates relative viscosity—the "thinness" or "heaviness" in mouth feel of the wine. A 12% alcohol wine, fermented completely dry, should have a Balling of −1.8 to −2.2, depending on nonsugar solids remaining in the wine. Note that this is a minus reading, which means that wine density is actually less than water at 0.0° Balling. As will be seen in the following text, the Balling test is the perfect tool for determining fermentation progress and controlling finished wine sweetness levels. A set of four Brix–Balling hydrometers for the ranges indicated earlier should cost less than $50.

A quicker and more convenient method of measuring Brix is by the use of a refractometer, although considerably more expensive than the hydrometer method. It operates by measuring the light refracted through a drop or two of grape juice on a prism. That light is directed to a scale which is read as one holds the refractometer up to a bright source of light. A refractometer is particularly good for testing grape sugars in the vineyards. This instrument cannot be used to measure Balling. Reliable Brix refractometers can be budgeted at about $150 or so.

Every vintner will face the prospects of adjusting the amount of sugar in juice, must, and wine. Grape juice or must measuring at 23° Brix (roughly a percentage of sugar solids) will ferment to about 12.3% alcohol by volume. The calculation is made by multiplying the Brix reading from the hydrometer by 0.535, which is the amount of alcohol which results from each degree of Brix. It is readily apparent that Brix is an essential test in calculating the precise amount of sugar addition, if any, to achieve the desired alcohol level from fermentation.

Methods for the operation of Brix–Balling hydrometers and refractometers are found in Appendix B. A table of sugar solids in solution for each degree of Brix is also provided in that section.

As in the wine cellar, sanitation in the laboratory is absolutely essential. After each use, each instrument should be thoroughly rinsed at least three times with distilled or deionized water and then placed upright in a storage rack to dry.

Residual Sugar

The Balling scale cannot be used to determine whether or not a "dry" wine is absolutely free of remaining fermentable sugar. It is important to know this prior to wine bottling in order to ensure that yeasts will not commence a secondary fermentation.

One simple and inexpensive method to analyze for residual sugar is by the use of Dextro-chek® tape. A piece of the tape is dipped into the wine sample, removed, and allowed to dry. Fermentable sugar, if any, will be revealed by a color change of the tape, which is then compared to a graduated scale included in the kit.

Total Acidity

This is a measurement of tartness. It is difficult to make definitive judgments of how tart each lot of juice or must is by taste alone. Sweetness has a "masking" effect which blunts the sensory effect of tartness. Notice that the term "sour" was not used, as this has to do with wine spoilage, such as the formation of vinegar.

Most table wine total acidity measures in the range 0.500–0.750 g/100 mL—from bland to tart. This numerical value can also be considered a rough expression of total acidity percentage.

The total cost for buret, buret stand, sodium hydroxide solution, and several 250-mL Erlenmeyer flasks should not exceed $100 and is money well spent to assure acidity levels are consistently balanced.

Appendix B provides the proper method for total acidity analysis, as well as a table which can help in easily determining results.

Volatile Acidity

Most of the acids formed by spoilage bacteria are volatile and can be isolated by using a Cash Volatile Acid distillation apparatus. Commercial vintners rely on this instrument to monitor the extent of volatile acid formation in wine, and thus the extent of spoilage bacteria activity. The ATF [27 CFR, Part 4, Section 4.21 (a)(1)] limits volatile acidity in commercial white wines to 0.120 g/100 mL, whereas red wines are set at a 0.140 g/100 mL maximum.

The Cash apparatus complete with condenser and attending glassware can exceed $400. Novices are encouraged to seek some assistance in individual operational instruction before attempting to operate the device.

Appendix B provides the proper method of using a Cash apparatus in the determination of volatile acidity and a table which can help in easily determining results.

Sulfur Dioxide

Another essential test is that for sulfur dioxide (SO_2), the agent used for preserving wines from further yeast and/or bacterial activity. Dry table wines having lower pH levels can be protected with as little as 30 mg/L. Wines having residual fermentable sugar and higher pH levels may require more than as 100 mg/L. The milligrams per liter expression is often made as parts per million (ppm), which is virtually the same value.

The instrumentation and materials needed to perform the Ripper method used in this analysis costs less than $100. The standardized iodine reagent, however, degrades rather quickly and should be kept refrigerated.

A better analysis for sulfur dioxide in wines is by the vacuum aspiration method. The glassware and other materials needed to set up this apparatus is a bit more expensive, but the results are more precise.

Analytical procedures for the Ripper method of determining both free and total sulfur dioxide are provided in Appendix B.

Sensory Evaluation

The investment in instrumentation for this analysis is simply a good set of durable wine glasses. A dozen or so restaurant-grade, all-purpose, "tulip-shaped" glasses are recommended. They should be identical so that wines can be compared next to each other. Good quality wine glasses should not exceed $5 each in cost. Finer grade glasses can exceed $50 each.

Essential to meaningful sensory evaluation of wines is an experienced palate. Sensory education can be the single largest expense incurred in tooling up for wine analysis. One of the best ways to dilute this cost is by finding ways to participate in tasting functions. Chapter meetings of the American Wine Society, Les Amis du Vin, and other wine groups usually include wine evaluations. Serving at wine competitions is another excellent way to develop expertise. Local wine retailers and restaurateurs may be sources from which to learn of promotional tasting programs. Groups of friends can certainly share on their own.

There are many different wine evaluation scales in use around the world. In the United States, two major formats are used most frequently: the 20-point scale and the 100-point scale.

Consumers may consider wine judgment as that limited to a very narrow set of personal likes and dislikes. Commercial wine judging expertise, however, requires a mindset which can separate personal preferences from actual flaws and virtues. See Appendix B for sensory wine analysis procedure.

Laboratory Refrigerator

Access to a refrigerator, including a freezer unit, is essential in the commercial wine laboratory. Sometimes, reliable, low-cost, used units can be found. Most analytical solutions, indicators, and reagents require refrigerated storage. Dried yeast should be kept frozen in order to maximize viability and storage life. Yeast cultures on agar or in broth should be refrigerated.

WINERY EQUIPMENT

There is virtually no end to the amount of equipment which can be employed in the modern commercial wine facility. It is this part of winemaking which can be approached with a number of rewarding philosophies. Some vintners pursue the challenge of expending the least possible cost for equipment—of "making do" and employing all manners of clever creativity. Conversely, others enjoy the pride of building and equipping what amounts to a rather "high-tech" winery. Most beginning wine producers find themselves somewhere between these extremes.

There are several equipment needs which are essential to making wine. They can be relatively expensive. Some winemakers reduce this burden by sharing equipment costs with friendly competitors.

Scales

It is necessary to fully document the net weight of all grapes and other fruit received at the winery, whether from estate vineyards and orchards or from independent suppliers.

ATF inspectors will require this information as a starting point in which to trace wine production histories. Some of the new digital scales, with 4 ft × 4 ft platforms and 5000 lbs. of capacity can be found for less than $3000. (See Fig. 8–7)

Crusher–Destemmer—Must Pump

Fruit must be "crushed"—a gentle cracking of the skins and pulp (not the seeds) sufficient for the pressing operation to be effective and efficient. Disintegrated seeds and stems will cause the resulting wine to be bitter and astringent due to excessive extraction of phenolic compounds.

Sometimes vintners are fortunate enough to find grower–suppliers who will sell fruit already crushed and destemmed. Be prepared to pay a premium for this service, however. The cost for a good quality crusher-stemmer-must pump (Fig. 8–8)

Fig. 8–7 Electronic scales.

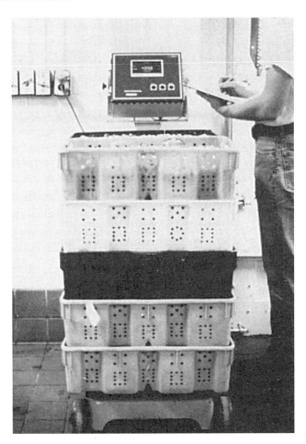

Fig. 8–8 Crusher–stemmer–must pump.
Courtesy: Prospero Equipment Corporation.

should be budgeted at about $1500 for a 1-ton/h unit. The heavy-duty work which these machines endure each vintage will wear the food-grade paint off steel hopper and shaft surfaces. These should be fully refinished immediately following the last crush each year. Larger-capacity machines, fabricated in stainless steel, will cost much more.

For some red wines, a portion of stems remaining in the must can be desirable. The best types of destemming devices are equipped with pumps permanently installed at the bottom of the receiving hopper. A crusher–destemmer–must pump generally costs progressively upwards of $1500 for higher-capacity machines built from stainless steel.

Press

No other piece of equipment is more symbolic of the wine art than the press. A basic piston-type basket press having a quarter-ton must capacity generally costs less than $1000 but requires considerable labor to operate.

Such presses operate by virtue of a wooden plate, or "piston," forced down-

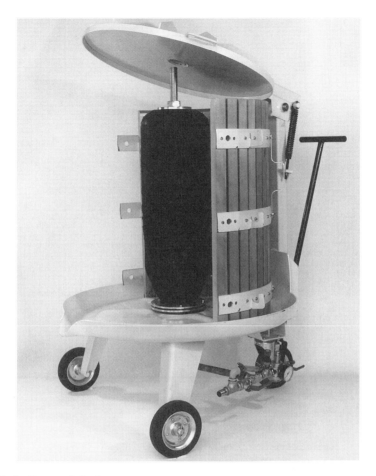

Fig. 8–9 Pneumatic basket press.
Courtesy: Prospero Equipment Corporation.

ward by the press head turned down the screw shaft running vertically through the center of the basket. The juice or wine is extracted out through the spaces left between the basket slats. Keep the shaft well lubricated with a food-grade vegetable grease. For another several hundred dollars of investment, the piston can be augmented with hydraulic pressure generated by a hand pump—well worth the price.

Better yet are vertical and horizontal pneumatic basket presses (Fig. 8–9) in which heavy-duty rubber "balloons" are inflated inside to exert pressure on the must against the press walls. These cost a bit more and are, again, well worth the price.

Best of all are "tank presses," which function with the balloon principle but with gentle action within stainless steel baskets and minimize extraction of pulp

Fig. 8–10 Stainless steel tank press.
Courtesy: KLR Machines, Inc.

and seed particles (Fig. 8–10). The cost of the smallest of these can exceed $10,000.

Pumps

Very small wineries can get along with centrifugal plastic-head pumps so small that they can be easily carried around the cellar. These cost less than $200 each and usually have a maximum capacity of about 10 gal/min. Greater capacities and efficiencies are found in larger stainless steel pumps mounted on rubber wheels. Depending on size and options, these generally range from $1000 to $3000 each.

Hose

The hose needed for transferring crushed white grapes (must) from the crusher–stemmer–must pump to either the press, or red must to fermenters, will need to be at least 3 in. in inside diameter (ID). This should be a good quality, food-grade hose of a length 10–15% longer than the longest anticipated distance for must transfer. For a fine quality 3-in. must hose, budget at least $15 per foot.

Transparent or translucent dairy-type hoses are recommended, as they are relatively inexpensive and durable if properly maintained. The ability to see liquid in movement through the hose has obvious advantages in operation. A 1-in. ID hose is recommended in the small winery. Rarely is there a need for less than 100 ft of

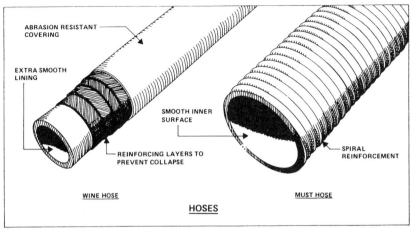
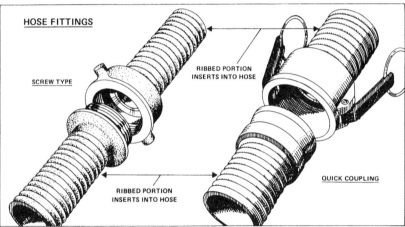

Fig. 8–11 Illustration of typical winery hoses and fittings. *Courtesy: Sebastiani Vineyards.*

such hose. Typically, two sections of about 6 ft or so are fitted as "sucker" hoses for the inlet of the pump, and two longer sections are cut as "delivery" hoses for the outlet of the pump. For a fine quality, 1-in. food-grade transparent transfer hose, budget at least $6 per foot.

In all cases, female fittings should be made at the end of the hoses, not on tanks, filters, or other units. The rationale for this is that the female fittings can better withstand hoses continually being dropped or hit on hard surfaces. Male fittings, whether threaded or adapted, will soon dent and foul, wearing to the point where leakage is difficult to control. Figure 8–11 illustrates typical winery hoses and fittings.

Tanks, Barrels, and Other Bulk Wine Vessels

There are all sorts and manner of containers used for making wine. Of critical importance is the material from which they are made. The best are constructed from stainless steel (Fig. 8–12), although glass carboys are often used for very small lots. Carboys should be budgeted at about $15 each.

Stainless steel and glass containers can be cleaned effectively with sufficient warm water rinsings to wash out any juice, wine, or solids which remained when emptied. This should be followed by the addition of $1/4$ teaspoon of potassium meta-bisulfite per gallon of rinse water and then drained dry.

Polyethylene can be acceptable if used only for short-term processing needs, such as for fermentation. Extended use has given rise to controversy over the development of off-flavors in wines. Used poly drums (55-gal capacity) can be found at very reasonable prices, sometimes at less than $10 each.

Although there is plenty of wine romance in using kegs and barrels in winemaking, they are also very difficult to employ properly. Wood is organic and can suffer deterioration, even ruination, from spoilage microbes. Wooden vessels must be kept moist, as they will shrink from drying, creating leaks and losses. Wood-aging essentials are discussed in Chapter 4.

Mild steel, iron, copper, aluminum, ceramics, and other materials are unacceptable. The danger is in the potential for extracting metals and minerals which can ruin wine clarity and flavor.

Fig. 8–12 Stainless steel drums.

Containers for fermenting white wines should be closed so that the only opening is that on which a "fermentation lock" (see Chapter 9) is attached. Red wine fermenters should have removable lids so that the must "cap" can be "punched" down easily several times each day.

It is recommended that the small winery facility be designed with containers which can be easily moved in order to optimize usage of space seasonally. A 5-gal glass carboy weighs about 50 lbs. when full, which is generally about the largest vessel one wants to consider moving without any mechanical assistance. Stainless steel vessels from 100 to 1000-gal capacity should be mounted on legs for ease in use and relocation around the winery.

Following fermentation, all wines should be processed and stored in vessels which are completely filled, a practice which keeps air from oxidizing wine. White wine fermenters can often serve double duty as good wine processing and storage/aging containers.

Key ATF Tank Regulations

27 CFR, Part 4, Section 24.167   *Tanks.*

(a) *General.* All tanks on wine premises used for wine operations or for other operations as are authorized in this part will be suitable for the intended purpose. Each tank used for wine operations will be located, constructed, and equipped as to permit ready examination and a means of accurately determining the contents. Any tank used for wine operations not enclosed within a building or room will be enclosed within a secure fence unless the premises where the tank is located are enclosed by a fence or wall, or all tank openings are equipped for locking and are locked when used for wine operations and there is no proprietor's representative on the wine premises, or the regional director (compliance) has approved some other adequate means of revenue protection. All open tanks will be under a roof or other suitable covering.

27 CFR, Part 4, Section 24.168   *Identification of tanks.*

(a) *General.* Each tank, barrel, puncheon, or similar bulk container, used to ferment wine or used to process or store wine, spirits, or wine making materials will have the contents marked and will be marked as required by this section.

(b) *Tank markings.* [Fig. 8–13]
  (1) Each tank will have a unique serial number.
  (2) Each tank will be marked to show its current use, either by permanent markings or by removable signs of durable material; and
  (3) If used to store wine made in accordance with a formula, the formula number will be marked or otherwise indicated on the tank.

(c) *Puncheon and barrel markings.* Puncheons and barrels, or similar bulk containers over 100 gallons capacity, will be marked in the same manner as tanks. A permanent serial number need not be marked on puncheons and barrels, or similar bulk con-

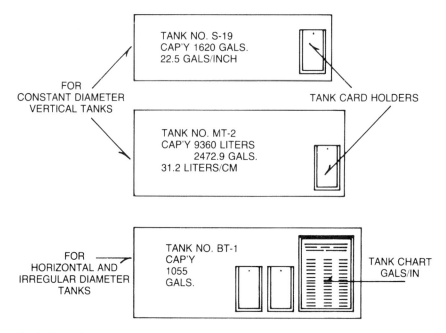

Fig. 8–13 Winery tank markings.

tainers of less than 100 gallons capacity, used for storage, but the capacity will be permanently marked.

27 CFR, Part 4, Section 170 *Measuring devices and testing instruments.*

(a) *Measuring devices.* The regional director (compliance) may at any time require proprietors to provide at their own expense equipment for ascertaining the capacity and contents of tanks and other storage containers, and scales and measuring devices for weighing and measuring wine, spirits, volatile fruit-flavor concentrate, or materials received and used in the production of wine. Where winemaking materials or other materials used in the treatment of wine are used immediately upon receipt on wine premises, or received and stored on bonded wine premises in original sealed shipping containers with a stated capacity, the quantity shown on the commercial invoice or other document covering the shipment may be accepted by the proprietor and entered into records in lieu of measuring the materials upon receipt.

Fermentation Locks

These are devices which allow carbon dioxide gas to escape from fermenters, but do not allow air, dust, and other contaminants to enter into the vessel. (See Fig. 8–14.) Air exposure can be devastating to wine—a source for oxygen required by vinegar bacteria and also for oxidative "browning."

Fig. 8–14 Portable poly all-purpose tanks with fermentation lock. *Courtesy: Prospero Equipment Corporation.*

Perhaps the most effective fermentation locks are inexpensive devices in which water floats a small plastic cylinder inside a larger one. As CO_2 gas is generated, it is allowed to bubble out through the water. Empty sparkling wine bottles, with sloping shoulders, are often placed neck down in fermenter bung holes. Carbon dioxide gas pressure will build in the tank until sufficient to lift the bottle out but not allow air back in.

Refrigeration

In that much of the flavor in white wines remains in the skins after pressing, it is obvious that very careful controls be placed on conserving the flavor extracted in

Fig. 8–15 Semi-truck refrigeration units adapted to winery cold-room refrigeration. *Courtesy: Thermo King Corporation.*

the resulting must. Fermenting white grapes in contact with the skins will serve to extract greater intensities of fruit flavor, but it also will leach out undesirable phenols. Consequently, vintners should equip their wineries with either a refrigerated room or jacketed tanks in which fermenting white wines generating heat can be cooled.

This can take the form of a specially designed room with adequate insulation cooled with an appropriately sized semi-truck unit (Fig. 8–15) or can be a custom-designed facility with remote compressor and condenser.

Many new wineries are generally designed with jacketed stainless steel tanks which can be used for fermenting, blending, processing, and storage. (See Fig. 8–16.) Chilled water or alternative refrigerants can be circulated through these jackets for pinpoint temperature control.

Filters

To have a "finished" look, commercial wines should be brilliantly clear. Ill-conceived "straining" operations often do more harm than good.

Vintners may choose a cartridge filter equipped with pump and hoses on a handy portable stand for filtrations of up to 100 gal. The entire unit usually costs about $500 with hard, durable, plastic cartridge housings; one of stainless steel costs a bit more. This type of filter employs disposable cartridges and is easy to use. As a bonus, the pumps and hoses can also be detached for use in transferring wine from one container to another. (See Fig. 8–17.)

For larger-quantity filtrations, a plate-and-frame filter using cellulose pads

Fig. 8–16 Dimple-jacketed stainless steel tanks.
Courtesy: Paul Mueller Company.

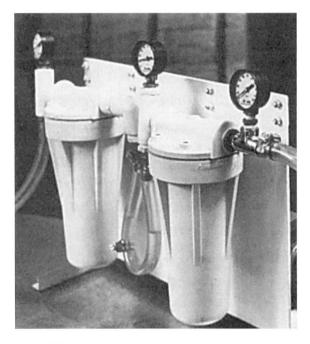

Fig. 8–17 Cartridge filter.

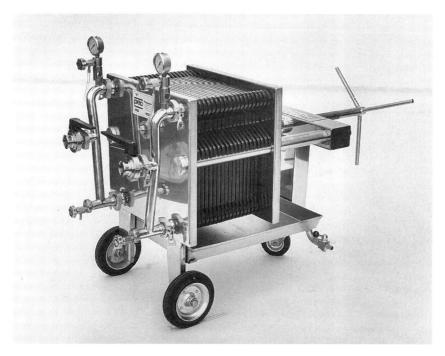

Fig. 8–18 Small winery plate and frame filter.
Courtesy: KLR Machines, Inc.

may be the best choice (Fig. 8–18). These are priced according to the number of plates and the type of materials used in construction. A stainless steel filter with ten 16 ft × 16 in. plates should be budgeted at about $4000.

Coarse filtration using diatomaceous earth (DE) in filters, which are much akin to swimming pool "sand"-type filters, serves to remove suspended particles in young wines long before they are naturally participated. This is particularly recommended in white and blush wines, in which optimal fresh fruit character is desired. This type of filtration can be achieved by employing DE plates and frames in appropriate filter types. There are also "dosing"-type filters which are specially designed only for DE filtration.

Fillers

Manual four- and six-spout, stainless steel, gravity fillers costing from $600 to more than $1000 can suffice most small wineries well into the future. These devices operate best with a constant supply of wine from a tank above, which is regulated by a float-level inlet valve. (See Fig. 8–19.)

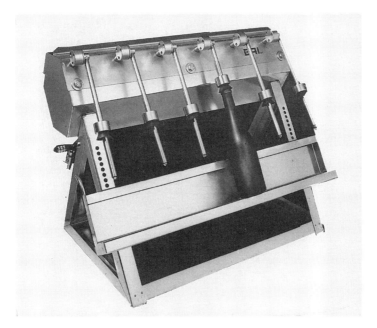

Fig. 8–19 Small winery manual filler.
Courtesy: Prospero Equipment Corporation.

Corkers

Wines can, of course, be packaged in bottles and sealed with screw caps. Corks are expensive, but their value in maintaining a resilient seal, in displacing headspace air in bottles, and in aesthetic appearance are far more marketable than caps.

A floor- or bench-type (Fig. 8–20) corker at less than $200 can work adequately over the short term, but a semiautomatic corking machine, ranging upward of $3000 will be required in the growing winery.

Most corkers operate by a set of "jaws" in which the cork is squeezed. A piston then drives the cork down into a bottle securely positioned under the jaws.

Labelers

The start-up winery can do well in the short term by using a manual label "pasting" machine which range in cost from $400 to more than $1000. These devices spread glue evenly over the backs of paper labels which are then applied directly to the bottles by hand. Various types of jigs can be devised in order to ensure that labels are placed at a uniform place on each bottle. If this uniformity is not closely adhered to, bottles on the shelves look uneven and create a poor product image.

It should be noted that haphazard label application (i.e., labels which are

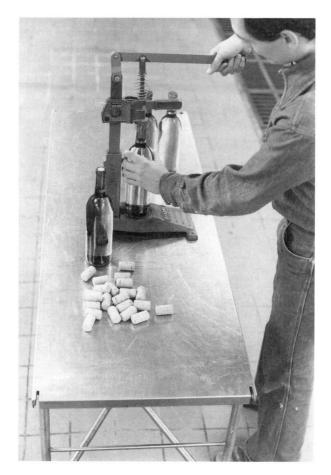

Fig. 8–20 Bench-model manual corking machine.

crooked, wrinkled, corners lapped over, etc.) illustrates to the consumer the vintner's same lack of care and pride for what is **inside** the bottle.

There are very few inexpensive semiautomatic labeling machines which operate dependably. Most successful vintners go directly from manual label pasting to an automated labeling machine (Fig. 8–21), which may cost upward of $7000.

MATERIALS

The philosophy of applying materials in commercial winemaking generally parallels that of investing in wine instruments and equipment. There are instances when

Fig. 8–21 Semiautomatic bottle labeler.
Courtesy: Prospero Equipment Corporation.

equally good arguments can be made for either the frugal or the fancy, and considerations in between. It would be difficult, however, to justify considerable expense for equipment while skimping on materials, and vice versa.

Try to minimize material inventories with the ideal being to have on hand only what is needed for each season. Some materials, such as potassium meta-bisulfite, lose strength over extended storage periods. Even in constant freezer storage, yeasts will loose viability. Inventory control is cost control. This is another area in which cooperating producers can find economies in sharing cost burdens. Larger unit purchases often significantly reduce unit prices.

Potassium Meta-bisulfite

Also known as KMS, this compound is of absolute necessity in small winery operations. Potassium meta-bisulfite is an inexpensive white powder which serves both as an antiseptic and an antioxidant. Sodium bisulfite is equally effective, but the potassium form is much more popular. With good acidity strength (pH 3.30 or below), 1 oz. of highly active KMS should result in about 30 parts per million (ppm) of free SO_2 in 100 gal of must, juice, or wine. KMS solutions are essential in cellar sanitation.

Be careful when using sulfites. The active ingredient is sulfur dioxide (SO_2), which is toxic and corrosive. Because KMS loses strength rather quickly, especially in warmer storage conditions, it makes sense to buy KMS from sources which are sure to have a fresh supply. Keep it in tightly sealed glass or poly containers in cool, dry storage.

Pectic Enzymes

Grape cell walls contain several major types of pectins which can reduce press yield efficiency, as well as cause further difficulty in resulting wine clarification and filtration. In the making of jellies, pectin is often added as an aid in solidifying fruit juice. In the making of wine, the opposite problem exists, and commercial winemakers often employ pectic enzymes, sometimes referred to as *pectinase,* to musts prior to pressing. Due to high cost and the short shelf life of most pectic enzymes, most vintners buy only what is necessary for the upcoming vintage season.

Yeast

Be very careful in selecting a source for yeast. It is tempting to take cultures offered by friendly vintners having the best of intentions. The danger exists in the inadvertent transfer of spoilage organisms along with the yeasts. It is generally acceptable to freeze clean yeast cultures for use the following year. However, there are typically fewer problems and better results from fresh yeast supplies purchased each year just before the vintage season.

Although much ado is currently published in consumer magazines about using natural yeasts, cultured wine yeasts (*Saccharomyces cereviseae*) are recommended for beginning commercial winemakers. Strains such as *Champagne, Epernay, Montrachet, Prisse de Mousse,* and *Steinberg* are offered in dehydrated pellets, which is the most convenient and efficient form to use. One heaping teaspoon of dry yeast pellets are intermittently stirred in a cup of warm (110°F) water for about 30 min before adding to a 3-gal starter batch of must or juice. Each starter batch of 1 gal should be sufficient to inoculate juice and must lots of 100 gal.

Fining Agents

One of the better clarification agents currently used is a proprietary compound called Sparkolloid®. Prepare in batches which can be easily handled. To each gallon of boiling water, slowly add up to 2 oz. of Sparkolloid while constantly stirring. One gallon of slurry is generally sufficient for 100 gal of wine—divide or multiply as necessary. While still hot, **slowly** add directly to the wine and gently stir. Again, **be careful,** as hot liquids can easily break cold glass. High grades of gelatin (over 100 bloom) can be used in a similar manner.

Fine red wines are often fined with egg whites. The whites of just two eggs, beaten to a froth (not stiff) are usually sufficient for a 50-gal barrel of young red wine.

One of the most common fining agents is bentonite, a clay which swells enormously in water and has a huge adsorptive capacity. Preparation requires extreme care to avoid clotting. Prepare by **very slowly** (it clumps very easily) adding up to 2 oz. of bentonite per gallon of constantly stirred boiling water. The resulting suspension slurry can be added directly to the young wine. One gallon of slurry should be fully sufficient for 100 gal of wine. Larger batches of slurry can be made up beforehand and stored cold for up to several months prior to use.

Make laboratory trials to determine the minimal amount of finings which can be employed, as most clarification agents reduce both color and flavor. Difficult wines may require increased dosages of finings. Sparkolloid, gelatin, and bentonite can be kept for several months if held in cool, dry storage.

Kieselsol has widely replaced bentonite in Europe and increasingly so in the United States. First, a light gelatin treatment (gelatin readily dissolves in water) is made and then Kieselsol can be added directly to the wine. It quickly coagulates to embrace any phenol degradation products which have resulted from the gelatin fining made previously. The important advantage of Kieselsol over bentonite and Sparkolloid is that it will not noticeably reduce color and flavor unless they are already precariously unstable.

Ascorbic Acid

More commonly known as vitamin C, ascorbic acid is sometimes added in amounts of 3–5 oz./ton of white grapes, which have a high propensity to brown from oxidation. The chief use of ascorbic acid is made as an additive to white wines just prior to bottling, in order to add "crispness" to acidity and freshness to overall character and to enhance resistance to browning.

Citric Acid

Citric acid is often added at a rate of 1–2 oz./100 gal after wines have been fined and stabilized, but before filtering. It is an organic acid which helps to further stabilize wines from forming precipitates. Citric acid solutions are also used in "neutralizing" filter media and in cellar sanitation.

Sorbic Acid

For wines finished with residual sugar, sorbic acid is highly recommended in order to help inhibit secondary fermentation. Sorbic acid is an effective inhibitor of yeasts (not bacteria or molds) at minimum levels of 200–300 mg/L. Wines having higher pH, lower total acidity, and weaker alcohol levels generally require a greater dosage. The ATF limit (27 CFR, Part 4, Section 24.246) is 1000 mg/L for commercial wineries.

Bottles

There is considerable history involved with the shape of wine bottles. The original Roman shape was rather cylindrical, with an abrupt shoulder reduced down to a narrow neck. This was the principal influence of bottles as we know them today. (See Fig. 8–22.) As the Roman Empire progressed to the west, into Spain, Portugal, and Bordeaux in southern France, the bottle remained in this traditional shape. But the occupation of northern Europe over several hundred years was to evolve a new tradition—a more gradual slope in the shoulders, as now expressed in the bottles from Burgundy, Champagne, and the Rheinland. Chapter 4 provides further background on the essentials of selecting wine bottles. Commercial prices range from $4 to $8 per case of twelve 750-mL wine bottles.

Table 8–1 provides an indication of what is generally considered appropriate for some of the more popular varietal table wines.

The most common wine bottle size is 750 mL, or 3/4 L which replaced the 1/5 gal, or "fifth" (757 mL), when the ATF changed to standardized metric measurements. See 27 CFR, Part 4, Section 4.37 for ATF bottle contents regulations.

Fig. 8–22 Traditional bottle shapes.

Table 8–1 Suggested table wine bottles.

Bottle shape	Varieties and types
White Wines	
Bordeaux	Aurora Blanc
	Niagara
	Sauvignon Blanc
Burgundy	Catawba
	Cayuga
	Chardonnay
	Delaware
	Seyval Blanc
Rhine (Hock)	Gewürztraminer
	Johannisberg Riesling
	Missouri Riesling
	Vidal Blanc
Red Wines	
Bordeaux	Cabernet Sauvignon
	Chambourcin
	Chancellor
	Ives
	Marechal Foch
	Merlot
	Norton
Burgundy	Baco Noir
	Concord
	Gamay
	Isabella
	Pinot Noir

Following is a list of common wine bottle sizes and names:

Bottle name	*Volume*
Split (or quarter bottles)	187 mL
Half-bottle	375
Bottle	750
Magnum	1.5 L
Jeroboams	3.0
Imperial	6.0

Splits are used extensively in the transportation and hotel industries, but are usually difficult to justify for small winery production due to the high cost of bottling and packaging. In addition, 187-mL bottles have a relative short shelf life. Half-bottles are designed chiefly for restaurant use when only one or two people are dining. Magnums, or double-bottles, are used primarily for festive occasions when eight or more guests participate. Some vintners fill a small percentage of production into half-bottles and magnums; the major share of inventory goes into the popular 750-mL size. Jeroboams and Imperials are very expensive to buy and difficult to handle.

The use of soft drink bottles, spirits bottles, jugs, and the like is not recommended. They can be used successfully but create an impression of carelessness and questionable ethics.

New bottles are usually transported in paperboard cartons. Cartons not in use can easily be broken down and folded for storage. When needed, they can be restored quickly with durable reinforced packing tape.

Most important is that bottles are clean before filling. All wine bottles, even new ones, need to be thoroughly cleaned. Used bottles should be scrubbed both inside and outside with a stiff bristled brush, followed with at least three rinsings of scalding hot water. If detergent is used, four or five rinsings are recommended. Cleaning can be made easier by rinsing bottles as soon as they are emptied. After draining, place them upside down in a paperboard container and seal with tape. Cases of empty bottles can be stored in most any place which is free from moisture, insects, rodents, and out of harm's way.

Corks

Corks are made from the bark of the cork oak tree, *Quercus suber,* and can cost up to $0.15 apiece, especially those in the higher-quality grades.

Corks are available in different sizes and in different finishes, as well as different grades (See Fig. 8–23.) Bottles are molded to a number of varying diame-

Fig. 8–23 Various cork types and sizes.
Photo: David Ferguson.

2" branded   2$^1/_4$" branded vintage-dated   1$^3/_4$" branded vintage-dated logo   1$^1/_2$" plain   1$^1/_2$" synthetic   1$^1/_4$" agglomerate

ters—for which matching cork diameters are needed. Typical cork dimensions are 9 mm in diameter by 1.75 in. long. Shorter corks can be used for wines which are expected to be in the bottle only several months; longer corks are necessary for reds designed for several years of bottle aging.

The best corks have designations such as "Extra Fine Quality" or "Grade AAA" (those having minimal natural imperfections). They are chamfered (edges rounded off), dedusted, coated with a thin coat of paraffin and silicone, and sterilized. These specifications add a few more pennies to cost, but added sealing integrity and fewer application problems are results well worth the expense. Figure on high-quality corks costing from $0.12 to $0.15 apiece.

Cellukork™ and SupremeCorq™ plastic cork manufacturers seem to be surmounting the initial problems of difficult driving and even more difficult pulling. There are indications of a rapidly improving technology, such as the virtual immortality of their ethylene–vinyl acetate co-polymer (EVA) formulation, improved lubricants, variable resiliencies, and even the possibility of scavenging headspace oxygen and SO_2. With prices remaining significantly less than natural corks, there seems to be a trend of more and more vintners moving toward increasingly larger use of these synthetic closures.

One of the newest developments is the "hybrid" cork, which is designed to bring together the virtues of natural cork with the dependability of plastic molded together to form a uniform closure which does not leak and protects against mold formation. See Chapter 4 for more information relating to corks and other closures.

Capsules

Capsules do not add any type of pilfer-proof seal to a wine bottle. Capsule value is totally aesthetic but, nevertheless, important. They provide a "finished" appeal to the final product and express the degree of pride and craftsmanship which a vintner has given his wine.

One popular type of capsule is shrunk on the bottle using a hot air gun similar to a hair dryer. Firm plastic capsules need no application machinery and are relatively inexpensive, although they are questionable in appearance when compared to the "roll-on" types, which provide a higher-quality image. Various capsule types are presented in Fig. 8–24.

All capsule types are offered in a number of colors and decor schemes. Prices range from less than $0.01 apiece in larger quantities, to more than $0.03 for high-grade custom capsules. See Chapter 4 for more information relating to capsules and bottles which are designed for no capsule application.

Labels

Labels disclose the winemaker, wine grape variety or type, the vintage year, and other important information. The ATF has exacting rules relating to label state-

210 *Getting Started*

Fig. 8–24 Various capsule types and sizes.
Photo: David Ferguson.

ments and decor which are strictly enforced in the wine industry. Every commercial wine label and label change must be submitted for approval using ATF Form 5100.31. Only the vintage date can be changed on a wine label without ATF authorization.

There is no single more important item in the wine business than the quality of the wine label. For beginning wineries especially, a case can be made for label quality being more essential than even the wine quality itself. The label is the item on which potential consumers focus; this forms their first impression. The reader is encouraged to review the discussion on label design given in Chapter 4.

The application of glue to labels is too slow and messy without at least a label pasting machine. Semiautomatic labeling machines can cost several thousand dollars. Straight-edge paper labels cost less and work best in these types of small winery labeling machines. Pressure-sensitive labels are not recommended in the start-up winery until experience can serve to deal with the difficulties of adjusting position and removal.

Cases

Bottles are sold by manufacturers by the case in a plain case with appropriate "contents" markings as standard procedure. Custom printing can be made, but most start-up vintners opt for the plain cases. ATF regulations for identity can be met by simply applying the same labels to each end of the case as are on the bottles inside. Some vintners prefer to apply various types of stamps to identify what is in each case.

CHAPTER 9

WHITE TABLE WINES

Dry white table wines are easy to make but challenging to make consistently well. They require delicate handling to avoid oxidation (browning) and careful attention to protect and develop delicate flavors throughout formative processing.

GRAPE VARIETIES

Chardonnay, Cayuga, Sauvignon Blanc, and Seyval Blanc make whites which are "vinous" in character (herbal and vegetable flavors) and are recommended for dry white wines. Varieties such as Aurora Blanc, Cayuga, Gewürztraminer, and Johannisberg Riesling are typically "floral" (honey and spicy flavors) and are recommended for both dry and semisweet wines. Varietals such as Catawba, Delaware, and Niagara are "fruity" (flavors of melons and raisins) and are recommended for both semisweet and sweet wines.

Whatever varieties are selected, they should obviously taste good and have the same type of flavor which is expected in the resulting wine. Harvest the first batches of white grapes on the slightly green side—which may be in the pH range 3.10–3.20. Adjust to earlier or later harvest, as may seem prudent, thereafter.

One variation on this theme are wines made from purposely overripened grapes—processes which have been pioneered in Germany. The goal is to maximize sugar content and develop added floral and fruit flavor components in Johannisberg Riesling and other varieties which are cultivated there. Success depends on high total acidity levels (over 0.70 g/100 mL) and an agreeable environment. As grapes mature past optimal ripeness, there becomes an increasing dependency for cold temperatures to inhibit the formation of molds. One mold, the celebrated *Botrytis cinerea,* or "noble mold," actually aids in the flavor development process. This microorganism permeates grapes skins, allowing berry water to evaporate out and concentrate sugar and flavor components retained in the fruit. Less experienced commercial vintners interested in trying their hand at late-harvest wines should experiment with small lots beforehand.

INCREASING WHITE WINE FRUIT FLAVOR INTENSITY

The most important factor in extracting optimal fruit flavor is that of closely monitoring pH during fruit maturation. Grape flavors change during ripening. Slightly underripened grape flavors, generally in the pH range of about 3.20–3.30, may tend to be masked with high total acidity—something akin to the green-apple flavor found in many cool climate Johannisberg Riesling wines. The same Riesling allowed to mature in warmer weather may reach a pH range of 3.30–3.40 or higher in which the typical terpene flavors of apricot and peach are readily distinguishable. A higher pH encourages the growth of bacteria and wild yeasts. Consequently, preservation of fruit flavor will be dependent on the winemaker's ability to conduct a straightforward cool fermentation of clean juice in sanitized stainless steel vessels.

Many vintners add pectolytic enzymes to musts held for up to several hours prior to pressing. Whereas this can contribute significantly to fruit flavor extraction, it can also result in a partial disintegration of grape skins. Consequently, reduced pressure is needed during the pressing operation.

Clarity cannot be a critical factor in the sensory evaluation of fermenting juice or must. During the peak of a normal fermentation, yeast cell population may reach more than 1 million/in.^{3} of liquid. Consequently, winemakers expect a heavy degree of turbidity due to this heavy suspension of yeast cells.

Thus, one of the most important elements of increasing fruit flavor potential is through reduction of solids resulting from crushing and pressing operations. The general idea is that fewer solids provides fewer substrates from which complex flavors develop and mask fruit flavor. In this case, ensure that crusher rollers are not cracking seeds and creating bits of skin and stems so as to minimize these sources for bitter phenols. On the other hand, some pulp solids are needed to ensure that there is sufficient retained nitrogen and key vitamins necessary for yeasts to perform a complete fermentation.

Suspended and precipitated solids are generally comprised of grape pulp, fragments of stems, skins, and seeds; sometimes, colloidal materials are also imparted from the must under heavy pressure. If left in the juice, these gross lees solids can release distasteful compounds formed by enzymatic breakdown and oxidation. Such solids are also related to colloidal formations which may require difficult filtrations diminishing and/or degrading fresh fruit flavors. Ideally, pressed juice should be held in cool 50°F stainless steel tanks for at least 24 h prior to commencing the fermentation regimen. The juice should then be racked (separated) from the settled gross lees into another tank prior to inoculating with yeast.

Some white wines such as *sur lies* Chardonnay and Seyval Blanc are designed to increase flavor resulting from extended exposure to secondary grape and autolyzed yeast lees following fermentation which is generally conducted in oak barrels. It is common to find sur lies terms ranging up to several months in duration, with a *batonage* (stirring with a clean stainless steel rod) several times per week. As would be expected, this practice is filled with potential microbiological and oxidative peril. An especially tenacious quality control program is required to ensure that this technique does not go awry.

Some vintners are finding more flavor by pressing some portion, as much as half, of their grapes in whole-cluster form. Champagne cuvée (table wines made before secondary fermentation) wines are totally pressed from whole clusters. Lower gallon-per-ton efficiency can be expected, although stems can serve as a press aid and a source of higher tannin content. It is important not to relate whole-cluster pressing with whole-cluster fermentation. Whereas some "nouveau" red wines benefit from whole-cluster fermentation, in whites this can extract bitter leucoanthocyanins which oxidize to a yellow pigmentation and a "leathery" tannic flavor.

Whole berry fermentations are also made; this requires a special type of destemming machine. Individual research is required in order to ascertain what extent, if any, that whole berry fermentation may improve the final product. It is common to see whole uncrushed berries survive more than 500 psi in the pomace of basket-type presses. Some vintners counter this by exerting excessive pressure which can extract undesirable bitter seed and stem phenols.

Yet another method for increasing white wine fruit flavor intensity is the retention of some portion of juice in cold storage. This is sometimes referred to as a *muté* and is added to the wine later, but before clarification and stabilization. If the variety and source of the muté is different than the wine in which it is blended, care must be exercised that the character of the finished product is not distorted. There are specific ATF regulations pertaining to this in 27 CFR, Part 4, Section 4.23.

HARVESTING

Machine-harvested fruit will have broken skins and some juice already extracted along with bits and pieces of leaves, stems, and other foreign matter. Consequently, grapes harvested by hand into clean containers are preferred. Expect a yield of about 130–150 gal of finished wine to result from each ton of grapes.

CRUSHING–DESTEMMING (FIG. 9–1)

Chill grapes, if necessary, as soon as possible after harvest. A temperature in the 45–55°F range is ideal. If there is excessive MOG (material other than grapes), set up a movingbelt *triage* line in which people are assigned the task of removing leaves, sticks, and other trash, along with unripe or spoiled berries. Record net

Fig. 9–1 Illustration of crushing–destemming operation.
Courtesy: Sebastiani Vineyards.

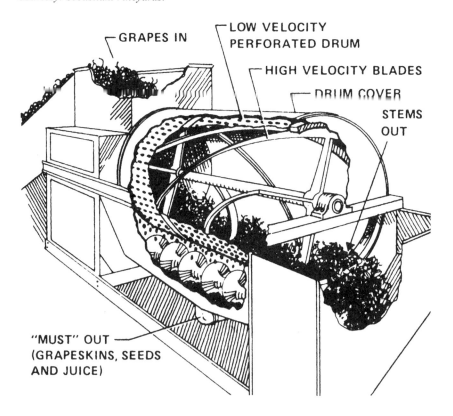

weight and rinse grapes (hand harvested only) with clean, cold water. Sulfur dioxide should not be added to prime-quality fruit, as it increases the extraction of bitter phenolic compounds from seeds and skins. However, if the fruit is suspect to degradation from some combination of delay after harvest, heat exposure, microbe infection, pest invasion, and so forth, it may be best to consider a sulfur dioxide application of from 40 to 60 ppm.

PRESSING

One of the most common errors in pressing juice from grapes is that of over-pressing (i.e., attempting to gain efficiency by long, high-pressure regimens). Although this succeeds in maximizing quantity, it is achieved at a reduction in quality. Free-run juice is the highest quality; the next best juice fraction is extracted with gentle pressure. After that, the winemaster should carefully evaluate the juice being extracted. Once the juice starts to clear, there is a good chance that undesired bitter and astringent phenols are being released from the skins and seeds.

The juice should be immediately transferred to cold temperature (45–55°F) for about 24 h. During this time, undesired suspended solids from grape pulp and seeds will precipitate. Juice is then racked (decanted) from this lees (sediment) and transferred to a fermenting vessel.

Pomace (skins, pulp, and seeds remaining in the press) should be immediately removed from the winery, as it quickly attracts insects and other pests. A good option for handling pomace is to transport it to a well-designed and maintained compost facility.

BRIX (SUGAR) ADJUSTMENT

Brix in white grape juice typically measures between 14° and 23°. Native *Vitis labrusca* varieties will generally ripen in the lower end of this range, *Vitis vinifera* at the higher end, and the French-American hybrids in mid-range. There are exceptions.

Each degree of Brix will convert to about 0.535% alcohol during fermentation. Consequently, to achieve 12% alcohol by volume, the white grape juice would require approximately 22.5° Brix. Thus, a natural Brix of 18.1° Brix would need to be adjusted with sufficient sugar to achieve a 4.4° increase in Brix.

The addition of sugar is called *chaptalization*. In some cases, grapes are excessively high in acidity, leading to both sugar **and** water being added—referred to as *amelioration*. Both of these adjustments have historically been controversial in fine wine circles.

Chaptalization and amelioration increase the resulting quantity of juice. This relegates a need for vintners to fully interpret the ATF regulations in this regard. They read as follows:

27 CFR, Part 4, Section 24.177 *Chaptalization (Brix adjustment).*

> In producing natural grape wine from juice having a low sugar content, pure dry sugar or concentrated juice of the same kind of fruit may be added before or during fermentation to develop alcohol. In producing natural fruit wine from juice having a low sugar content, pure dry sugar, liquid sugar, or concentrated juice of the same kind of fruit may be added before or during fermentation to develop alcohol. The quantity of pure dry sugar, liquid sugar or concentrated juice added may not raise the original density of the juice above 25 degrees Brix. If grape juice or grape wine is ameliorated after chaptalization, the quantity of pure dry sugar added to juice for chaptalization will be included as ameliorating material. If fruit juice or fruit wine is ameliorated after chaptalization pure dry sugar added under this section is not considered as amelioration material. However, if liquid sugar is added to fruit juice, the volume of water contained in the liquid sugar will be included as ameliorating material.

A simple mathematical formula can be used to calculate granular sugar additions:

$$\text{Gallons of juice} \left(\frac{\text{Desired Brix} - \text{Existing Brix}}{100 - \text{Desired Brix}} \right) = \text{Pounds of sugar to add}$$

Consider the following example:

$$1743 \text{ gal} \left(\frac{22.5 - 18.1}{100 - 22.5} \right) = 99 \text{ lbs. of sugar addition}$$

Sugar additions are, however, illegal in some states, including California, and deficiencies in Brix must be made with the addition of concentrates. Consider the following formula:

$$\text{Gallons of juice} \left(\frac{\text{LSR} - \text{LSE}}{\text{LSC} - \text{LSR}} \right) = \text{gallons of concentrate to add}$$

where LSR = pounds of solids per gallon required, LSE = pounds of solids per gallon existing in the juice, and LSC = pounds of solids per gallon existing in the concentrate.

Consider the example of 1743 gal of white grape juice which has 18.1° Brix. Grape concentrate at 68.0° Brix is to be added in a sufficient quantity to raise the entire blend to exactly 22.5° Brix. First find LSR, LSE, and LSC from the Brix Table provided in Appendix B.

$$1743 \text{ gal} \left(\frac{2.054 - 1.623}{7.586 - 2.054}\right) = 135.8 \text{ gal of concentrate}$$

However, native *Vitis labrusca* grapes usually ripen with total acidities well beyond the 0.900 g/100 mL which may be considered maximum tartness. Consequently, winemakers may opt to ameliorate with both sugar and water in order to dilute this acidity. A well-designed lot origination form (Fig. 9–2) will permit

Fig. 9–2 White wine lot origination form in service.

```
                              LOT ORIGINATION
DATE             9-24-95              CULTIVAR  CHARDONNAY
LOT NO.          541                  REMARKS:  rec'd 9-22- 74°
SERIAL NO.       11-95                triaged for slight
FERMENTER NO.    F-4                  botrytis and MOG, no
NET GALLONS      1,814                KMS, 3oz P-enzyme
NET TONS         11.22                6 press loads with
GALLONS PER TON  161.7                rice hulls - juice
pH               3.30   T.A.  .915    chilled to 47° in 5-7
BRIX             19.1   ALC.   -      for 48 hrs - 58 gals lees
EXT.              -                   destroyed on 9-24

AMELIORATION =    5  %     @ 22.5° BRIX   = 12  ALCOHOL

  1,814        GALLONS START    = 1,909  RESULTING TOTAL
   .95         INVERSE OF                GALLONS OF PRODUCT
               AMEL. PERCENTAGE

  1,909        TOTAL            @22.5° BRIX  (2.054 LBS/GAL)= 3,921  TOTAL
               GALLONS                                               LBS.

  1,814        START            @19.1° BRIX  (1.720 LBS/GAL)= 3,120  START
               GALLONS                                               LBS.

    95         AMELIORATION                             801   ADDITION
               GALLONS                                        LBS.

    59         SUGAR AS               x .074 (GAL/LB)=   59   SUGAR AS
               GALLONS                                        GALLONS

    36         WATER
               GALLONS

 Pasteur       YEAST ADDITION = RATE OF   1   LBS PER  1,000 gals
REMARKS:       dry yeast mixed in 15 gals of water @ 100°
```

vintners to accurately calculate sugar/water additions and simultaneously maintain a good record of increases in gallonage.

Total acidity in white grape juice should ideally measure between 0.65 and 0.75 g/100 mL or, roughly, between 0.65% and 0.75% acidity. In the more delicate white varieties, such as Chardonnay and Seyval Blanc, it may be best to leave a slightly high acidity unadjusted. Juice from the floral and fruity varieties can benefit by ameliorating high acidity with the judicious addition of clean potable water.

A good rule of thumb is to optimize, not maximize, amelioration. Calculate water additions by first dividing the desired total acidity by the naturally existing total acidity. Second, using the *inverse* of the result as the percentage of water to add to the juice. The inverse is that number required in addition to an *obverse* decimal number needed to result in 1.00. The following is an example of 100 gal of juice analyzed to contain 0.96 g/100 mL total acidity:

1. 0.85 [desired total acidity (T.A.)] divided by 0.96 (existing T.A.) = 0.89
2. Inverse of 0.89 = 0.11 (percent water addition to juice)
3. 100 gal of juice multiplied by 0.11 = 11 gal of water

The inverse for a desired T.A. of 0.80 in the example would be 0.17 (0.80 divided by 0.96 = 0.83). For a desired T.A. of 0.75, the inverse would be 0.22 (0.78 divided by 0.96 = 0.78), and so on.

It is essential that winemakers fully understand the ATF rulings relating to amelioration.

27 CFR, Part 4, Section 24.178 *Amelioration.*

(a) *General.* In producing natural wine from juice having a fixed acid level exceeding .500 gm/100 ml, the winemaker may adjust the fixed acid level by adding ameliorating material (water, sugar, or a combination of both) before, during and after fermentation. The fixed acid level of the juice is determined prior to fermentation and is calculated as tartaric acid for grapes, malic acid for apples, and citric acid for other fruit. Each 20 gallons of ameliorating material added to 1,000 gallons of juice or wine will reduce the fixed acid level of the juice or wine by 0.01 gm/100 ml (the fixed acid level of the juice or wine may not be less than .500 gm/100 ml after the addition of ameliorating material).

(b) *Limitations.* (1) Amelioration is permitted only at the bonded wine premises where the natural wine is produced.
(2) The ameliorating material added to juice or wine may not reduce the fixed acid level of the ameliorated juice or wine to less than .5 gm/100 ml.
(3) Except for wine made exclusively from loganberries, currants, or gooseberries, the volume of ameliorating material added to juice or wine may not exceed 35

percent of the total volume of ameliorated juice or wine (calculated exclusive of pulp). Where the starting fixed acid level is or exceeds .769 gm/100 ml, a maximum of 538.4 gallons of ameliorating material may be added to each 1,000 gallons of wine or juice.

NOTE: The author has converted ATF "grams per liter" to "gm/100 ml" as the latter is more commonly used throughout the U.S wine industry. Europeans more often use grams per liter, or g/L.

ACID ADDITIONS

Lower grape acidities are encountered more frequently in warmer climates such as central California than those grown in more cooler regions of the United States. Tartaric acid additions to increase total acidity can be calculated from Table 9–1. Stir in all additions and test for Brix and total acidity once more to ensure that the desired levels have been achieved. Post additions in the appropriate record.

YEAST INOCULATION

Refer to the discussion of yeast in the "Materials" section of Chapter 8, so as to become familiar with media and rates used for juice inoculation. Follow the manufacturers' instructions for cultured wine yeast preparation, or refer to the text addressed to dry yeast pellets. Inoculate juice with cultured wine yeast starter and gently stir. Clean yeast preparation containers, stirrers, and other associated instruments with scalding water and drain dry.

At one time, winemakers inoculated by simply taking a few gallons of another fermenting white wine, or lees sediments, and adding straightaway to another lot

Table 9–1 Total Acidity Table for Tartaric Acid Additions

Total acidity (g/100 mL)	Per-gallon addition for 0.600 g/100 mL	Per-gallon addition for 0.700 g/100 mL
0.450	5.7 g	9.5 g
0.500	3.8 g	7.6 g
0.550	1.9 g	5.7 g
0.600	—	3.8 g
0.650	—	1.9 g

of juice. This practice is now known to be a potential source for encouraging and perpetuating wild yeast and bacterial infections, as well as other problems.

FERMENTATION

Clean fermenters made of glass or stainless steel are ideal for white wine fermentation, processing, and storage. One variance is to ferment the juice from *vinous* white grapes in oak barrels to achieve an added "toasty–nutty" flavor complexity.

Fermenters should be filled to about 70% of capacity, which will allow room for foaming to occur without fouling the fermentation lock. Each fermenter should be properly fitted with a clean lock and placed in a cool dark room or cellar which is freely ventilated. Ideal temperature is in the 55–60°F range.

IMPORTANT NOTE: FERMENTATION IS DANGEROUS! IF THE FERMENTER VESSEL IS NOT PROPERLY VENTED WITH A FERMENTATION LOCK, THE CONTAINER MAY EXPLODE. FURTHER, CO_2 GAS GENERATED FROM FERMENTATION CAN BE ASPHYXIATING IN CLOSED, POORLY VENTILATED AREAS. EMPLOYEES, VISITORS, AND ESPECIALLY CHILDREN SHOULD BE CAREFULLY SUPERVISED IF ALLOWED IN LOCATIONS WHERE ACTIVE FERMENTATIONS ARE TAKING PLACE

It may be a day or two, perhaps several, for fermentation to commence so that regular bubbles can be observed in the lock along with thin spots of foam forming on the surface. If a delay of more than 1 week occurs, gently heat the juice up to 65–70°F until fermentation starts and then return it to the cooler temperature. Such delays are generally attributed to deficiencies in free nitrogen and/or poor yeast viability.

No two fermentations are exactly alike. Three fermenters of the same juice may take three different lengths of time to finish fermenting. At 60°F, however, a Balling below $-1.5°$ should be achieved within 3 weeks or so.

STUCK FERMENTATION

A "stuck" fermentation is one which has not run the full course—a fermentation which has stopped even though a good share of the fermentable sugar remains. This is generally caused by having added too much KMS at the crusher and/or having fermented at temperatures which are radically low or high. Conversely, too little potassium meta-sulfite (KMS) may permit certain yeast-inhibiting bacteria to grow in the must. Grapes having been heavily treated with fungicides in the

FERMENTATION CONTROL RECORD

TANK NO. F-4   LOT NO. 541   VARIETY CHARDONNAY

DATE	TIME	TEMP.	BALL.	ALC.	T.A.	V.A.	C	C	N	T	
9-25	7:00am	55°	22.2	—	.870	—	✓	✓	✓	✓	
9-26	7:30am	57°	21.6				✓	✓	✓	✓	
9-27	7:30am	58°	20.0	—	.855	—	✓	✓	✓	✓	gassy
9-28	7:00am	59°	18.0					✓	✓	✓	
9-29	7:00am	60°	15.7	—	.833	.012	✓	✓	✓	✓	-temp
9-30	8:00am	62°	13.1	ran cooling jacket for 2 hrs							
10-1	7:00am	54°	11.5	5.1	.773	.018	✓	✓	✓	✓	
10-2	7:30am	57°	9.5								-temp
10-3	7:30am	59°	7.2	ran cooling jacket for 2 hrs							
10-4	7:00am	57°	5.5		.743	.018	✓	✓	✓	✓	
10-5	7:30am	59°	4.1				✓	✓	✓	✓	
10-6	7:00am	60°	3.0	9.0	.728	.024				✓	-temp
10-7	7:00am	61°	1.9	ran cooling jacket for ½ hr							
10-8	7:30am	59°	1.0			.024	✓	✓	✓	✓	
10-9	7:30am	59°	0.5	10.9	.720	.024	✓	✓	✓	✓	
10-10	7:30am	59°	0.0				✓	✓	✓	✓	
10-11	7:00am	60°	-0.4	11.3	.713	.024	✓	✓	✓	✓	
10-12	7:30am	60°	-0.7				✓	✓	✓	✓	
10-13	7:00am	60°	-1.1	11.7	.713	.024	✓	✓	✓	✓	
10-13	10:00am	racked to TK S-2, 38 gals lees destroyed, work order no 191-95									

REMARKS: no cracked seeds in lees. TK F-4 needs a new manway gasket

Fig. 9–3 White wine fermentation control form in progress.

vineyard can also be a problem source. More recently, blame has been placed on diseased grapevines, such as those infected with *Phylloxera,* which are unable to uptake essential nutrients from the soil.

There is no simple remedy for stuck fermentations. If the problem can be positively diagnosed as excessive SO_2 content, then vigorous agitation may help—at the risk of oxidation. An appropriate temperature adjustment and/or the addition of a modest nitrogen supplement, such as diammonium phosphate (DAP), up to 150 parts per million, may also be effective. See 27 CFR, Part 4, Section 24.246 for specific types and limits authorized by the ATF.

FIRST RACKING

As soon as fermentation bubbles have rescinded to a very slow repetition, the new white wine should be racked (decanted) off the *gross lees* (grape particles and dead yeast sediment) into clean storage vessels.

One variant of this is traditional to the white Burgundy wines made in France which remain *sur lies,* or "on the lees," for several months after fermentation is completed. Ideally, this results in the development of complex "cheesy–earthy" flavors. Without proper care and expertise, however, the lees can also provide plenty of growth nutrients for spoilage organisms. Another variation is to arrest fermentation in order to conserve some predetermined level of residual sweetness. This is usually performed by chilling and then cold storage at about 30°F, along with a coarse filtration. This treatment will slow yeast viability and remove most viable yeast cells.

The new wine should be racked into containers which can be filled with perhaps only a slight amount of air space. This void will allow the final few CO_2 bubbles to escape without fouling the fermentation trap, which is continued in use after the first racking. It helps to determine which containers will be used beforehand so that the new wine does not require handling over and over again. Continue fermenting at the same 55–60°F temperature. Clean and sanitize empty fermenters and racking equipment thoroughly. Post racking data and resulting gallonage in the appropriate record.

SECOND RACKING

After bubbling has completely stopped (it may be only a day or two after the first racking up to perhaps a week or two), repeat the racking process. New white wine should test with only a trace of residual sugar. If so, the tank should be filled completely in order to expel any air, and a loose cap or plug should be used to cover the second-rack containers instead of the fermentation lock.

An addition of 1 oz. of KMS per 100 gal should increase the sulfur dioxide content in the young wine by about 20–30 ppm (parts per million). This is recommended to help further protect against spoilage microorganisms and oxidation. Continue storage in a cool, dark, cellar-like environment.

From this point on, throughout the life of the wine, it will be necessary to ensure that it is stored in containers which are completely full. Avoiding air contact upon the wine reduces its exposure to oxygen which is a component for oxidative browning and the growth of spoilage organisms.

BLENDING

The general rule of thumb in blending is to associate the different virtues of individual wines into a resulting wine having a more broad and satisfying appeal. As an example, grapes harvested a bit too green may have made a wine which has superb acidity but is shy in flavor. Another lot may have been harvested a bit overripe and the resulting wine is overpowering in flavor, but weak in acidity. Laboratory-scale trials will reveal the best blending ratio.

Wines should be blended before stabilization. The many factors involved with stability sometimes result in stable wines blended together to form an unstable wine. See Chapter 4 for an in-depth discussion of blending considerations. Each blend will, of course, give cause for entries in appropriate records.

FINING AND STABILIZATION

Young white wines (2–4 months old) typically have a hazy translucent appearance, caused principally by proteins and other organic residues. Most new wines also have excessive potassium and tartrate ions in solution. This eventually crystallizes as potassium bitartrate, or "cream of tartar."

Gelatin and kieselsol or bentonite clay and Sparkolloid® mixtures are recommended for clarification, or "fining," of the haze. A typical application is about 1 oz./100 gal of wine. Description and preparation of these materials is found in Chapters 4 and 8.

After the finings have been added, the wine should be "topped up" to ensure each container is completely filled, sealed, and allowed to rest for 4–6 h at normal room temperature. Cold stabilization should then commence. This is performed by storing the wine at a temperature of 26–28°F for at least 3 weeks. This time can be significantly reduced by the addition of specially prepared potassium bitartrate crystals—the "seeding" of particles which serve as nuclei in the formation of new crystals.

The refrigeration capacity to maintain detartration temperatures is expensive, both in equipment and energy. One alternative is allowing ambient winter cold into winery storage cellars, which can work fine as controls are in place to counteract severe cold temperatures. Freezing wine can radically distort flavor profiles and, worse, expand to rupture containers. Frozen water pipes are, of course, yet another jeopardy.

It is easy to confuse tartrate crystals with filter media when wine samples are viewed under a light microscope. The preparation of slides with various crystals and filter media can help to develop an expertise in using microscopy as an effective quality control device. See Figs. 9–4 through 9–7.

Potassium bitartrate *argols* (crystals) resulting from detartration can easily clog drains; they should be removed into a solid waste disposal. Thoroughly clean empty clarification and stabilization containers. Ensure that all clarification agents used are posted in the record—along with the temperature(s) and duration of cold storage for stabilization. Table 9.2 provides a source which may be useful in determining the cause of instability problems.

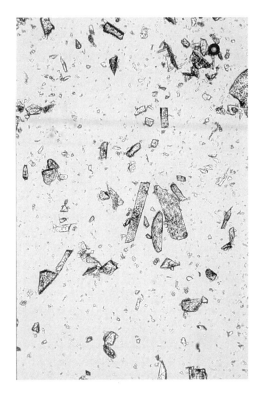

Fig. 9–4 Light microscope image of potassium bitartrate crystals.

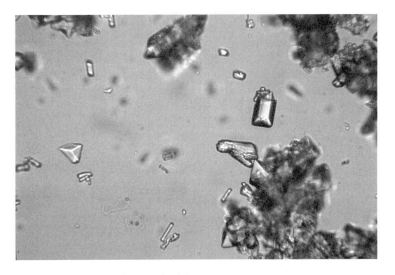

Fig. 9–5 Light microscope image of calcium tartrate.

White Table Wines 225

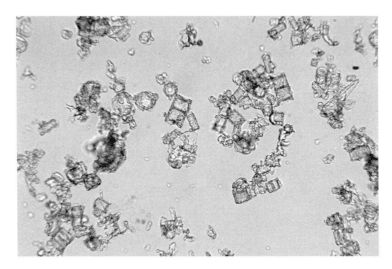

Fig. 9–6 Light microscope image of diatomaceous earth filter media.

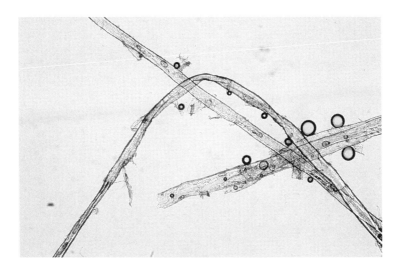

Fig. 9–7 Light microscope image of cellulose pad filter media.

Table 9-2 Determination, Causes, and Recovery of Positive Instability in Wines

Designation of turbidity or separated material	Cause	Prevention	Recovery
Organic copper turbidity	Copper contact of beverages (spray medium, copper piping, poorly silver plated or tinned fittings or equipment).	Avoidance of copper contact. Tinned copper pipes are often defective and should be replaced if possible with stainless steel.	About 90% of the existing copper is eliminated by fermentation. Addition of sulfurous acid is eventually able to reducing copper turbidity to a soluble stage. (Total SO_2 must not be in unfermented beverages be above 80 mg/L.)
Calcium sulfate sediment	This turbidity occurs in brandy which has been diluted with calcium-containing water.	Avoidance of calcium-containing water for dilution purposes. Employment of softened water.	Cold storage and filtration, or ion-exchange treatment (IR 120)
Winestone sediment	Rapid cooling or higher tartaric acid content, or brief cask storage.	Control of tartaric acid content before bottling. At 0°C, about 1.5 g/L tartar is soluble in the beverage (with 10% alcohol by volume)	Cold storage of wines and filtration or do tartaric acid determination and as far as possible on this basis carry out chemical deacidification.
Calcium tartrate sediment	During short observation period after chemical deacidification. Perhaps employment of calcium containing filter sheets.	Wash filter sheets in tartaric acid solution! Observation time before racking off from deacidification sludge should be at least 4 weeks.	Cold storage of beverages and filtration.
Copper sulfite turbidity	Copper containing beverages with simultaneous high total sulfurous acid content (over 120 mg/L).	Prevent copper uptake! Remove copper piping and replace with stainless steel or hose.	Reduce copper concentration by blending; aeration of beverage to bring about disappearance of SO_2 (insoluble cuprous compounds thereby go over into soluble cupric compounds)

Table 9-2 (continued)

Designation of turbidity or separated material	Cause	Prevention	Recovery
Ferric phosphate turbidity	High iron content (More than 6–8 mg/L)	Prevent iron uptake! Lacquer bare iron parts.	Remove iron by Aferrin fining. Lower the iron content by blending with iron-poor beverage. Gelatin fining in connection with stronger dosing with sulfurous acid.
Pigment–tannin sediment	High pigment and tannin content.	By longer storage of harsh and highly colored beverages in casks, sedimentation in bottles can be largely avoided.	Gelatin fining with 20 g/100 L and dosing with sulfurous acid up to a content of 25–30 mg/L free SO_2 helps to remove an excess of tannin and leads to stabilized beverage.
Iron–tannin turbidity	High iron and tannin content.	Avoid iron uptake! Lacquer bare iron parts.	Gelatin fining for removal of excess tannin. Blend with iron and tannin-poor beverages. Citric acid addition (50 g/100 L in wine; 200 g/100 L in fruit wine.)
Biological turbidity	Yeast and bacterial activity.	Sterile bottling as preventative measure. Sanitary operations!	Tight filtration (EK filter) and dosage with sulfurous acid; pasteurization.
Thermolabile protein turbidity	Unfermented beverages, especially grape juice, contain under certain circumstances thermolabile protein, which, after heating (pasteurization) and cooling lead to turbidity.	(a) Pasteurization done twice, filtration between times. (b) Bentonite treatment	Beverages with protein sediment are pasteurized, cooled and filtered. Repeat heating leads to no new turbidity.

Source: Tanner, H. and Vetsch, U. 1956. How to characterize cloudiness in beverages. Am. J. Enol. 7, 142–149.

AGING

There is a wide variance in the manner in which oak aging is employed in winemaking. Some white wines, especially the *vinous* ones described at the beginning of the chapter, can significantly improve from a very carefully controlled regimen of oak aging. The desired effect is a bit of "toasty vanilla" character which can be a fine addition to bouquet and flavor.

Wines made from the floral and fruity varieties are often better without any oak aging complexity in the bouquet and flavor profile.

Oak aging in commercial wineries is usually performed in expensive barrels having 55 gal of capacity. Smaller containers such as 15–30-gal kegs can be used effectively but are usually constructed from inferior types and grades of oak. The smaller the keg, the more wood surface area exposure there is per unit volume of capacity—and the faster the wine will respond in aging. Wooden containers require special cleaning and maintenance in order to be effective. See Chapter 4 for a full discourse on barrel aging.

Oak chips and oak extracts can also be used effectively, provided they are employed properly. Obviously, these do not add the charisma in the cellar as do barrels.

Wine should be evaluated at least once weekly during wood aging in order to monitor development of bouquet and flavor. As the desired effect appears, evaluate more frequently in order to avoid the wine becoming too "oaky."

Detailed wood aging history is a particularly valuable record to maintain for future reference.

BALANCING AND PRESERVATION

Wines should be balanced and preserved just before bottling. To *balance* a wine is to bring harmony to sweetness/acidity ratios. What is perfectly balanced to one person may be out of balance to another. Most avid wine consumers prefer dry wines, especially with food. Some people prefer a touch of sweetness, whereas others enjoy wines which are very sweet. Table 9–3 presents some generally recognized levels.

Cane sugar or citric acid can be added to achieve the desired balance for each wine type made. ATF regulates sugar additions in 27 CFR, Part 4, Section 24.179 and acid additions in Section 24.182.

Take small samples and test various laboratory-scale additions first. From this testing, the "ideal" prescription of sugar and/or acidity can be calculated for the entire lot. Stir thoroughly as sugar and acid additions do not readily dissolve in cold wine.

Table 9-3 Typical White Table Wine Sugar–Acid Balance Levels

Balance level	Degrees balling	Total acidity (g/100 mL)
Tart	−2.0	0.700
Dry	−2.0	0.600
Semidry	−1.0	0.600
Semisweet	0.0	0.600
Sweet	≥2.0	0.600
Insipid	≥2.0	<0.500

Preservation can be achieved with a final addition of KMS, usually at a rate of 1 oz./100 gal for dry wines at a pH of 3.30 or under.

Wines having residual sugar, or added sugar, should be preserved with a KMS addition of at least 2 oz./100 gal and even higher doses for pH levels greater than 3.30. Be certain that the KMS supply used is fresh and active. A failure to halt yeast action can result in secondary fermentation and exploding bottles. As a further safeguard, a dosage of sorbic acid is also recommended. The following formula can be used in calculating the proper addition:

$$\text{mg/L of sorbic acid desired} \left(\frac{150}{112}\right)(3.8)\ \text{wine (gal)} = \text{mg of sorbic acid required}$$

Take the following 100-gal example where, 350 mg/L sorbic acid is desired:

$$350 \left(\frac{150}{112}\right)(3.8)(100) = 178{,}120\ \text{mg}$$

Dividing the 178,120-mg result by 1000 results in 178 g of sorbic acid addition required—or about 6¼ oz. ATF limits sorbic acid additions to a maximum of 300 mg/l is 27 CFR, Part 4, Section 24.246.

Post all sugar/acid, KMS, and sorbic acid components used for balancing and preservation on the appropriate record.

FILTRATION

Brilliantly clear wine has a visual appeal, but the cost of filters and filter media are serious cost items. The most common filter used in small wineries is a cartridge type fed by a small pump. This involves a comparatively small filter investment but a high cost in cartridge media, which is typically fabricated from various cellulose and other fiber materials. Although these materials may be advertised as "inert," they can impart unwanted flavors into the wine and/or absorb desired flavors from the wine. Consequently, thorough rinsing with water, in which 5 g/L each of KMS and citric acid is dissolved, should be pumped through each cartridge before commencing subsequent wine filtrations. If both white and red wines are to be filtered, filter the white(s) first.

Although the savings may be tempting, it is not recommended that cartridges be "cleaned" for reuse. Spent cartridges contain heavy concentrations of microorganisms trapped inside the media and should be disposed of immediately. Disinfecting, soaking, and other contrived attempts cannot achieve reclaiming cartridges for commercially dependable results.

Once a small winery has grown to produce more gallonage than cartridge filters can handle, it is time to consider a plate-and-frame type which uses cellulose pads for media. These plates are standardized 16 in. × 16 in. and setups can exceed 40 pads for larger lots. It is an obvious waste to equip the winery with far greater capacity than foreseeably necessary. Pad media availability is diverse—from several microns of coarse porosity to less than 1 micron nominal, which can remove most yeasts and larger bacteria cells.

Although commercial membrane filtrations (Fig. 9–8) can remove all yeasts

Fig. 9–8 55-Plus Beverage Monitor
Courtesy: Millipore® Corporation, Bodford, MA.

and bacteria from wines, this technology is available only to those wineries able to afford the many thousands of dollars needed for equipment and media. For all intents and purposes, filtration is purely cosmetic—repeated filtration does not remove winemaking flaws but will diminish fruit bouquet and flavor. See Chapters 4 and 8 for additional information regarding wine filtration.

After filtration operations are completed, the filter, tubing, pump, and other equipment used in filtration should be cleaned with scalding hot water and drained dry.

BOTTLING

The bottling operation is usually one of the most festive operations in winemaking but should be performed with serious attention to the exceptional care required for sanitary practices. Figure 9–9 is a schematic of a bottling line operation.

It is recommended that Chapters 4 and 8 referring to bottles are reviewed before commencing bottling operations. New bottles should be rinsed at least once with scalding hot water. High-pressure, "jet" spray devices work best. Used bottles should have been cleaned as soon as they were emptied, and stored upside down. Under the best of storage conditions, however, used bottles may need scrubbing with a brush to clear out foreign matter. Used bottles are, of course, recyclable, but the expense of cleaning is often more than the cost of new.

New wines will undergo an abrupt change in flavor and overall character due to the rigors of preservation, final filtration, and bottling—a condition referred to as "bottle shock." This state usually passes in a few weeks and the wine returns to near its original condition.

Most white wines are best up to only several years of aging. Some people feel that the older a wine, the better. Not true. Whereas some red wines may improve with several decades or more of aging, they, too, will reach an optimal point in life and then diminish in quality.

White wines should be aged in a cool, dark, cellar-like environment, free from vibration and drastic temperature variances. If space is at a premium, place white and blush wines closest to the floor, where it is generally cooler. Ideal storage temperatures are consistently maintained at 55–60°F.

PACKAGING

Some winemakers prefer to apply capsules and labels right after bottling and corking. This can work fine as long as the bottles are stored in cases or some other protective container.

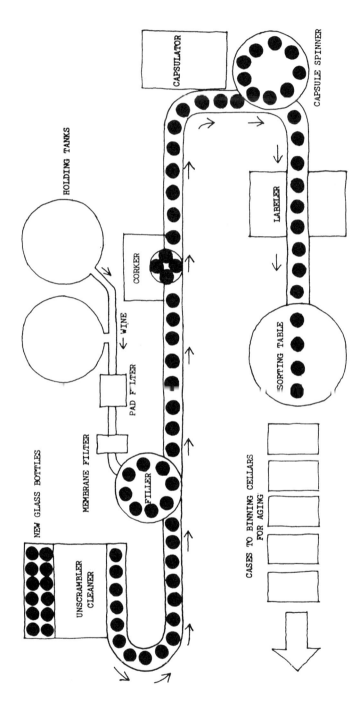

Fig. 9–9 Schematic of bottling line operation.
Courtesy: Sebastiani Vineyards

Bottles laid to age in cellars, however, can collect cobwebs, dirt, mold, and grime. Rinsing each bottle with warm water will help in the successful application of packaging materials. Hard-water rinsings will generally leave unsightly mineral spots unless each bottle is dried.

Apply glue to fully cover the back of each label with as thin a layer as possible. Identically position labels on bottles, as varying heights of labels create a haphazard impression. Many vintners prefer labels produced with an adhesive coating already applied. These can work very effectively in the hands of experienced personnel. Once in contact with a bottle, self-adhesive labels are difficult to adjust for position or to remove. See Chapters 4 and 8 for further information relating to packaging materials and equipment.

Once packaging is applied, bottles can be carefully inserted in cases so as not to scuff the labels. Some vintners choose to wrap bottles in tissue paper before casing. This is obviously an extra expense and can be argued both pro and con as to consumer preference.

Cases can then be stored in a cool (60–65°F) dry warehouse (Fig. 9–10). Be sure paperboard cartons are stored on a pallet or some other support which will protect them from wet floors.

Fig. 9–10 Wine warehouse.
Courtesy: Bouchard Pere et Fils.

BULK WINE BOOK INVENTORY CARD

Lot. No. 541	Type-Variety CHARDONNAY	Class -14%	Color WHITE

REMARKS:

Date	Racked From Tank	W.O.	Gallons	Treatment-Disposition	Racked To Tank	W.O.	Gallons	PHYS INV	ALC	BALL	T.A.	V.A.	FSO2
10-13	F-4	191	1,814	racking	S-2		850	✓	11.9	-1.9	.683	.024	—
					S-3		850	✓	12.0	-2.0	.690	.024	—
					1bbl		50	64 gals juice lees destroyed					
10-17	to wine acct			analysis	S-2		850	✓		-2.1		.024	36
	to wine acct			add KMS	S-3		850	✓		-2.0		.024	36
	to wine acct				1bbl		50	✓		-2.1		.030	32
10-24	S-2	199	850	racking	S-6		1,500		12.0	-2.1	.683	.024	20
	S-3		850		4 bbls	(1YR AM OAK)	212	18 gals wine lees destroyed					
	1bbl		50		2cbys		10						
11-8	S-6	211	1,500	fining	S-2		850	✓	1lb bentonite 1lb KHT crystals				
	4bbls		212	and	S-3		850	✓	1lb bentonite 1lb KHT crystals				
	2cbys		10	stabilize	2cbys		10	✓	12gals lees destroyed				
11-9					S-2	analysis	850	28°	12.1	-2.1	.668	.030	—
					S-3	analysis	850	29°	12.1	-2.0	.660	.024	8
12-15	S-2	244	850	filter w/	15bbls	(NEW FR. OAK)	890	50° ✓	12.1	-2.1	.638	.030	—
	S-3		850	coarse	14bbls	(1YR FR. OAK)	812	51° ✓	12.0	-2.0	.630	.024	—
	2cbys		10	pads	3cbys		15	✓	13 gals filter loss				

BULK WINE BOOK INVENTORY CARD

Lot. No. 541	Type-Variety CHARDONNAY	Class -14%	Color WHITE

REMARKS: continued from reverse side

Date	Racked From Tank	W.O.	Gallons	Treatment-Disposition	Racked To Tank	W.O.	Gallons	PHYS INV	ALC	BALL	T.A.	V.A.	FSO2
12-22	1cby			top up	4 bbl	SAMPLES		56° ✓				.024	—
1-6-96				top up	4 bbl	SAMPLES		54° ✓				.024	—
1-13				top up	4 bbl	SAMPLES		52° ✓				.024	—
2-6				top up	4 bbl	SAMPLES		51° ✓				.030	—
2-21				top up	4 bbl	SAMPLES		51° ✓				.024	—
3-5	(-1cby used)			top up	4 bbl	SAMPLES		51° ✓				.030	—
3-20	1cby			top up	4 bbl	SAMPLES		52° ✓				.030	—
4-3				top up	4 bbl	SAMPLES		51° ✓				.024	—
4-17				top up	4 bbl	SAMPLES		52° ✓				.030	—
5-3				top up	4bbl	SAMPLES		54° ✓				.030	—
5-21	(-1cby used)			top up	4bbl	SAMPLES		56° ✓				.030	—
6-5	15bbls	99-96	870	adjust				58°	12.1	-2.1	.638	.030	—
	14bbls		812	and	added 2 lbs citric acid, 1½ lbs KMS								
	1cby		5	filter	BT-1	.7 MICRON PADS	1680	59°	12.1	-2.1	.645	.030	48
6-5	BT-1	100	1680	bottle	697cs	+7 bottles		64°	12.1	-2.0	.645	.030	40
6-5	transferred			697 cs and 7 bottles to cased goods acct									
6-5	this acct			closed									

Fig. 9–11 White wine bulk inventory form in service.

White Table Wines

CASED GOODS BOOK INVENTORY							
Lot No. 541	Vintage 1995		Type-Variety CHARDONNAY			Size 750 ML	
Class ~14%	Color WHITE	Minimum Inventory	Bottle Type BURGUNDY DEAD LEAF GREEN PUNT		Closure NAT WAXED 13/4" STERILE		Other
Date	Cases In	Cases Out	Balance on Hand		Ref No.	Inv.	Remarks
6-5-96	697cs 7bot		697cs	7bot			
7-10		1 bot	697cs	6bot	for analysis	✓	bottle sick
8-14		1 bot	697cs	5 bot	for analysis	✓	bottle sick
9-3		1 bot	697cs	4bot	for analysis	✓	bottle sick
9-4	released	1 cs 4bot	696 cs		family use		
9-5		10 cs	686 cs		retail shop		R-94-96
9-6		50 cs	636 cs		Royal Dist.		P.O. #44707
9-6		40 cs	596 cs		Jones Bev.		P.O. # JB-791-96
9-6		40 cs	556 cs		Castle Whsle.	✓	P.O. # C-1766-W
9-10		50 cs	506 cs		Filmore Dist.		P.O. # 944-6
9-10		30 cs	476 cs		Phillips Bev.		P.O. # PB-6497
9-11		40 cs	436 cs		Morgan Dist.		P.O. # 7533
9-11		10 cs	426 cs		retail shop	✓	R-97-96
9-12		10 cs	416 cs		O'Brien Rest.		P.O. # 196-96
9-12		4bot	415 cs	8bot			4 bottles broken
9-12		8bot	415		family use		
9-12		30 cs	385		Barnes Whsle.	✓	P.O. # BW-64170

Fig. 9–12 White wine cased goods inventory form in service.

RECORDS

Figure 9–11 and 9–12 present a typical white wine processing history using the record format presented previously.

CHAPTER 10

RED TABLE WINES

It generally takes more time to bring red wines to the point of bottling than whites. Although fermentation should take place in only a week or so and the clarification/stabilization process is more simplistic, the aging of heavy-bodied reds to maturity is more complex.

GRAPE VARIETIES

Cabernet Sauvignon, Chambourcin, Chancellor, Merlot, and Norton make reds which are "vinous" in character (herbal and vegetable flavors), dense in color, heavy bodied and require long terms of aging. Varieties such as Baco Noir, Gamay, Pinot Noir, Marechal Foch, and Zinfandel, are typically "berry–cherry" in bouquet and flavor, moderate in densities of color and body, and can mature in just several years of bottle age. Native American reds such as Concord, Isabella, and Ives make "fruity–grapey" wines which are typically bold in every sensory aspect and can often be marketed when just 1 year old.

Most red grapes should be harvested at peak ripeness or just slightly on the overripe side, which would generally be in the pH range 3.30–3.40. This allows for full color development and will encourage later malolactic fermentation, if so desired. The buttery–cheesy flavor component from this bacterial fermentation usually has a negative effect on wines made from the native American red varieties; consequently, these may be harvested just slightly on the green side in the pH range 3.20–3.30.

As with making white wines, the very first consideration is for sanitation. Reds are fermented in open-top containers at higher temperatures, making the exposure to spoilage organisms an even greater consideration. It is absolutely essential that equipment and containers be thoroughly cleaned with scalding hot water and drained dry before securing from use.

INCREASING RED WINE FRUIT FLAVOR INTENSITY

Close observation of the ripening process each vintage season can provide essential data for anticipating the precise time for harvest in order to achieve the red wine flavor profile desired. Much of this scrutiny can be determined by simply tasting representative berry samples. The flavor identified in grapes should be captured into the wine made from them. Analysis of pH, Brix, and total acidity data should be recorded and correlated with optimal maturity levels for future years.

Immature red grapes in the pH range 3.10–3.20 typically make wine which is rather vegetative in flavor. Generally higher pH levels, in the 3.30–3.40 range, are associated with more intense fruit flavor.

One method commonly employed in Europe is the *saignée* process. This involves separating free-run and/or lightly pressed juice from red grape musts immediately after crushing. This initial fraction is usually cool-fermented separately for blush wines or for making blending wines. The drained must is then fermented in the normal red wine manner. This obviously results in a greatly reduced juice portion in which color and other solids are dissolved during extraction by fermentation. Fully ripened grapes are needed for the saignée in order to reduce high concentrations of acidity. Generally, shorter on-skin time exposure is made in order to avert excessive levels of phenol extraction.

Another method of increasing red wine fruit flavor is with the *maceration carbonique* process which is closely identified with the famous "Beaujolais nouveau" wines. Closed fermenters are partially filled with crushed grapes and then filled with whole clusters. Carbon dioxide generated by the fermenting crushed portion is held under a modest pressure in the tank until the desired color and flavor extraction is achieved—followed by immediately pressing the entire lot. Carbonic maceration generates dense purple color extraction and lower concentrations of tannins. Unless this typical "nouveau" character is desired, further blending is needed to achieve good overall balance.

HARVESTING

Hand-harvested grapes are preferred. Broken grape skins resulting from mechanical harvesting at higher pH levels encourages *apiculate,* or "wild," spoilage yeasts to grow before a cultured yeast strain can be inoculated. Expect a yield of about 140–150 gal of finished wine from each ton of grapes.

CRUSHING–DESTEMMING

Commence the winemaking process as soon as possible after harvest. Grapes received in the 75–80°F temperature range are ideal. Remove leaves and trash along with unripe or spoiled berries in the same *triage* process as described for white wines. Record net weight and rinse grapes (hand harvested only) with cold water.

Reds, unlike whites, may benefit from some portion of the stems being retained in the must—adding *tannin* (complex phenolic compounds) which may contribute a bit of appealing astringency and extend aging potential. Perhaps up to one-half of the stems might be considered for retention in vinous grape musts; less in the berry–cherry flavored varieties and little or none in fruity–grapey reds. Experiments can be conducted to learn optimal stem/must ratios.

Clean the crusher thoroughly with scalding hot water and crush whole bunches of grapes for the stem-retained portion and destem the balance. Crusher–destemmer devices are discussed in Chapter 8. After crushing operations are completed, clean empty grape containers, crusher, and destemmer thoroughly with scalding hot water and allow to drain dry.

Crushing should take place so that the resulting must is deposited directly into a clean fermenting container. Red must fermenters should have a "jacket" around the outer wall in which cold water can be recirculated to lower the rapid temperatures generated by active fermentation. (See Fig. 10–1.) Even so, fermentation may take place in only a few days, as yeasts are very active in the ideal elevated temperature environs of red grape musts.

BRIX (SUGAR) ADJUSTMENTS

Brix in red grape juice typically measures between 14° and 23°. Native *Vitis labrusca* varieties will generally ripen at the lower ends of this range, *Vitis vinifera* at the higher end, and the French-American hybrids mid-range. There are exceptions.

Each degree of Brix will convert to about 0.535% alcohol during fermentation. Consequently, to achieve 12% alcohol by volume, the red grape much would require approximately 22.5° Brix. A natural Brix of 17.6° Brix would thus need to be adjusted with sufficient sugar to achieve a 4.9° increase in Brix.

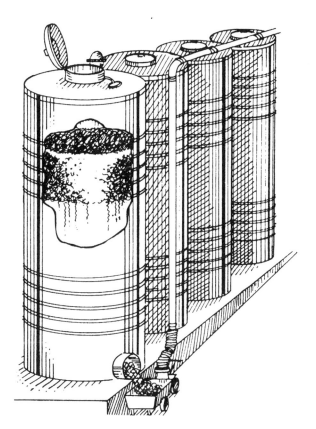

Fig. 10–1 Jacketed red wine fermenting tanks with floating "cap" of skins. *Courtesy: Sebastiani Vineyards.*

As mentioned previously, the addition of sugar is called *chaptalization*. In some cases, grapes are excessively high in acidity, leading to both sugar **and** water being added—referred to as *amelioration*. Both of these adjustments have historically been controversial in fine wine circles. ATF regulation, 27 CFR, Part 4, 24.177 relates to Brix adjustments.

The major difference between adding sugar to white juice and red must is that the winemaker does not have a finite measurement of the juice contained in must which has not yet been pressed. It is thus necessary to conservatively estimate the amount of juice for a sugar addition. Depending on grape variety, ripeness, and press efficiency, history will usually determine juice potential to be within the range of 160–180 gal/ton. Once the fermented red wine is pressed out, a precise determination can made for final adjustment if so desired.

The same mathematical formulas for calculating sugar and concentrate additions used for white juice in Chapter 9 can be employed for red musts. A well-designed lot origination form will permit vintners to accurately calculate sugar/water additions and simultaneously maintain a good record of increases in gallonage.

ACID ADDITIONS

Total acidity in red musts should ideally measure between 0.65 and 0.75 g/100 mL or, roughly, between 0.65% and 0.75% acidity. Most red wine styles are constructed on medium to heavy-bodied consistency. It may be best to attempt no adjustments on musts having only a slightly high acidity as compared to the thinning which would result from ameliorating that tartness with water. The calculation and methodology for red wine additions are the same as provided in Chapter 9 for whites.

Stir in all additions and test for Brix and total acidity once more to ensure that the desired levels have been achieved. Post additions in the appropriate record.

YEAST INOCULATION

The rationale for selecting and preparing red wine yeast starters are identical for white wines as outlined in Chapter 9. At one time, it was rather common practice to utilize yeast sediment remaining in the bottom of red wine fermenters or some actively fermenting red juice as an immediate starter culture for inoculation of the next batch of crushed must. Research indicates that neither of these practices may be recommended. Wild yeast and/or bacterial infections are easily spread, and significantly more undesired fermentation by-products are generated by perpetuating yeast cultures.

FERMENTATION

The fermentation philosophy for red wines is much different in red wines as compared to whites. Whereas it is critical to maintain cooler temperatures and restrict exposure to grape stem, pulp, and seed particles, it is the opposite for reds. Color and flavor extraction from the red grape skins usually is ideal at 80–90°F. Figure 10–2 shows the method of taking the temperature of a red wine fermenter.

As mentioned previously, some portion of grape stems left during the crushing and destemming operation may be preferred by some winemakers in developing added red wine complexity. As will be mentioned later, postfermentation maceration of red wine in contact with the skins may be another source of desired complexity in the finished product.

Fig. 10–2 Taking temperature of a red wine fermenter.
Courtesy: Wine Institute.

IMPORTANT NOTE: CAUTION! DUE TO THE ELEVATED TEMPERATURES AND TUMULTUOUS FERMENTATION OF RED WINE MUSTS, AN ACCELERATED AMOUNT OF CO_2 GAS WILL BE GENERATED. HEAVY-DUTY VENTILATION SYSTEMS AND AIR MARKS CLOSE AT HAND ARE CRITICAL TO THE PREVENTION OF ASPHYXIATION. UNASSIGNED PERSONNEL AND VISITORS, ESPECIALLY CHILDREN, SHOULD NOT BE ALLOWED IN LOCATIONS WHERE ACTIVE RED WINE FERMENTATIONS ARE TAKING PLACE.

As fermentation commences, grape skins will rise to the top of the must, forming a "cap" over the juice below. In order that maximum color is extracted from the skins by the alcohol being generated by fermentation, the cap should be regularly pushed back down into the juice—an operation referred to as "punching." Punching also serves to discourage aerobic spoilage organisms which may be obtaining oxygen over the cap. Using a clean wooden, plastic, or stainless steel paddle, the cap should be punched at least three times daily. Some winemakers prefer to "pump over" rather and punch. This operation is the simple attachment of a pump and hose to the bottom of the fermenter in order to deliver wine over the top where it is sprayed evenly across the surface of the cap (Fig. 10–3) and, as with punching, should be performed at least every 8 h during fermentation.

Fig. 10–3 Illustration of pumping over the cap.
Courtesy: Sebastiani Vineyards.

Fermenter temperature should be checked at least twice daily and adjusted by running hot or cold water as needed in the jacket around the outside of the fermenter. If there is no jacket, a handy device for this is an old garden hose which can be punched with small pinholes every inch or so. The hose is plugged at one end and strapped around the upper outside wall of the fermenter. Increase pressure from the hose bib slowly until an even flow of cold water covers the tank wall. Although the cold water provides some cooling, it is the evaporation of the water which creates the major cooling effect. Fermenting red wine musts under temperature adjustment should be agitated continually in order to avoid extreme heat in the center of the vessel.

One of the newest, and most expensive, innovations in red wine fermentation is the rotating stainless steel fermenter (Fig. 10–4) which is very efficient in the amount of time and energy needed to gain optimal skin exposure to the fermenting wine and minimal oxygen exposure to spoilage organisms.

Stuck (incomplete) red wine fermentations are rare and most of these due to temperatures having allowed to exceed 100°F. If the problem can be positively diagnosed as excessive heat, then immediate cooling is an obvious treatment. Light agitation to replenish some oxygen for remaining viable yeast cells may also serve to resume fermentation. The addition of a light dose of food-grade diammonium phosphate (DAP), up to 2 oz./100 gal will help replenish free nitrogen required by yeasts.

Fig. 10–4 Rotating stainless steel fermenters.

Post inoculation date, temperature checks, heating/cooling and other pertinent data in the appropriate record.

PRESSING

There are several major rationales in determining the precise time for pressing new red wine from the must.

One variance is that of color extraction. Color intensity is primarily a function of pigment development in the berries during a given growing season and the pH at harvest; it is not directly correlated with extended skin contact. Most red pigments are fully extracted by the time Balling reaches 5.0 or so. This level is recommended as a beginning standard for pressing berry–cherry varietal musts.

Yet another variance is that of limited skin contact–in the range of 10.0°–5.0° Balling. Pressing fruity–grapey musts at this midway stage of fermentation renders wines with a more simplistic profile of flavors—uncomplicated by the extraction of tannins and other complex compounds.

MACERATION

Maceration of fermented red wine must simply means holding it in the fermenter for some period of time, usually 10–30 days, after fermentation has been completed. This should not be confused with *maceration carbonique* for "nouveau" wines, which is very different than red wine must maceration.

Red wines designed for long life, such as the heavily structured Meritage reds, must have sufficient tannin to slow the oxidation process known as aging. One approach is in determining the amount of tannin desired to be extracted from the skins and seeds. Bordeaux chateaux often keep Cabernet Sauvignon and Merlot musts "on the skins" up to several weeks of extended *maceration* time after fermentation has been completed. A close watch needs to be maintained during this time to guard against white "islands" of spoilage bacteria developing on the surface of the wine. Any suspicion of such activity justifies immediate pressing.

Maceration can easily be excessive, due primarily to the chemical makeup of the tannins extracted from the must. This is very complex, as tannins are polyphenolic polymers formed by oxidation and condensation reactions. Tannins from red grapes harvested on the immature side generally tend to form as coarse and bitter compounds. Conversely, the more ideal, somewhat astringent tannins, those associated with complimenting the aging process, are developed in grapes which have fully matured—even perhaps just slightly overripened. Fermented must can be gravity fed to the press from fermenters designed for that purpose. Otherwise, fermented must should be transferred to the press with a portable must pump such as that illustrated in Figure 10–5.

Fig. 10–5 Portable must pump with hopper and auger.
Courtesy: Scott Laboratories.

FREE-RUN AND PRESS WINE

Free-run red wine from the press (Fig. 10–6) is usually the best quality. The press wine should be darker in color due to the extraction of pigments from the skins under pressure. The free-run and press-run sublots should be held separately for later blending considerations.

If the red wine has fermented to less than $-1.0°$ Balling, it can be placed into clean containers, filled, sealed with fermentation locks, and stored in a dark 55–60°F chamber. If above $-1.0°$ Balling, the containers should be filled to about 80–90% capacity until the $-1.0°$ Balling level is reached, and then racked to full containers equipped with fermentation locks. If the estimated gallonage of juice in the red must was reasonably accurate, then the level of alcohol developed from fermentation should not vary much from what was calculated. Providing the yeast population is still active, increases in alcohol can be made by a sugar adjustment and calculated by using the ratio $1.0°$ Brix = 0.535% alcohol.

RACKING

Young red wines will generally finish bubbling within a few days after pressing and should indicate no more than a trace of residual sugar using the Dextro-Chek®

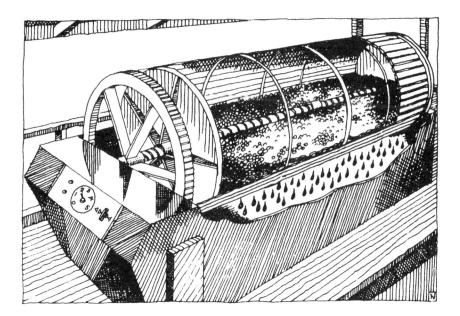

Fig. 10-6 Free-run juice falling from a Vaslin-Type rotating basket press. *Courtesy: Sebastiani Vineyards.*

tape. If so, racking can be made into a clean vessel filled completely in order to expel any air, and a loose cap or plug is used to cover the second-rack containers instead of the fermentation lock.

An addition of 1 oz. of potassium meta-sulfite (KMS) powder per 100 gal should increase the sulfur dioxide content in the young wine by about 20–30 ppm (or mg/L). This is recommended to help further protect against spoilage microorganisms and oxidation. Continue storage in a cool, dark, cellar-like environment.

From this point on, throughout the life of the wine, it will be necessary to ensure that it is stored in containers which are completely full. Avoiding air contact upon the wine reduces its exposure to oxygen, which is a component for oxidative browning and the growth of spoilage organisms.

BLENDING

As discussed in Chapter 4, the general rule of thumb in blending is to associate the different virtues of individual wines into a resulting wine having a broader and more satisfying appeal. A typical blending motive for white wines is to achieve a delicate expression of varietal fruit flavor expressed upon balanced acidity with little tannin astringency. Vinous reds are generally blended to achieve the opposite—bold flavors, heavy body, dry of sweetness, and moderate tannin. Berry–cherry reds may be a bit lighter in overall expression and fruity–grapey reds may approach white wine blending rationale.

Wines should be blended before stabilization. The many factors involved with stability sometimes result in stable wines blended together to form an unstable wine. The new blend will cause a new record to be started. Wines taken for the blend should be deducted from their appropriate records.

MALOLACTIC FERMENTATION

As mentioned in Chapter 3, the controlled growth of select bacteria species (*Lactobacillus* and *Leuconostoc*) can serve to transform malic acid, which has a rather apple-like character, to lactic acid, which has a pronounced "buttery–cheesy" (diacetyl) flavor. Such bacteria are thus known as malolactic and their fermentation process also includes a small reduction in total acidity and a small increase in acetic acid, as well as some CO_2 gas. The virtue of malolactic fermentation in vinous and berry–cherry red wines is to add an interesting nuance of buttery character. As a rule, fruity–grapey reds are not improved by this treatment.

The inoculation of a viable malolactic bacterial culture in a young red wine should be only when the pH is above 3.35 and the temperature is in the range 65–70°F. Be forewarned that this condition also invites the growth of acetic bacteria and other spoilage organisms.

This can be a frustrating process, as the activity of malolactic bacteria can vary significantly and is difficult to anticipate. Because these organisms require only a small amount of oxygen, this fermentation can be conducted in nearly full containers equipped with fermentation locks. Close observation remains critical, however. Inspect several times daily for the emergence of CO_2 bubbles in the fermentation lock and the development of a buttery–cheesy odor.

FINING AND STABILIZATION

The high level of phenolic compounds in red wines is usually sufficient to reduce protein suspensions. Consequently, young red wines (2–4 months old) are typically "bottle bright" and have no need for clarification. During barrel aging, red wines usually precipitate unstable tartrates without the use of cold storage. Because of this, many commercial vintners choose not to stabilize red wines in cold storage.

Egg-white fining is a common procedure employed by the finer vintners of red wine. Several egg whites beaten to a froth, not stiff, and gently stirred into 50 gal of wine serves to soften the harshness of young wines having an abundance of tannins. Alternative methods of red wine fining can be found in Chapter 8.

BARREL AGING

Oak aging for red wines generally transcends a much longer time in wood to achieve the desired "toasty–vanilla" character in bouquet and flavor than for whites. The vinous red varieties may take up to several years in 50-gal barrels in a cool cellar environment. (See Fig. 10–7.) Berry–cherry varietals may need a year or so, and perhaps no wood aging is best for the fruity–grapey reds. Remember that aging progresses faster in smaller oak containers.

All wooden containers will lose some wine contents—mostly due to assimilation and seepage. Each keg and barrel should be carefully examined at least monthly in order to 'top up' with wine. Generally only a few tablespoons, certainly less than a cup, is needed. This procedure refills the 'ullage' (air space) with a similar type wine so that oxygen is not available for bacterial growth. If a white 'film' of bacteria or wild yeast growth appears on the surface, the wine used for filling should be treated with KMS at a rate of about 1 teaspoon per gallon. This should inhibit the microorganisms and allow aging to continue.

Fig. 10–7 Visitors observing red wine barrel aging. *Courtesy: Sterling Vineyards.*

Some commercial winemakers prefer to "wet-bung" barrels after filling in order to reduce the jeopardy of bacterial infection entering barrels through the bunghole. It is a procedure in which barrels are filled with bungs securely driven and secured. Barrels are then rolled to the side so that the bungs are situated at about "2 o'clock" so that wine constantly covers the bung hole during the aging period. Subsequent ullage is created in an aseptic atmosphere which resists spoilage.

An in-depth discussion of barrel aging is provided in Chapter 4.

BALANCING AND PRESERVATION

Red wines are rarely finished with any residual sweetness and, thus, red wine sugar/acid balance is not generally an issue unless total acidity is excessively high or low. Red wines at these extremes may be best used in blending.

Dry red wines can be preserved with a final addition of KMS, usually at a rate of $1/2$ oz./100 gal for dry wines with a pH at 3.30 or lower.

FILTRATION

Red wines often clarify so well that filtration is not needed or desired. Some degree of sediment is expected in older bottles of red wines anyway—the reason that fine crystal wine decanters exist.

Careful racking, fining, and barrel-aging regimens can result in perfectly clear red wines. Under these circumstances, there may be justification to consider no filtration. This will conserve color and flavor. Should no filtration be decided on, it is recommended that the label on this finished wine indicate that the wine was "unfiltered" so that consumers can expect some sedimentation.

Chapters 4, 8, and 9 provide complete information relating to common filter types, filter media, and filtering techniques.

Low-porosity membrane filtration is not recommended for red wines, as extensive color pigmentation may be lost.

BOTTLING AND CORKING

Refer to Chapter 8 in order to select the proper bottle shapes in accordance with red wine traditions. The bottling procedure for reds is identical to that for whites, as discussed in Chapter 9.

Red wines such as "nouveau" and fruity–grapey types, as well as others designed for consumption in the short term, can be bottled and corked in much same manner as for white wines. (See Fig. 10–8.) Heavier reds having greater tannins for longer terms of bottle aging are usually given longer corks, 2 in. or more, in order to extend sealing potential.

Fig. 10–8 Compact single-unit bottling and corking line.
Courtesy: Prospero Equipment Corporation.

Bottle Aging

Greater amounts of color and other phenolic compounds extracted during fermentation are correlated with longer bottle-aging times—from several weeks to several years. Some commercial vintners routinely bottle-age red wines for 1 year prior to market release. The ideal environment is 50–60°F, dark, and free of vibration. (See Fig. 10–9.)

Bottles can be placed in horizontal tiers or upside down to keep corks moist. If paperboard cases are used, be sure to keep them on a pallet or some other support to avoid wet floors. As time passes, regular sensory evaluations should be made to monitor aging progress, and tasting notes posted in the appropriate record.

Fig. 10–9 Red wine bottle aging.
Courtesy: Bouchard Pere et Fils.

	LOT ORIGINATION	
DATE	9-14-95	CULTIVAR CHAMBOURCIN
LOT NO.	521	REMARKS: machine
SERIAL NO.	8-95	harvested - ±1% MOG
FERMENTER NO.	F-12	must temp 74°F out
NET GALLONS	EST. 984	of crusher - added
NET TONS	5.79	50 ppm KMS during
GALLONS PER TON	EST 170	crush - yeast added
pH	3.35 T.A. 1.028	immediately in F-12
BRIX	18.8 ALC.	
EXT.		

EST.
AMELIORATION = __10__ % @ __22.0__ BRIX = __11.5__ ALCOHOL

__EST 984__   GALLONS START = __1,093__ RESULTING TOTAL
   .90   INVERSE OF   GALLONS OF PRODUCT
              AMEL. PERCENTAGE

__EST 1,093__   TOTAL   @ __22.0__ BRIX   (__2.004__) LBS/GAL= __2,190__ TOTAL
              GALLONS   LBS.

__984__   START   @ __18.8__ BRIX   (__1.691__) LBS/GAL= __1,664__ START
              GALLONS   LBS.

__109__   AMELIORATION   __526__ ADDITION
              GALLONS   LBS

__39__   SUGAR AS   x .074 (GAL/LB)=   __39__ SUGAR AS
              GALLONS   GALLONS

__70__   WATER
              GALLONS

__PDM__   YEAST ADDITION = RATE OF __1__ LBS PER __1,000 gals__

REMARKS: yeast in 4 gals. of water @ 105°F

Fig. 10–10 Red wine lot origination form in service.

PACKAGING

Unless bottled red wines are placed in some type of sealed container for bottle aging, it is difficult to escape at least some sort of dust or grime during that time. As a rule, a buildup of cobwebs, dust, mold, and other matter needs to be washed off with a careful rinsing with warm water. Sometimes more than one rinsing is necessary. Hard water can leave unsightly mineral spots unless each bottle is dried.

FERMENTATION CONTROL RECORD

TANK NO. F-12   LOT NO. 521   VARIETY CHAMBOURCIN

DATE	TIME	TEMP.	BALL.	ALC.	T.A.	V.A.	C	C	N	T	
9-14	9:00 am	74°	21.2			.012	✓	✓	✓	✓	+temp
9-14	10:00 am	circulated hot water in F-12 jacket for ½ hr.									
9-15	9:00 am	82°	20.0	pumpover 10 min		.012	✓	✓	✓	✓	
9-15	2:00 pm	83°		pumpover 10 min							
9-15	9:30 pm	84°		pumpover 10 min							
9-16	9:30 am	86°	16.0	pumpover 10 min		.018	✓	✓	✓	✓	gassy
9-16	2:30 pm	87°		pumpover 10 min							
9-16	9:30 pm	87°		pumpover 10 min							
9-17	9:00 am	88°	11.6	pumpover 10 min		.024	✓	✓	✓	✓	OK
9-17	2:30 pm	88°		pumpover 10 min							good color
9-17	10:00 pm	89°		pumpover 10 min							
9-18	9:00 am	88°	8.1	pumpover 10 min		.024	✓	✓	✓	✓	OK
9-18	2:30 pm	89°		pumpover 10 min							
9-18	9:30 pm	88°		pumpover 10 min							
9-19	9:00 am	87°	5.9	pumpover 10 min		.030	✓	✓	✓	✓	OK
9-19	2:30 pm	87°		pumpover 10 min							
9-19	10:00 pm	86°		pumpover 10 min							
9-20	8:30 am	84°	3.8	pressed		.036	850 gals free-run to 5-7 +cooled 150 gals to 3 (2 yr) am oak ttles				
9-21	8:00 am	65°	2.5			.036	✓	✓	✓	✓	
9-22	8:30 am	66°	1.5			.036	✓	✓	✓	✓	
9-23	8:00 am	66°	0.7			.036	continued on reverse side of card				

FERMENTATION CONTROL RECORD

TANK NO. 5-7 and 3 bbl/s   LOT NO. 521   VARIETY CHAMBOURCIN

DATE	TIME	TEMP.	BALL.	ALC.	T.A.	V.A.	C	C	N	T	
9-24	9:00 am	66°	0.2			.036	✓	✓	✓	✓	
9-25	8:30 am	65°	-0.4	11.1	.893	.030	✓	✓	✓	✓	
9-26	9:00 am	66°	-1.1			.036	✓	✓	✓	✓	
9-27	8:00 am	66°	-1.5			.036	✓	✓	✓	✓	
9-28	9:30 am	65°	-1.7			.036	✓	✓	✓	✓	
9-30	8:00 am	64°	-1.9			.030	✓	✓	✓	✓	
10-2	9:30 am	63°	-2.0	11.9	.885	.036	✓	✓	✓	✓	rack
10-4	11:00 pm	64°	-2.0	racked to TK 5-8. 14 gals juice loss. Destroyed work order no. 184-95							

Fig. 10–11 Red wine fermentation control form in service.

Information relating to packaging equipment, materials, and techniques is found in Chapters 4, 8, and 9.

Once packaging is applied, bottles can be carefully inserted in cases so as not to scuff the labels. Some vintners choose to wrap bottles in tissue paper before casing. This is obviously an extra expense and can be argued both pro and con as to consumer preference. Cases can then be stored in a cool (60–65°F), dry warehouse. Be sure that paperboard cartons are stored on a pallet or some other support which will protect them from wet floors.

RECORDS

Figures 10–10 and 10–11 present an example of a typical red wine lot origination format along with a typical fermentation history.

Once young red wines are fully fermented, the bulk wine and cased goods record-keeping procedures are similar to those illustrated for white wines in Chapter 9.

CHAPTER 11

BLUSH TABLE WINES

Making pink table wines brings together some elements from both white and red winemaking procedures. Consequently, much of the chapter will refer to Chapters 9 and 10.

GRAPE VARIETIES

Color hue and intensity is one of the most important criterion in "blush" wines—a delicate pink pigmentation instead of the generally heavier color intensity found in most traditional "rosé" wines.

Both blush and rosé wines require grapes having pigmentation properties that are stable and resist oxidation. Orange hues generally indicate an excess of bound tannins in relationship to the amount of anthocyanin pigmentation in solution. One of the best methods of achieving a true blush color hue is to add a pectolytic enzyme right after crushing and then allow the pigmentation to leach out in the must from 24 to 36 h before inoculation to commence fermentation.

Most of the varieties noted as vinous in the Grape Varieties section of Chapter 10 are good choices for blush and rosé wine product. Other reds, such as Zinfandel and Concord, can also achieve superior results.

256 *Blush Table Wines*

Harvest should take place when grapes are slightly underripened, with the same general quality control parameters as outlined for white table wines in Chapter 9.

CRUSHING–DESTEMMING, SUGAR AND ACIDITY ADJUSTMENTS, AND YEAST INOCULATION

The same general procedure and manner outlined for red table wines in Chapter 10 can be employed for processing blush and rosé wine. Minimal use of sulfite is recommended in order to reduce bleaching out pink color hues.

It is imperative that grape containers be thoroughly cleaned after each use in order to minimize the possibility of contamination. An effective custom-made grape lug rinser is illustrated in Figure 11.1.

For heavily pigmented red grapes, such as Cabernet Sauvignon and Zinfandel,

Fig. 11–1 Plastic grape lug rinser.

some vintners simply drain the juice from crushed must through a stainless steel screen prior to fermentation (Fig. 11–2). The remaining must is then fermented on the skins for red wines, which will, of course, be much more concentrated in color, flavor, acidity, and tannins, although a bit less strong in alcohol due to much of the sugar having been removed with the loss of free-run juice. This separation process is sometimes referred to as *saignée*.

PRESSING AND FERMENTATION

The principal concern for determining the precise time for pressing is ensuring that the desired amount of color is extracted from the must. (See Fig. 11–3.) Allow for

Fig. 11–2 Draining juice from red grapes on a screen.
Courtesy: Sebastiani Vineyards.

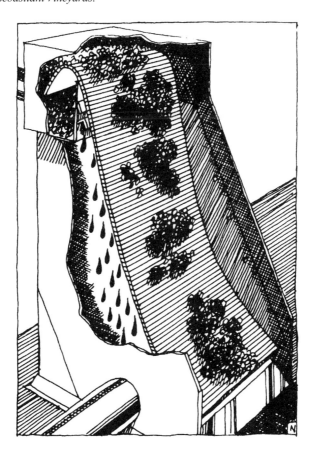

258   *Blush Table Wines*

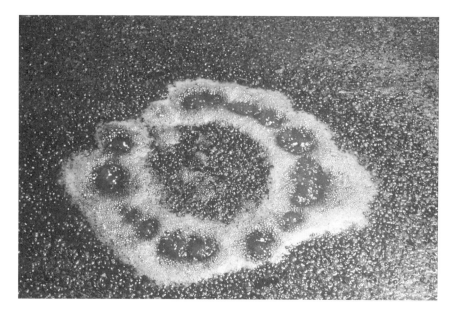

Fig. 11-3   First indication of red must fermentation.
Courtesy: Food and Wines from France.

a little extra color development, as some pigmentation will be lost in later processing functions. Some red grape varieties release color more quickly and more intensely than others, requiring close attention to the development of hue and density.

Conduct pressing and fermentation operations in the same manner as discussed for white table wines in Chapter 9. No juice racking should be necessary, but cool fermentation temperatures are particularly important in avoiding development of orange hues due to oxidation.

RACKINGS

These can also be performed in the same procedure as prescribed for white table wines. Blush wines do not generally improve from *sur lies* retention, as the "yeasty" flavor component which results does not typically enhance the fresh fruit flavors desired in most pink wines. The racking of young blush wines should be made to clean cool storage jacket stainless steel tanks such as those illustrated in Figure 11.4.

BLENDING

Adjustments in color can be made by appropriate white wine or red wine additions as determined by laboratory-scale blends.

Fig. 11–4 Jacketed stainless steel fermentation, detartration, storage, and blending tank room with diatomaceous earth filter.

CLARIFICATION AND STABILIZATION

Depending primarily on the amount of skin contact during fermentation, blush wines may or may not have sufficient phenolic extraction to reduce suspended proteins. Hazy light blush wines should be clarified and stabilized in the manner of white table wines, whereas clear darker rosé wines can be treated as reds.

AGING, BALANCING, AND PRESERVATION

Blush wines are usually best when fresh and young, rarely benefiting from any type of extended aging program. Balance and preserve the same as discussed for white table wines in Chapter 9, ensuring that residually sweet wines are treated with both potassium meta-sulfite and sorbic acid.

FILTRATION

Blush wines often create a dilemma when it comes to being filtered. Pink hues in young blush wines are generally dull, without the sheen of brilliant clarity. On the other hand, significant color loss can be realized during filtration. This is the pri-

Fig. 11–5 Inverted bottle spray washer–rinser.

mary reason why reference was made above in the section on pressing to allow a little extra color to develop during fermentation.

Proceed with filtering in the same manner as described in for white table wines in Chapter 9.

BOTTLING, CORKING, AND RECORDING

All of these blush and rosé table wine operations can be performed in the same manner as described for whites. An effective custom-made bottle rinsing device is illustrated in Figure 11.5.

Apart from allowing a few weeks to recover from "bottle shock," the delicate and elusive character of fresh, uncomplicated pink table wines usually renders bottle aging an unnecessary and undesirable element of processing.

CHAPTER 12

FRUIT AND BERRY WINES

Edited by: Dan Archibald
Publisher, **Fruit Winemaking Quarterly**

One of the many attractive aspects of making fruit and berry wines is that many species ripen during the spring and summer months—well before the vintage season commences for grapes, allowing for more efficiency and variety in the winery facility. The opportunity to add different types of wine to the product portfolio is also an appealing prospect for many vintners.

In developing an understanding of fruit and berry wines, it should be remembered that, commercially, these products are typically viewed as including everything other than wine made from grapes. Fruit wines offer a far greater array of flavors than grapes and can stretch the imagination across a far wider field for how wine can or should taste. Fruits and berries, like grapes, will vary in color, flavor, and acidity profile, depending on the soil and climate in the region where they are grown and the methods of horticulture applied in orchards. Thus, the same care and consideration in the selection of varieties and grades must be given to fruits and berries in order to achieve top-quality wine results.

Although general methodologies are much the same as for grape musts, one should be extremely careful when applying pressure to fermented berry and fruit musts. Consistency is often very loose and may flow between the press staves without efficiently separating liquid from solids. A stainless steel or poly colander or mesh nylon bag may be of help. Also, even moderate pressures can extract very astringent and bitter phenols from the pits, pulp, seeds, and skins. Taste the press wine frequently to detect excess bitter phenol extraction.

As a rule of thumb, fermentation temperatures and processing regimens for most berry and fruit wines can proceed in the same general manner as suggested for grape wines in Chapters 9–11. Use white, red, or blush table wine procedures as appropriate to the color of the berry and fruit wines being made.

While exceptions exist, most berry and fruit wines are rather short lived, certainly as compared to red vinous grape wines. Consequently, long-term wood and bottle-aging regimens are not recommended.

APPLE WINE

Although ancient, medieval, and renaissance life was replete with apples in Europe, apple trees first came to America from England, quickly becoming a staple in the diet of the New England colonists. By 1776, the apple had made its way westward into the Finger Lakes region of upstate New York; many of the plantings were made by the Mohawk and Seneca Indians. Today, apples are one of the most widely cultivated fruits across the northern United States and Canada.

Apple wine blending is an essential function in composing the desired product type and style. Varieties of apples vary in their aromatic and flavor character, their sugar content, as well as their astringency and acidity. (See Table 12–1.) Much of the characteristics presented in Table 12–1, of course, is subject to variability due to the growing climate and soil, ripeness level, and treatment provided during growing, harvest, and storage. As with grape wine, the variety of apple selected, and its state of ripeness and condition is all-important to the level of quality which can be achieved in the final product.

It is generally agreed that the best apple wines have a near colorless hue with brilliant clarity, moderate bouquet intensity, and sugar/acid balance a bit on the tart side. This can be readily achieved by insisting on fresh fruit harvested slightly on the green side of ripeness. A "triage," or hand removal of rotten apples, is highly recommended.

Table 12–1 Relative Aroma and Acidity Table for Apple Varieties

Full Aroma	Low Acidity	High Acidity
Delicious	Baldwin	Granny Smith
Golden Delicious	Cortland	Greening
Jonathon	Delicious	Jonathan
McIntosh	McIntosh	Northern Spy
Rome	Rome	Winesap

It is imperative that pressing **immediately** follow crushing. One has but to recall the last time an apple was eaten in order to understand how quickly apple pulp will commence to brown from oxidation. Some of the most successful apple winemakers sprinkle 4–6 oz. of malic acid (citric acid is not permitted by 27 CFR, Part 4, Section 24.182c) and potassium meta-sulfite (KMS) evenly over each ton of apples going into the crusher in order to inhibit oxidation and the growth of spoilage microorganisms. An addition of pectolytic enzyme is also encouraged, in accordance with the manufacturer's recommendations.

Juice temperature should be adjusted to about 55°F as soon as it is taken from the press with an immediate inoculation with approximately 1% fermenting white wine yeast starter. Apples are generally deficient in sugar to reach a sufficient alcohol level. Sugar additions can be calculated in the same manner as described in Chapter 9.

Fermentation should be maintained at 55–60°F; rackings, clarification, and the balance of apple wine processing can also be patterned similar to white table wine production from grapes.

BRAMBLE BERRY WINES*

Most bramble berries (berries which grow on thorny bushes) are deficient in natural sugar to reach a sufficient alcohol level after fermentation. Sugar additions can be calculated in the same manner as described in Chapter 9. Most of the berry pulp will liquify during the enzymatic and fermentation reactions.

Only prime fresh or frozen, fully ripened, food-grade berries are recommended. Commercial experience has shown repeatedly that poor-quality berries, or those distressed from poor transit and storage, make equally poor wines. Spoiled berries should be culled by triage and the remaining fruit gently crushed so as not to crack seeds which can release astringent phenolic compounds.

Well known for their high acid content, the major challenge to develop a berry wine with good balance is of utmost importance to the winemaker. Without this balance, most berry wines, particularly raspberry, are excessively acidic.

One of the simplest methods of reducing the acidity level is by amelioration with water and/or sugar—although this can also have the negative effect of producing a wine with light body and thin consistency. Because flavor is the most important single factor on which the quality of a fruit wine depends, the winemaker should strive to find the best balance of fruit, water, and sugar. ATF rules for these functions are found in 27 CFR, Part 4, Sections 24.178 through 24.182.

Inoculate with Prisse de Mousse yeasts and ferment in much the same manner

---

*Adapted from Dan Archibald, "All About Raspberries and Gooseberries," *Fruit Winemaking Quarterly*, Volume 2, No. 3, Summer 1992, and Edwin Patterson "Experiments and Experiences with Raspberry Wine Making," Fruit and Berry Wine Symposium Proceedings, May 1992.

as for light, fruity red table wines as discussed in Chapter 10. If needed, adding some of the amelioration water in the form of ice can help to control fast-rising fermentation temperatures.

Particular attention should be placed on monitoring any development of volatile acidity. Berry wines seem to be particularly conducive to this type of spoilage. Some producers buy frozen berries in order to suspend any bacterial action until just before fermentation. In any case, berry wine fermenters should be closed except for fermentation locks, and frequent volatile acidity analyses should be run in the laboratory.

Crushed berries typically ferment rather quickly, usually in just several days. Temperature should not be allowed to exceed 70°F and fermentation should achieve at least $-1.5°$ Balling before pressing.

The best results in pressing wine from fermented berry must are generally obtained from gently and slowly pressing the mass in mesh bags. This helps to strain out the seeds and other pulp which can later cause astringency and/or bitterness. Heavy rates of rice hulls, often two or three times that used for other fruits, can help achieve optimal wine extraction from the press.

Some berry winemakers add up to 3 lbs. of bentonite per 1000 gal of newly pressed wine in order to immediately compact the lees during the final throes of fermentation. This will permit optimal efficiency in the first racking, which should be into storage at 30–40°F. Cold storage in combination with a moderate sulfite addition will help to avert malolactic fermentation—to which berry wines are very susceptible. A second racking should also be made in cold storage.

After several months of continual cold storage, the new berry wine may be sufficiently clarified for a coarse filtration, which is frequently difficult and slow, requiring a larger filter capacity than for most other wines.

BLUEBERRY WINE*

Blueberries are versatile in that they can make equally good dry red table wines and sweeter dessert wines.

It is recommended that fully ripened, even a bit overripened, blueberries be considered in order to achieve good color and flavor. Some of the best blueberries for wine are those which commercial packers discard as "floaters" in the washing process. These are overripened and generally perfect for winemaking when carefully sorted. As only a small percentage of the berries sorted each day are floaters, it may require several trips to the packing house to get a sufficient quantity. Alternatively, they could be collected and frozen for later delivery.

---

*Adapted from Douglas Welsch, "Making Better Wines from Blueberries," Fruit and Berry Wine Symposium, May 1992.

Blueberries have a delicate skin with tiny seeds in a semisolid pulp. Most crusher–destemmer–must pump machines used for grapes work adequately for crushing blueberries. Frozen berries should be thawed before commencing this operation.

Add 50–75 ppm of sulfite to berries in the crusher and pump must to a holding tank where a pectic enzyme should be added at two to three times the recommended rate for grapes. Allow the crushed berries to set for 2–4 days in contact with the skins in order to extract flavor and color into the juice. Then press, settle our juice solids, ameliorate up to 30% of the resulting product, inoculate with active cultured yeast, and ferment as discussed for white table wines in Chapter 9.

Crushed blueberries will generally have a Brix of 7°–9°, a total acidity of .800 to .900 g/100 ml, and a pH of between 3.10 and 3.30. Brix adjustment up to about 22° is recommended for dry wine, and up to 24° or 25° if the resulting wine is to be residually sweet. The amelioration may be as high as 30% percent with sugar and water in order to reduce the bold character of the fruit and the astringency from the skins.

One of the most common difficulties encountered in making blueberry wine is in fermentation, as it is typically deficient in a nitrogen source for yeast growth. Diammonium phosphate (DAP) should be added in the full amount of the 8 lbs./1000 gal as permitted in the ATF regulation 27 CFR, Part 4, Section 24.246. The recommended yeast strains are FERMIVIN® or PASTEUR® in order to best transcend the full term of fermentation. The temperature should be maintained at 55–65°F, during which it may take several weeks to complete the fermentation. If the wine is designed to be residually sweet, fermentation can be stopped by chilling to 30°F or so and adding a light treatment of sulfite.

The resulting wine should then be clarified, stabilized, and coarse filtered as for white table wines. Some styles of blueberry wines benefit from brief periods of oak aging.

CHERRY WINE*

As with grape wines, the best cherry wines are made from top-quality fruit, ideally a bit overripened to maximize color and sugar. Brown rot, a common problem in overripened cherries, will ruin cherry wine and must be avoided.

Frozen Montmorency pitted cherries are recommended, as the pits being removed will reduce the almond flavor which results from fermenting in contact with the pits. Most varieties of sweet cherries have a very unstable color.

Cherries should be crushed in a grape crusher–destemmer–must pump ma-

---

*Adapted from Douglas Welsch, "Making Better Wines from Cherries," Fruit and Berry Wine Symposium, May 1992.

chine with the crusher rollers separated so as to minimize cracking of the pits. Sulfite in the 50–75-ppm range can be added at the crusher and the must pumped to a red wine fermenting vessel. Inoculate with yeast which accentuates fruit character, such as Epernay®, and ferment cool—below 55°F if possible.

Punch down the cap at least twice daily and press after 4–6 days of skin contact using a gentle pressing action in order to avoid damaging pits or macerating skins. The skin color is readily extracted during fermentation. Pressing as soon as possible will help to reduce the bitter herbaceous skin character and almond flavor from the pits. Be careful if using a bladder or membrane press, as broken pits have sharp edges and may cause some damage. Immediately chaptalize the new wine with sufficient sugar to increase the alcohol to 12–13% alcohol by volume. No water additions are recommended, as cherry wine color and flavor are typically rather delicate. Continue fermentation at the cool temperature. The color pigments in young cherry wines tends to brown unless treated with 2–3 lbs./1000 gal of Polyvinylpolypyrrolidone (PVPP) during the latter stages of fermentation. Because of high acidity levels in cherry wines, most fermentations are stopped at 0.0° Balling, or even before.

Most cherry wines tend to be protein unstable and should be fined with 4–5 lbs. of bentonite per 1000 gal. In some cases, even more bentonite may be needed. Run a lab trial first to determine the ideal level of bentonite treatment. Excessive use of bentonite diminishes color and flavor in wine. The wine is then finished along the same general practices and procedures as discussed for white table wines in Chapter 9.

Cherry wine does not age well. Hence, it should be made, processed, bottled, and sold well within a 1-year production–marketing cycle.

PEACH WINE

The first major hurdle in the production of peach wines is in the removal of the pit. This is usually too large to go through a grape crusher–destemmer–must pump without damage to both the pits and the machine. The pits must be removed in order to avoid the astringent and bitter phenolic compounds which are extracted during fermentation. Consequently, access to a special pitting machine is needed, or else its a matter of performing this arduous task by hand.

Pitted peaches can be run through a grape crusher with the addition of 50–75 ppm of sulfite. The peach must/puree may not be liquid enough to pump with centrifugal pumps; if so, it would need to be transferred by a progressive cavity pump, or by hand, to a clean fermenter where an addition of about twice the normal pectic enzyme recommended for grapes is indicated. As peaches are deficient in the essential nitrogen for yeast growth, diammonium phosphate (DAP) should also be added at a rate of 4–6 lbs./1000 gal—up to the ATF limit of 8 lbs. if necessary.

Inoculate with Epernay® yeast to accentuate flavor extraction and ferment cool—at less than 55°F if possible. As fermentation lags below 0.0° Balling, a light sulfite addition can be made and the wine immediately racked to cold storage at about 30°F. It can then be fined with 2–3 lbs. of bentonite per 1000 gal, more if determined necessary by laboratory trials, in order to achieve protein stability. Continue cold storage until clarity sufficient to permit at least a coarse filtration is achieved.

Balance acidity and sweetness as desired. Citric acid additions are permitted up to a maximum of 5.8 lbs./1000 gal in the ATF regulations (27 CFR, Part 4, Section 24.246). Sweetening additions are regulated in Section 24.179 of the same regulations. The wine is then finished along the same general practices and procedures as discussed for white table wines in Chapter 9.

Peach wine does not age well. Hence, it should be made, processed, bottled, and sold well within a 1-year production–marketing cycle.

STRAWBERRY WINE*

Strawberries are unusual in that they can be frozen and still make superior wine. Whether fresh or frozen, the berries should be fully ripened and unsound berries removed, a tedious triage (sorting and culling) process but essential because a small percentage of rotten fruit can have a profound negative effect on wine flavor.

Fresh strawberries do not need to be crushed before pressing, although frozen berries need to be thawed beforehand. Berries should be carefully placed in a mesh bag in order to contain the stringy pulp and the tiny seeds while the press slowly extracts the juice and free pulp. Prior to pressing, a treatment of about 3 g ($^1/_2$ teaspoon) of citric acid per 100 lbs. of berries plus an addition of a pectolytic enzyme in the manufacturer's recommended dosage should be made.

Juice from the press should ideally be very pulpy, as the pulp is needed for sufficient flavor extraction during fermentation. The temperature should be adjusted to about 55°F and immediately inoculated with 1% or 2% of actively fermenting yeast starter.

Strawberries, like bramble berries, are generally deficient in sugar to reach a sufficient alcohol level. Sugar additions can be calculated in the same manner as described in Chapter 9. Most of the pulp will liquify during the enzymatic and fermentation reactions.

Fermentation should be maintained at 55–60°F, with rackings made every week or so until the new wine is dry of sugar or nearly so. The last racking should

---

*Adapted from Dan Archibald "All About Strawberries," *Fruit Winemaking Quarterly*, Volume 4, Issue 3 July 1994.

be made into a blending tank in order for color to be adjusted with natural strawberry concentrate. The color extracted from most varieties of strawberries is unstable and highly susceptible to browning during fermentation. The desired addition of concentrate for color and sweetness should be made, thoroughly mixed, and immediately transferred to storage at about 28–32°F until sufficient clarity is achieved for filtration and bottling. An additional racking or two may be necessary in cold storage until final clarity.

The balance of strawberry processing can be made in the same manner as outlined for white table wine production from grapes in Chapter 9.

DRIED FRUIT WINES

The best raisin wines can approach a similarity to some of the "late-harvest" types made from overripened grapes—dark straw–gold wines with a soft apricot–caramel character.

Although choices are usually limited, the freshest possible raisins are best, providing the most natural nutrients for yeast growth. Avoid using "golden" raisins, as these are generally processed with high concentrations of sulfur dioxide which can inhibit fermentation.

The sugar content of most raisins ranges between 70° and 80° Brix and must be diluted with water to about 22° Brix prior to inoculating with wine yeast. High sugar concentrations in musts have a dehydrating property which stresses yeast activity.

Precision water addition calculations are difficult due to the presence of skins and other unfermentable solids in the raisins. A typical water addition is about three times the **weight** of the raisins. For example, 10 lb. of raisins at 75° Brix could be calculated to need 30 lb. of water—or at 8 lb./gal, approximately $3^3/_4$ gals. Warm water will help to dissolve sugar solids from the raisins more quickly. Gently knead and mix raisins in the water to help create a raisin "must."

Measure Brix and total acidity, adjusting each as outlined in Chapter 10. Inoculate with wine yeast and continue with the same general red wine procedure.

Most raisin wines are finished with a bit of residual sweetness—in the typical "late-harvest" style. Be sure that adequate potassium sorbate is used as a yeasticide to inhibit secondary fermentations.

The following specific ATF regulation applies:

27 CFR, Part 4, Section 24.202 *Dried fruit.*

In the production of wine from dried fruit, a quantity of water sufficient to restore the moisture content to that of the fresh fruit may be added. If it is desired not to restore the moisture content of the dried fruit to that of the fresh fruit, or if the moisture content is not known, sufficient water may be added to reduce the density to 22 degrees Brix. If the dried fruit liquid after restoration is found to be deficient in sugar, sufficient pure dry sugar may be added to increase the total solids content to 25 degrees Brix. After addition of water to dried fruit, the resulting liquid may be ameliorated with either water or sugar, or both, in such total volume as may be necessary to reduce the natural fixed acid level of the mixture to a minimum of 5.0 grams per liter; however, in no event may the volume of the ameliorating material exceed 35 percent of the total volume of the ameliorated juice or wine (calculated exclusive of pulp). Pure dry sugar may be used for sweetening. After complete fermentation or complete fermentation and sweetening, the finished product may not have an alcohol content of more than 14 percent by volume nor may the total solids content exceed 35 degrees Brix.

MEAD

Mead is wine made from honey. There are many mead "recipes," and variation far exceeds standardization in the manners and methods employed in its making.

The same general methodology can be employed as outlined for white table wines in Chapter 9 commencing with "Sugar Adjustment."

One critical variable is, of course, the sweetness level of the honey to be used—usually in the general range of 60°–70° Brix. This is difficult to measure directly by hydrometer and many home meadmakers will add warm water at a rate of twice the volume of honey and then take a Brix test. From this result, a final water adjustment can be calculated to pinpoint the Brix required to achieve the desired potential alcohol.

Honey is highly deficient in natural acidity. Consequently, an addition of about 25 g of citric acid per gallon of diluted honey is recommended. The lack of sufficient yeast nutrients should be supplemented with a dosage of Yeastex 61® in the amount of 6 g/gal.

Fermentations may be difficult to get started, resulting in the need to consider the use of an unpreserved fruit juice as an active starter medium.

In order to discourage stuck fermentations, which are common in meadmaking, fermentation should be conducted in the 65–75°F range.

Mead is finished both dry and sweet. Close attention should be given to the following specific ATF regulation:

27 CFR, Part 4, Section 24.203   *Honey Wine*

In the production of wine from honey, a quantity of water may be added to facilitate fermentation provided the density of the mixture of honey and water is not reduced below 22 degrees Brix. Hops may be added in quantities not to exceed one pound of each 1,000 pounds of honey. Pure dry sugar or honey may be added for sweetening. After complete fermentation or complete fermentation and sweetening, the wine may not have an alcohol content of more than 14 percent by volume nor may the total solids content exceed 35 degrees Brix.

CHAPTER 13

MARKETING

Adam Smith, an 18th-century economist and philosopher wrote:

> Consumption is the sole end and purpose of production; and the interest of the producer ought to be attended to only so far as it may be necessary for promoting that of the consumer.
>
> *An Inquiry into the Nature & Causes of the Wealth of Nations*
> [1776] Vol. 1, Book 1, Chapter 8

RATIONALE

Whereas the excitement and pride of releasing the first vintage is fulfilling, it is a grave mistake to depend on the public sharing that enthusiasm once all the opening celebrations are over. It is at this point that many new wineries commence the failure process. Consequently, it is of essential importance that a comprehensive marketing plan be in place for generating a flow of revenue.

Some vintners view marketing as nebulous and secondary in the entire scheme of things. Winemakers spend money on equipment, labor, and other physical tools to grow the grapes or make the wine, but they hesitate to allocate the proper budget to marketing. Marketing has to be an integral and active part of the winery business for a winery to survive, much less thrive. Without proper, consistent product marketing, the wines produced will be unknown in the marketplace. Thus, wineries should make an equally appropriate investment in the marketing portion of their business, as they do to their vineyard and cellar operations. Consider marketing an activity which must be performed regularly and routinely to attract new customers and reinforce relationships with existing customers.

In marketing, repetition is key. Perhaps the best example of this is the repetition of advertising. However, in any marketing venture, it is essential to remind the public of the product name and its benefits, on a number of occasions, via a number of mediums, in order to make an impression. It can take as many as seven to nine contacts to make an impression or make a sale. If the target audience is being reached in a number of different media over and over again, there is a greater likelihood that a given product will come to mind when they are receptive to that experience.

Most of the wine sold in the United States is in supermarkets or wine stores. There is comparatively little demand for direct wine availability in winery retail shops. Conversely, it is wine sold at retail which is the principal revenue from which small vintners depend. As a result, vintners must create an ambiance at their wineries which provides a wonderful memory—the winery experience. Consumers are not merely at a winery to buy wine; they are there to be educated and entertained.

Given all the ideals of modern marketing, vintners face a special set of circumstances, in that wine marketing is heavily regulated across the United States by both federal and state governments. The federal rules are found in 27 CFR, United States Code of Federal Regulations enforced by the Bureau of Alcohol, Tobacco, and Firearms (ATF). State statutes are administered by the alcohol beverage authority operating in each capital city.

MARKETING PLAN

An elemental aspect of this is recognizing that "marketing" is not just another word for "sales." Marketing consists of five fundamental functions: public relations, advertising, sales promotion, sales, and market feedback.

The manner by which these are allocated funding and support is called the "marketing mix." A marketing plan with a delineated budget will provide a guide for everyday decisions made as contingencies arise over the course of time. It is important to "plan your work and work your plan."

This process should start long before the doors of a new winery are opened for business. The marketing plan commences even before site selection for locating a winery and deciding which types of wines will be produced. It is prudent to consider the following marketing questions:

Who are the projected customers?
What types of wines do they drink?
Where do they live?
What other attraction would they visit in the locale?
Why would they come to a winery, instead of buying a standard table wine off the shelf of the local grocery?
How will they arrive at the winery?
How much do they normally pay for wine?
How much are they willing to pay for wine from a winery?
What kind of entertainment do they like?
What will they do while they are at the winery?
How much of the budget will be allocated to marketing?
How will results be measured?

Essential in the marketing plan is what the winery will provide for visitors and potential consumers. The winery will succeed only if it meets the customer's needs—not vice versa. Failure to recognize consumer expectations can mean the difference between success and failure for a small business.

Having a marketing plan will give provide focus and direction to the winery business. Issues to address in the marketing plan include the following:

Mission statement
Current market conditions, including competitive analysis
Winery description
Opportunities and obstacles
Measurable, attainable objectives (specifics on market share, profits, and sales volume)
Strategies for reaching each objective
Tactics for carrying out each objective
Action calendar specifying action to be taken in each area of the business from production to marketing activities
Evaluation of the program

The winery name will undoubtedly be the first impression others have of the business. The name should identify the producer, the products, and the consumer benefit. Avoid names which are difficult to pronounce and have negative or ambiguous connotations. Bypass names which could present limitations in the future. It is also highly recommended to do a legal name search after choosing a name and before opening.

As a vintner reviews this chapter and undertakes marketing projects, a foremost consideration in every marketing project should be a method of market evaluation. Devise a specific tracking mechanism into every project to measure the effectiveness of resources budgeted for each activity. Results from the marketing effort should be evaluated to ascertain how each part of the project met, exceeded, or fell below expectations. Some efforts will generate immediate results but others will not and will need to be evaluated based on their overall or long-term consequences. Market feedback will be covered in depth later in this chapter, but its importance to every facet of the plan is so crucial that immediate attention to the subject is imperative.

MARKETING BUDGET

The determination of a marketing budget should be needs-driven as opposed to resources-driven; that is to say, once an adequate marketing plan is decided upon, resources should made available to realize that plan. One guideline, set by *Vineyard & Winery Management* magazine, is that 35–40% or more of the winery budget should be initially allocated to marketing the winery. Once the winery has become established, plan to allocate at least 20% to marketing endeavors.

The following overview of marketing functions and channel activity should be supplemented with in-depth study. Major industry periodicals such as *Vineyard & Winery Management, Wine East, Practical Winery & Vineyard,* and *Wines and Vines,* continue to publish informative articles addressed to wine marketing concepts and issues. Public libraries have further resources on the subject of marketing.

PUBLIC RELATIONS

In new wineries, PR is generally the marketing segment which receives the most attention—primarily because clever vintners will find ways to maximize results with minimal costs.

Public relations consists of media relations in which print, broadcast, and other public media are pursued in order to help "get the word out." PR also includes trade relations in which restaurateurs, retailers, wholesalers, and other licensees are pursued for business alliances.

Winery events such as art shows and music programs may also be conducted under the umbrella of public relations. Winery newsletters, videos, and so forth are also placed in this area. Many vintners also place the winery tasting room and retail shop in this responsibility, as well. Tourism plays an important role for the small vintner in many cases. Therefore, it is prudent to establish a good public relations program with tourism in mind.

Tourism

Tourism is a key part of selling the small, family winery experience to the public. Tourism is second only to health care as the largest employer of people across the United States.

Good tourism bureau and association public relations includes offering the winery as a host to welcoming parties for guests coming to the locale. Perhaps specially labeled wines may provide an appropriate gift for dignitaries at special events in the community. Ascertain how the winery can best participate for everyone concerned. This may be as a contributor in a charity auction, or by setting up a booth for tasting, or as a presenter of the winery video at some gathering. Some revenue-generating events may be an opportunity for glass, bottle, or case sales.

Keep tourism offices brochure racks filled as a matter of routine. Offer tourism personnel copies of photos, slides, videos, and other visuals which may help the winery missions. Some tourism centers may permit setting up a special winery display. Share with them recent media attention given the winery and reinforce the winery's position of importance as a tourism attraction in the locale.

Traveler Motivation

Travel patterns are dynamic, requiring constant study of trends and impacts. Among the most important travel elements is weekend travel. Dual-income families, often with two different vacation times, are taking shorter, more frequent trips closer to home. Consequently, winery products and entertainment must be designed and marketed to provide solutions to wants and provide value as a leisure activity.

Consumers buy goods and services for expectations of benefits. The propensity toward travel/tourism is strongly correlated with the probability to purchase leisure goods and services, such as stereo equipment, theater and opera tickets, fashion goods, dining out, books—"wants" as compared to "needs" in basic economics. Wants goods and services are often associated with a high-profile image, wealth, sophistication, and greater quality. Incorporating these concepts into the winery visit experience is essential to realizing full marketing potential.

The winery experience covers the complete experience from the time travelers leave home to the time they return. The accessibility of the winery, its comfort zones, the mood generated by winery personnel, their recommendations of restaurants and hotels, and many other factors are all part of what will be identified with the winery and its experience.

Most travelers favor leisure-time activities that allow them to be socially interactive, to do something worthwhile, and to feel comfortable in their surroundings. They enjoy learning, varied experiences, and participating in activities which are fun and challenging.

Tourists visiting a winery should be identified in order to best design methods and manners in reaching them. Consider that local people within a 30-min drive will probably not be reached with brochures at highway rest areas. Conversely, people living more than a 2-h drive out into the region are probably not going to be reached with local advertising.

Winery Video

Videos can bring the experience of being up close and personal in the vineyards, crushing grapes, bottling, and other seasonal activities year round. Difficult concepts are made easier to understand with well-designed audio/visual presentations.

Although the advent of relatively inexpensive video cameras makes in-house productions affordable, most of these fall far short of what consumers are accustomed to seeing on their screens at home. Vintners make a serious error in assuming they can do just as good a job as the professionals in making a video program.

Professional videos are, however, very expensive items. It is common for the cost of a 10 min production to exceed $100,000. On the other hand, they can be an excellent way to deliver winery information to the public. A video can supplement a winery tour and totally replace the winery tour when the video is presented off-premises at various gatherings. Copies may be distributed to public libraries, wine shops, television stations, wine and food writers, travel writers, and tour companies, among other sources.

In order to optimize the viewer's attention span, a video should last no longer than 10 min. Concentrate on the most important points, making them clear and concise. Winery videos should express a relaxed, unhurried mood, with charm, character, and dignity. Avoid scenes which are loud, raucous, and nerve-wracking. As videos become increasingly obsolete, production should also include a CD-ROM version. This will provide the ability for delivery over the Internet—to millions of computer users across the world looking for information on a specific subject matter.

Motorcoach and Group Tours

Successful motorcoach marketing can generate significant added revenue and promotion to winery operation. This travel method is expected to increase as the large segment of people born in the late 1940s, the "baby boomers," become "empty nest" parents preferring less long-range driving.

Travel and community associations are usually searching for creative attractions to help attract motorcoach tours and group travel to their locales. Establish a good working relationship with tourism personnel on how to best serve ongoing mutual objectives. Supply them with brochures. Keep them informed of seasonal and special winery events by sending them newsletters and news releases. Consider cooperative advertising and opportunities in tourism and private publications.

Most importantly, ensure that the winery can accommodate the number of people in each group scheduled to visit the winery. Check and double check how the group will be received and how the tour will be conducted. Ensure that there is a properly trained staff, adequate facilities, and sufficient time allotted to the tour and for tasting room sales.

Wine Trails

Routes on which wine visitors travel among wineries have become popular in areas where vintners are located in a proximity which permits cooperation with each other. Neighboring wineries can produce joint advertising materials (Fig. 13–1) and conduct special events.

Wine trail events can be centered around theme activities: holiday wines, new vintage weekends, gourmet foods, or other themes in which the consumer receives a benefit as they travel among wineries. Some wine trails require advance sales or charge a per-person fee to cover promotional costs.

Another wine trail concept is that of a "passport club" using a wine trail brochure. The passport brochure is stamped or punched by each winery visited along the trail—such that achieving some number of visits wins a nice prize or incentive.

Fig. 13–1 Wine trail promotional pamphlets.
Photo: Dave Ferguson.

Networking

Even before the winery is opened it may be appropriate to host a gathering of other area vintners and trade association personnel in order to express how the new winery may hopefully fit in the industry and community.

Provide a place for other local wineries to place their brochures. New vintners often conceive of this as a conflict of interest. On the contrary, the notion of visiting several wineries can be far more enticing to people than making the trip to search out an isolated vintner. Having good things to say and do among other vintners in the locale will pay dividends in return. Some wineries within specific regions have had great success in cooperative marketing ventures.

Every vintner should fully understand all of the services and materials available from state wine trade associations in order to take full advantage of them. Every state which has a wine trade association structures it differently. Some are funded by members, whereas others are state supported or funded by a combination of sources. Participate in programs and extend a cooperative hand in assisting association goals. These organizations can be an integral source in getting new winery marketing off to an affordable effective start.

Vintners should also consider seeking appropriate appointments on community organizations in the winery locale. Offering the winery as the host for businesses, associations, service clubs, and other organized groups can be another manner in which residents are attracted. This activity fosters good relations and will pay dividends far into the future.

Vintners should visit restaurants, hotels, gourmet shops, and other local businesses on a regular basis. Discuss topics of interest and the possibility of projects which are of mutual benefit, such as sponsorships for a festival or support of a civic function. Similarly, invite the business staff to visit the winery so that they can enjoy the experience and be able to convey that to their customers. Pursue manners which permit the power of advertising to be shared in cost. Various packages containing admission, gift items, meals, and other enticements only add to consumer appeal.

Wine tastings as part of special gatherings in art galleries and antique shops create a particularly good synergy. Consumers feel a more social atmosphere, gallery proprietors create an entertaining experience, and vintners enjoy an association of other fine art along with their wines.

Education

Educational events give the small winery credibility, a promotional tool with media, and a less commercial agenda for attracting consumers. Adults today are looking for interactive educational experiences where they can meet others who enjoy the same interests. Vintners may want to teach classes at community centers, continuing education classes at colleges and libraries, or on the winery premise. Class topics can range from vine culture or winemaking to every aspect of wine appreciation. Public education may include a regular wine or winery column to be published in the local newspaper, or as a host for a radio or television show.

Media Relations

Media relations is quite simply the ability to build good relationships with print and broadcast media contacts. Study their work and develop projects which are "win–win" for both participants. This is a superb way to gain credibility with winery visitors.

Journalism Basics

Understanding the elements central to the business of journalism is key to successful relationships in working with media inquiry.

Accuracy ensures the fairest and most complete story on a subject. Make certain the information they need is complete and precise. It is important to be brief and clear and to center on the most important elements.

Reporters strive to show the reader what is happening and, as such, they rely on seeing events first hand. They frequently use anecdotes, quotes, and statistics to illustrate what they have seen. In addition, they always like to speak directly with the vintner, vineyard manager, or chief participant.

Newsworthiness includes breaking news (storms, fires, etc.) and features which are composed of human interest, or "soft" news. In addition, newsworthy activities can arise out of conflict, unique plots, and timely or seasonal events. It is essential that vintner stories identify a significant benefit to an audience. Depending on the subject matter, wine news can appear in a feature section, business news, a food column, a farm report, or the travel page. The challenge is that stories on consumer products such as wine are viewed suspiciously by the media because they are highly commercial. Trend stories diminish the commercial aspect or objections a reporter may have. For example, new vineyards planted in California and New York, as well as by the local estate winery "may signal future growth of wine as a part of the American diet." Present information in an entertaining and/or educational manner so that the idea dilutes the inherent commercial bias.

In order to support, refute, and confirm material presented in an article and to establish its accuracy, reporters will look for other sources such as trade magazine statistics or state experts to reinforce what they are reporting. Suggest resources or people within the wine industry for their use.

Audiences identify with success stories of people working hard, overcoming adversity, creating new products, contributing to the community, and so on. If the story is about a dramatic increase in sales, it is likely that someone's personal experience with selling will lead the article. These stories are not about ideas or

Fig. 13–2 Promoting awards and accolades.
Photo: Dave Ferguson.

things but about the people involved and the resulting human consequences. Of particular benefit are awards and accolades such as demonstrated in Figure 13.2 which can provide an interesting focus for media coverage.

News Releases

This is the standard method of first contacting members of the media. A good lead in a news release answers six basic questions of journalism in 30–35 words, or 2 sentences: Who, What, Where, When, Why, and How. This brevity improves visibility and readability and gets the message across immediately so the editor can quickly determine its newsworthiness. Good releases are less than one page in length, double spaced with 1-in. margins on all sides of the winery letterhead stationery. Provide the media with several interesting focal points in order to gain their interest for writing their own stories.

Always list a first and second contact name with day and evening phone numbers on the top of the news release for easy access. A headline which is clear, concise, and catchy has the best chance to win media attention. Avoid trade jargon and euphemisms, as they obscure clear understanding. Include any item which may make the reporter's response easier such as a simple map portraying winery location, along with phone, fax, and e-mail numbers. Include the winery mission statement developed in the business or marketing plan in every news release in order to reinforce brand identity. Eliminate misspellings and factual errors, reading individual sentences from end to beginning in order to catch any mistakes.

Basic grammar school English is all that is needed for the construction of an effective news release. Media personnel prefer simple words, short sentences, and brief paragraphs. The use of vivid nouns and active verbs are encouraged as long as they are consistent with the subject matter. Is it a "good" wine, or a "berry-filled red wine the shades of garnets and rubies?" Create an attractive vision which portrays the sight, smell, taste, and feel of the wine, or the festivity.

Television

Television demands a demonstration of the event or activity. Assist the video journalists by suggesting good shots or angles and by offering the use of props and necessary equipment. Ensure the tasting room and retail shop has good consumer activity.

Television news editors commit crews to cover various events in the community. The contacts for talk shows are producers or community affairs directors. One-page faxes are an especially good method for reaching the media with last-minute details or reminders. There are books published that list media in specific areas with addresses and phone and fax numbers, or simply call the switchboard and ask to whom to send a news release.

A dynamic, informed spokesperson makes a better interview for the broadcast media. Enthusiasm can be entertaining if its charismatic and not overdone. Speak conversationally when interviewed and give complete answers, as the questions often will not show in the story. Make eye contact with the interviewer and any objects being discussed. Avoid looking into the camera lens unless there is a "Tell the audience . . ." question posed by the reporter.

Media Formats—Timing and Deadlines

Daily newspapers must turn out a breaking story in a matter of hours to ensure it is timely. Major features in various sections can be planned 6–8 weeks in advance, particularly with traditional stories during the Holiday Season, Valentine's Day, the Fourth of July, and others. Avoid holding a promotion or releasing news on election day or tax day, unless the winery news is somehow interestingly associated with such events. Generally, the deadlines for daily newspapers run late afternoon for morning editions and late morning for afternoon editions. Of course, these times should be avoided. At any time, it is always a good idea to begin a conversation with any reporter with, "Do you have a minute?" If they are on deadline, ask what time would be best to call back.

Although print circulation is generally smaller, weekly community newspapers are widely read by the citizens of the area they serve. Moreover, they are so specifically targeted that a vintner can do well in them if trying to reach a particular geographic and demographic audience. The smaller editorial and reporting staffs of weeklies do present challenges in practicing firsthand reporting. Consequently, a well-written news release can sometimes be printed verbatim if it is created in a news format. The worst day to call a weekly with soft news is a day or two prior to the publication day.

Magazines are currently a rapidly advancing market with highly targeted audiences. Editorial calendars are planned a year or more in advance for major features and annual holiday or seasonal stories. Stories are often photographed the previous season to capture the appropriate background images. For event news, or short items of two or three paragraphs, 60–90 days is standard, although some have the flexibility to receive information in as little as 45 days.

Television talk shows are usually taped several weeks or more in advance and the community affairs director at the station is usually the chief contact. These programs are a great opportunity for building relations with station personnel for better programs in the future. In the case of a spontaneous news event, it is obviously necessary to call immediately.

Radio news in small communities, and some medium-sized urban markets, rely primarily on the wire services for breaking news reports. There can be opportunities to have local winery news events mentioned if the news director or public service director are contacted during nonbroadcast times to discuss such items. A fact sheet on the event is sufficient, and many can incorporate the event into their broadcast schedule as late as the day of the event, if necessary.

Call a few days after sending the release to offer additional information on the subject to which the news release pertained. That will give a reporter another reason to look over your release and further determine its value. Be prepared to quickly tell them the story again in 10 words or less. Also be prepared to follow up immediately with any requested information.

Many print media and some broadcast have listings of special events offered in the local and neighboring communities. They are typically well noted, and many readers plan their free time around the activities listed. Often the information must be submitted at least 2 weeks in advance in order to adequately facilitate preproduction needs.

Interviews

It is important to be well prepared for media interviews. Consider the purpose of the interview and its major points. Have some statistics and other key facts accurately memorized. Most reporters cover a range of subjects and cannot be an expert on particular areas, so keep explanations simple. Practice good faith; be honest and direct with the media. Correct any mistake immediately, or as soon as may be appropriate in the interview. Seek help in preparation for interviews—particularly in setting up "mock interviews" which pose difficult questions. One needs a positive, nondefensive response to, "Why did your Vignoles win a silver medal instead of gold? or "How can you justify a 15% increase in the price of your

Chardonnay?" Listen carefully to each question asked. Pause before answering in order to prepare thoughts. In every case, answer the question succinctly and then return the dialogue to the interviewer. Talking too much can cause confusion. Do not talk off-the-record. If there is something which should not be disclosed in public, it should not be disclosed in private.

Interviewing Techniques

An interview should proceed as an educational exchange. Reporters will often commence interviews with simple remarks in order to establish a "comfort zone" and friendly rapport for the upcoming interview. Vintners should be relaxed and well rehearsed but able to respond spontaneously.

Reporters may ask questions which are confusing or inappropriate. These types of questions can be dealt with by simply asking for them to be repeated. For example, "I'm not sure what you're asking," or "I'm not sure I understand your question" forces the reporter to rephrase and provides some time to think about a response. In classic paradox, silence is an important tool in the field of journalism. Reporters use lulls in dialogue to create uneasiness in the interview. Should this happen, allow them to break the silence. Avoid off-hand comments. Remain alert, professional and friendly even when the interview appears over. Interviews are not over until everyone in the reporting crew has left the scene.

Unfortunately, most everyone may, one day, face an adversarial role or a "lose–lose" question perhaps best left unanswered. Nevertheless, one of the worst responses is, "No comment." It is better to answer questions, otherwise the reporter is apt to look for the response from another source, which may provide biased and/or incorrect information.

Although it may be a bit distressing, the best response may be to simply "tell it like it is." Audiences are more apt to identify with someone willing to speak the plain truth. Be credible and respectful of the opponents. Praise the people with diverse viewpoints. Honesty is the best possible pathway through difficult questions and topics.

Visuals

Visual aids increase reader interest. Working with writers and editors of travel and wine/food periodicals will usually be far more successful if professional photographs and illustrations are offered.

If in-house photos are used, they should be produced with high contrast and sharp detail. Shoot pictures of interesting activity in progress. Keep subjects close together and backgrounds neutral so they will not distract from the action.

Do not use the same photos for publicity as for advertising. Always mail photos, slides, or other illustrations in a sturdy stiffened envelopes to help protect from damage. Photos provided to the press are rarely returned but including a self-addressed stamped envelope may help.

Photo formats preferred by trade magazines and newspapers are color 5 in. × 7 in. glossy prints. Color slides or transparencies are preferred for consumer magazines, brochures, and advertising. Identify subjects and activities pictured and the contact names with phone numbers. Attach this information on the back of each print with pretyped pressure-sensitive labels, or attach the print to 8 $^1\!/_2$ in. × 11 in. paper with the information on the paper. Do not staple prints or write directly on prints.

Here are some important media "do's":

Do tell the truth
Do state the impact quickly and clearly
Do vary sentence length and structure
Do use colorful words
Do read copy aloud
Do use authoritative sources to justify what is said
Do give directions to the winery
Do use photos when appropriate
Do tell of possible photo opportunities
Do follow up without being a nuisance
Do localize regional, national, or international stories
Do return phone calls promptly
Do respect media deadlines
Do follow up verbal conversations in writing
Do keep copy simple, brief, and accurate

Conversely, here are some essential media "don'ts":

Don't leave unanswered questions
Don't link advertising and editorial support
Don't confuse with jargon
Don't overstate the case

Media Kits

A media kit has useful information which reporters can store in their files for both current and future reference. These need not be expensive productions but should be well designed with salient information and illustrations. Standard file folders can work well. Pocket folders can also be employed with the winery logo, label, or name neatly printed on the tab. The following items are typical in winery media kits:

- Fact sheet containing statistical information such as vineyard acreage, winery production capacity, and number of employees
- Historical information of the estate
- Program schedule (if an event)
- List of participants connected (owner, winemaker, vineyard manager) with biographies and with day and evening phone numbers
- Brochure
- Question-and-answer sheet
- Visuals—photos, logo sheets, labels, slides, or color transparencies
- Maps and directions
- Charts or graphs
- Good copies or reprints of pertinent news articles—include publication name, date, and author(s)
- Newsletters
- Glossary of wine terms, particularly for nonwine media

Obviously, all of these would more than fill up a file folder. Be selective with essential materials which tell the winery story without repetitiveness and overwhelming volumes of materials.

A monthly "special visits" program during which VIP media and trade people are individually invited to the winery for a tour and luncheon is one of many effective ways to present new information and develop friendly contacts. Similarly, regular tour and tasting programs for clubs, associations, conventions, tour groups, and other gatherings can also help focus the public eye upon the winery.

SPECIAL EVENTS

Special events add to the collective impact of the marketing mix created by winery advertising and public relations. This rationale is closely associated with the postulate that small wineries often find that a large part of their mission is in providing good entertainment.

In spite of all the time-saving conveniences available today, people work more and have less time for leisure. Consequently, there is competition for what seems to be a diminishing amount of free time. Well-designed special events can encourage busy people to choose a winery visit. If the experience is good, they will return.

Special events generate publicity, awareness, new business, repeat business, customer loyalty, and word-of-mouth promotion. However, these do not single-handedly increase sales, or succeed without sufficient planning and resources, or guarantee profitability.

Special events (e.g., as shown in Fig. 13–3) can be enhanced by having logo products on hand for sale and prizes, offering percentage-off coupons for various items on sale, and mailing follow-up brochures and invitations relating to another winery event or a personal visit.

As with any marketing function, it is important to determine who the potential audience is before deciding on the type of event. Once this question is answered, planning can move on to targeting participants, such as home winemakers, "empty nest" couples, gourmets, locals, visitors within a 50-mile radius, novice wine drinkers, or experienced wine drinkers.

Project a reasonable set of goals for each event, such as the number of persons attending and dollars of wine and gift shop items sold. Decide beforehand how these goals are to be evaluated.

Ensure that timing does not conflict with other major area gatherings. Check to

Fig. 13–3 A "Wine Stomp" at a vintage festival.

see if there are other consumer events being held within a reasonable radius. Coordinating with that facility may be better than competing with their event. Having two or more major product influences, such as wine and food, could become synergistic in attracting people.

Small winery vintners have found success in offering barrel tastings after the holidays. Blind tastings bring out those who enjoy the challenge of detecting which is which. Component tastings help to sharpen consumer palates. Tasting events are perhaps the very best type of program in generating product and consumer loyalty.

One of the first comments about home winemaking demonstrations in a winery is that they are a conflict of interest. Not so. Home winemakers are typically among the most loyal customers, as they buy gift items at the winery and home winemaking supply items if they are offered.

A well-designed class or series of demonstrations offered by a vintner can be "instant expertise" in the mind of every wine enthusiast who attends. Special prizes and discounts on materials and gifts help to make these experiences frequent topics of discussion in amateur winemaking circles.

Wine cookery demonstrations, particularly if conducted by a popular food authority, are excellent attractions. Food and wine pairing events are also well received by people at every level of gastronomical expertise. If practicable, a food and wine event held among the tanks and barrels lends itself to an experience hard to forget. These, along with other types of demonstrations and shows, can be attractive promotional events for the media—effecting increased awareness and profits for clever vintners.

It is, of course, elemental that the timing of special events be made at appropriate times. Outdoor programs should be relegated to late spring, summer, and early fall seasons—with plenty of shelter available in case of inclement weather. Ensure there is sufficient space for parking and for the event itself. Clearly mark the facilities for entry, exit, parking, rest rooms, no admission, and so forth. It is a good idea to project facilities needs for up to double the number of people actually expected. Project the number of winery staff needed to take care of all the special needs associated with large gatherings of people. It is a good idea to have several supervisory people circulating around the facilities in order to maintain order and fill in any needed support. Inform local law enforcement and emergency units about large events well ahead of time. They may require that special personnel are assigned to the premises.

IMPORTANT NOTE: BE SURE TO CHECK ATF, STATE, AND LOCAL REGULATIONS AS TO THE SPECIAL PERMITS WHICH MAY BE REQUIRED TO CONDUCT PUBLIC TASTINGS, SALES, AND SPECIAL EVENTS.

It is, of course, essential that the public know about each special event held at a winery. Marketing research indicates that it takes seven to nine contact messages in order for consumers to take action. These contact messages include direct mail postcards (Fig. 13–4), fliers, advertising, media relations, winery and association newsletters, and posters along the winery tour route, in the tasting room, and in the windows of local businesses.

An example of a good campaign would be to place posters in the winery several months in advance of the event. Promote the event in the winery newsletter. Send direct mail postcards to appropriate mailing lists about 1 month prior—announcing the date and a phone number for more information. At about the same time, a press release should be sent to all the regional print media. Distribute posters to store windows in the area and send a release to regional radio and television. During the week prior to the event, advertisements should be run repeatedly on both radio and television. The following constitutes a special events checklist:

A mission based on consumer needs
An objective based on vintner needs
Professional security and emergency services
Target audience
Theme
Timing

Fig. 13–4 Postcard invitation to a winery special event.

Facilities
Personnel
Publicity
Admission fees
Staff training
Food
Entertainment
Drawings and raffles
Designated drivers

Newsletters

Newsletters are elemental to successful winery marketing. These provide ongoing awareness, new information, continued relationships, and generate media coverage. A winery newsletter should reemphasize the image and character of the winery by maintaining consistency in design and theme. (See Fig. 13–5.) If a winery has a country theme, its newsletter should not be designed in a contemporary format or vice versa.

Most consumers find a sense of personal identification with a winery when

Fig. 13–5 Winery newsletter formats.
Photo: David Ferguson.

they receive a newsletter. Newsletters are more credible than advertising or some forms of direct mail.

Determine the newsletter audience in order to reach those individuals who comprise the best targets for winery information. Add new media, community leaders, and customers routinely. Include a return postage-paid "subscription" card once each year to ensure that each addressee wants to receive future issues.

Budget newsletter expense for at least 1 year and then set a schedule for mailings—quarterly is the most common frequency. Assign responsibility for research, writing, photography, and art work, and set firm deadlines. Stick to the schedule and be consistent. This can make the difference between a newsletter that is a successful sales tool and one that is a haphazard, cost-prohibitive burden.

An effective newsletter should have an objective. It could be to increase sales in the retail shop or to educate customers. In every case, provide material in a variety of formats, such as question-and-answer, feature profile, interview, letters column, or a calendar of events. Newsletters can include a list of wines and gift shop items, along with an order form.

Each newsletter headline must capture reader attention. This may be the only part of the newsletter that some people will have the time to read. It is thus essential that headlines are attractive, accurate, clear, and deliver a meaningful message.

Make the newsletter easy to read. Font type and size should be a minimum of 10 points up to 12 points. A sans serif type is good for headlines, whereas serif is appropriate for the body. Italics should be used sparingly.

ADVERTISING

Everyone knows the inescapable power of advertising. Quite apart from television, radio, newspapers, magazines, and junk mail, modern advertising is painted on cars, trucks, buses, signs, and buildings. It is printed on clothing, pens, pencils, cups, bags, leaflets, and most any other item which can present an image to a potential consumer.

Apart from the essential consumer brand awareness provided by advertising is the ongoing value of advertising visibility of the winery by the media, tourism personnel, local business people, and government officials.

Although advertising programs can be expensive, the newly established commercial vintner should incorporate this cost into the marketing plan and budget.

The ATF enforces regulations specific to wine advertising. These are found in 27 CFR, Part 4, Subpart G *Advertising of Wine*. Every wine marketer should be familiar with these rules before attempting any advertising function.

Consider ways advertising costs can be shared with other wineries, hotels, restaurants, and other participants. Ask ad salespeople how other small businesses achieve their goals or if there are other small businesses which might want to share the costs. Seek grant programs to offset costs via local economic development agencies and tourism offices.

Brochures

Another advertising program common to new wineries is an economical and attractive brochure which can be distributed to hotel/motel lobbies and highway rest areas. (See Fig. 13–6.) There are companies that specialize in brochure distribution to tourism-related businesses. Many state tourism offices can provide a list of rest areas. A brochure will go where the vintner cannot. It will reinforce the winery image and bring customers in the door.

Professionalism pays. Home-made brochures often dissuade people from visit-

Fig. 13–6 Winery brochures.
Photo: David Ferguson.

ing the winery. This should not be confused with "down-home" appeal, which is a mode successfully employed by many vintners. What may look good to the eye and ego of a vintner may really be a warning in the eye of a potential winery visitor.

Direct Mail

This is an interactive system of marketing which employs various advertising media to effect consumer focus on sales, image, winery programs, and other events. If the target is well defined, direct mail can be one of the most profitable of the advertising media.

Repeated use of direct mail which is properly designed and timely can also help to build and reinforce long-term consumer relationships. Use a variety of direct mail formats (newsletters, postcards, etc.) to achieve your goals. The principle factor in design is to avoid as much as possible the "junk mail" image. Direct mail flyers should be attractive, colorful, upbeat, and to the point. The message should be clear, supported with action-oriented results.

Postcards enjoy high readership because recipients know they involve little time commitment for reading. It is simple to turn it over, quickly read the message and evoke interest. This type of direct mail is particularly useful when several consumer contacts are part of an ongoing program.

Figure 13–7 is an example of an interactive direct mail program. The mailing list is the key element in a direct mail campaign. There are three types of direct mailing lists: house lists, response lists, and compiled lists:

- A *house list* is comprised of people who have already done business with the winery. These lists will typically outperform the other two in sales. The guest book is the best source to develop a house list. Keep all direct mail lists updated and check with postal authorities to update addresses periodically. It is a good idea to send a test mailing to the winery so mail delivery time can be determined.
- A *response list* will include people who have requested information from various promotions or advertising.
- The *compiled list* is usually purchased from a targeted source. A good example of this would be the list of people in a given set of zip codes who subscribe to a consumer wine magazine. These can be both expensive and large in scope. Ensure that the target list is up to date and comprised of people who best represent the demographics and psychographics inherent with potential winery customers. Investigate whether or not other firms, especially other vintners, have used the list and, if so, when. Inquire as to any time limits existing on the list and if any discounts are available.

Direct mail should create an offer to consumers for something they want and at the same time invite them to take action. Once a decision has been made as to the targeted audience, the offer and the design of the materials will evolve from the customer needs list.

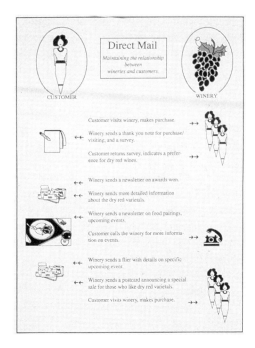

Fig. 13–7 Direct mail flowchart.

Database Marketing

This method advances in tandem with the growth of computer technology. Databases provide the storage, sorting, and tracking of information appropriate to every marketing need. As an example, consumers can be identified by their patterns of visiting and buying. They then can be targeted for future events, making them good prospects who will attend and buy again.

Billboards

One form of advertising pursued by vintners is billboards in key locations to notify motorists that a winery is ahead and how to reach it. The final decision for billboard rental should be made after an exhaustive study—preferably in association with an advertising agency. Although many tourists may arrive at the winery in peak months because of billboard advertising, the cost can be monumental. Expense can sometimes be reduced by sharing billboards with another business in the same location. Seasonal billboard programs are another possibility for trimming this cost. Some offer weekly rentals, which can build excitement for special events or seasonal activities.

An attractive sign in front of the winery is also considered advertising but is really more of a landmark which reassures first-time visitors that they have arrived at the correct location. A special paint job with the winery name and logo on the winery truck, however, will be seen all around the area. Vintners should ensure that all signage for the winery is in accordance with local codes.

Some state Departments of Transportation operate a highway signage program in which wineries can participate. These programs usually have requirements for eligibility and specifications that need to be followed, along with a fee schedule. Requirements may be based on attraction maintenance, parking facilities, rest rooms, accessibility, hours of operation, or visitor attendance. For all of these reasons, it is imperative to maintain an ongoing count of visitors arriving at the winery in order to measure the effectiveness of the program costs.

All billboards and signage should be designed and produced by commercial sign companies. Professional creativity will include the major elements necessary in order to attract the attention of potential visitors. Homemade signs and billboards often work to the disadvantage of vintners by sending the wrong message to the wrong people.

Depending on the locale in which each winery is situated, some advertising alternatives are more cost-effective than others. Attractive window displays in various businesses are one of many ways in which advertising can be employed with minimal cost. Specialty advertising for various community events, such as logo tee shirts at the county fair or lapel buttons at conventions, can also deliver the winery awareness. Once again, there should consistency in maintaining a positive winery image.

Internet

Marketing through the Internet is a trend which promises continued growth. A widely used tool on the Internet is the World Wide Web, which can take one virtually anywhere.

Commercial vintners can use this tool to set up a winery home page or can hire a service to add the winery to a wine mall. The information a winery could add to this page includes publications, events, winery history, wine descriptions, and a winery tour, complete with photos. The page can include an order form for consumers as well as e-mail capabilities. Make sure that the page is continually updated to offer the consumer ongoing benefits. Before selling on-line, vintners should ensure they are properly licensed and know to what states they can legally ship.

Wine trade publications are good sources for information relating to marketing on-line. Some have included a regular column for communicating this information on a regular basis.

Word of Mouth

According to many vintners, the word of one consumer to another is the best form of advertising. The visitor's experience through the winery tour, tasting, products, and, most importantly, its people determines to what degree word-of-mouth advertising will be made.

In addition, the marketing mix creates the climate for ongoing word of mouth. When the public sees a vintner's name, image, wine, or presence at events or in media, it creates small talk among acquaintances.

SALES PROMOTION

The generation of sales depends on the effectiveness of all the other elements of marketing which precedes the decision of buyers to act. It is exemplified in the winery retail shop as the traditional "monthly special" deal, such as offering a full case of 12 bottles priced at a regular cost of 10.

Sales promotion is involved when wines are displayed with medals they have won in competition or when "cut case" (cases stacked three high with the upper case cut open to reveal the bottles inside) displays bring attention to "close outs" and "releases." (See Fig. 13–8.)

Sales promotion in the general marketplace is far more costly than in the winery and should be carefully budgeted. Larger vintner firms often hire professional agencies to conduct sales promotion activities on their behalf in key markets. With clever ideas and good sense, small winery vintners can successfully create their own.

Wholesale sales promotion is exemplified by programs in which incentives are offered to encourage increased placement of wines in retail accounts. This generally involves a special "post-off" price over the duration of the promotion period, as well as attractive premiums which are awarded to winning sales personnel.

In restaurants, it is common to see winery displays in the entrance foyer and new "clip on" entries in the wine list. Offer table tents, wine list suggestions, and printing assistance if feasible. Education of the wait staff to promote wines to their clientele usually follows. Follow-up visits are necessary to regenerate enthusiasm and reward good performance.

Sales promotion in retail wine shops is exemplified by small cards called "flaggers" or "shelf-talkers" which are placed on the shelves in front of bottles. These draw special attention to customers by announcing some prestigious medal won in competition, a wine and food pairing, a high rating given by a notable wine writer, or some other type of accolade. Retail sales promotion includes the popular "coupon" refund programs, consumer contests, and other forms of attention-getting activity. Use caution to maintain the winery image.

Fig. 13-8   Cut case display.

SALES

The generation of sales depends directly on the effectiveness of all the other elements of marketing. It is the aggregate influence of these which will ultimately determine whether or not people will trade their dollars for a vintner's products.

Sales Personnel

Potential sales can easily be lost by individuals who are unqualified to sell. It is imperative that sales personnel are fully informed of their products and their position in the industry. Their demeanor and attitude must be totally positive. Fairness, honesty, and tenacious effort are other essential personal attributes. It should go without saying that good grooming and appropriate attire are also important.

Successful marketing is heavily dependent on opportunities pursued promptly. New retail stores and restaurants should be identified and approached far in advance of opening for business. Recognized opportunities for promising public relations, sales promotion, and advertising should be quickly relayed back to the management team for consideration prior to making the first sales call.

Niche Sales

It is obviously important to explore niche marketing for the new winery. Private labeling (special labels designed with restaurant or retailer names and logos) is a good example. Some vintners pursue private label business with local restaurants and retail licensees. Others provide special labels for business customers at holiday gift-giving time. It can be profitable for wedding receptions and other occasions, too. Remember, however, that each and every label must first be submitted on form ATF F 5100.31 and approved before being used.

Winery Tour, Tasting Room, and Retail Shop

Arriving visitors should be warmly received and welcomed. There should be a minimal wait in a comfortable area before departing on the tour. It should go without saying that the tour guides and the winery should be clean and neat. Poor housekeeping can make the best of wines seem tainted. Guides should speak clearly and conduct the tour so that it parallels the wine process—starting with where grapes are received and ending where the wine is bottled and warehoused.

Tours should move along with brief stops at each point of interest—20–30 min in total duration. If possible, it makes good sense for guides to continue on after the tour as the tasting room clerk. The tour provides a good chance for a friendly acquaintance to be established and helps to put visitors at ease while tasting and considering purchases.

Guides should have a pleasing demeanor and be willing to happily answer questions—particularly fielding "silly" questions as though they are perfectly appropriate. Visitors should never be embarrassed. Vintners should require guides to read wine industry news articles so they are "in the know" about important developments, along with information which may better help their tour and tasting delivery. Idle time for guides can be used in replacing stocks and housekeeping chores.

The tasting room and retail shop must be designed and decorated to achieve an ambiance conducive to progressive wine marketing. Professional assistance is highly recommended.

White and blush wines should be placed in a 50–55°F refrigerator under the counter. All wine bottles should be uncorked in front of visitors with a friendly flair. The clerks should check color, clarity, and bouquet to ensure the wine measures up to standards. Glasses (plastic will offend many wine enthusiasts) should be poured with just a taste—about 1-oz. portions are sufficient. Smaller (6–8 oz. capacity) tulip-shaped glasses generally display table wines best and are both easy to clean and store. (See Fig. 13–9 for various types of glass.)

During pouring, the neck of the bottle should not touch the lip of the glass. As each pour is made, the bottle should be turned a bit as it is withdrawn to prevent dripping. A plain white service towel may be held under the bottle as it is moved from guest to guest.

The small portions poured allow for swirling the wine in the glass, to better judge color and clarity, and to increase the surface area from which the precious bouquet vapors rise. Small portions also cost less. Bottled water, small bites of bland cheese, and unsalted crackers should be available for tasters to cleanse their palates between wines.

Invite questions during the tasting and become genuinely interested in the guests—where they are from, other winery visits they have made, favorite wines, and so forth. Not only is this part of their winery experience, but it is a good way to approach them in one or more surveys designed for some marketing objective. Point out one or two interesting attributes about each wine when it is served. Between wine samples, discuss a special promotion, recent awards, or any other meaningful information relating to wines available for sale, but avoid the "hard sell"; the wines being tasted will speak for themselves.

Almost everything free is attractive to consumers. In the small winery, free information, free parking, free tours, and free samples are typical, along with quantity discounts. It is important, however, that consumers recognize the value of these. Some vintners sell the wine glasses in their tasting rooms in order to receive the free wine sample, and refund the price against purchases made later. Some

Fig. 13–9 Various types of wine glass.

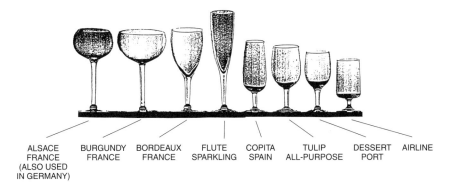

ALSACE BURGUNDY BORDEAUX FLUTE COPITA TULIP DESSERT AIRLINE
FRANCE FRANCE FRANCE SPARKLING SPAIN ALL-PURPOSE PORT
(ALSO USED
IN GERMANY)

consumers decide to keep the glasses anyway. Discounting and price cutting too deeply can cause concern about the value of the product.

Frequency programs recognize and reward the best customers. Some are designed to earn points on the number of visits and are others based on some level of volume purchased. Others have special rewards for being members of a winery club membership. All of these should address services, programs, and events such as barrel tastings, harvest suppers, new release premieres, and so forth which could be interesting to repeat visitors.

Customer Service

Genuine concern for every winery visitor is just as important as wine quality. It is not realistic to expect every customer to be totally focused on the vintner, the winery, and the products offered. Try to be patient and understanding with people who seem brusque and disagreeable. Some may never be won over to the good will expressed by a persistently friendly winery tour guide or tasting room clerk. But many will be, and most never forget that grand experience.

Knowledge is power—and confidence. Winery staff dealing with customers should be fully informed about the winery, its products, and other area attractions. When a potential customer is warmly welcomed, they will feel comfortable about asking questions. Some customers go back to the same winery over and over again because the staff is always cheerful, offers help, and understands individual needs. Recognize these needs and try to serve them. This can put their mind at ease and make the difference between whether sales are gained or lost and, more importantly, whether the visitor returns.

It should be remembered that the bottom-line function of the tasting room and retail shop is to generate revenue. Although some vintners are good at the sale, they fail to win "after sales." This refers to offering visitors an opportunity to receive notices of close-out sales, new vintage releases, and other events which may be of interest. This can be achieved by actively maintaining the *house list* described earlier—a most valuable resource on which each enthusiastic customer's name should be added. In some well-established retail shops, after-sales amount to more than 50% of total sales.

Gift Shop (Fig. 13–10)

Some states prohibit the sale of gifts or any other nonwine items in the retail facility. Alcohol Beverage Commission (ABC) regulations should be fully examined before including this important revenue source in the marketing plan.

Most market studies indicate that the following items, in order of importance, account for the most sales in winery gift shops:

- Logo tee shirts
- Logo wine glasses

Fig. 13–10 An active winery gift shop.

Logo corkscrews and cork pullers
Food items
Tip-top® pourers
Logo bottle bags and totes
Sparkling wine stoppers
Cookbooks
Novelty bottle stoppers
Shirts, hats, and other clothing
Vacuum wine resealers
Nonlogo wine glasses
Candle holders
Wine books
Locally made pottery
Logo napkins, coasters, and other table amenities

The seasonality of some items should be expected to be closely correlated to sales demand. For example, logo tee shirts and picnic items are usually big sellers during the summer. Conversely, bottle bags and totes, books, jams, and jellies are major gift items in demand just before the holidays. Although some consumers may spend an aggregate of more than $100 in a winery gift shop, higher unit prices of more than $40 per item seem to be comparatively less popular. Consequently, designer wear, comprehensive books, and crystal wine glasses at higher-end prices turn over relatively slowly.

In much the same manner as unpopular wines are replaced with new items which will hopefully sell better, the same goes with gift shop items. There is little profit, indeed, a cost burden, in allowing unsold merchandise to continue taking up space which could generate more profit with another product.

Try different manners of merchandising gift shop items and keep up with seasonal decorating themes such as Valentine's Day, Independence Day, Halloween, Thanksgiving, and, of course, the Holiday Season.

Offer drawings for logo items. These can be shipped to visitor homes and reinforce the winery experience. Provide employees with logo shirts, hats, name tags, and buttons with appropriate special messages to wear when working. Include gift shop items as part of the tasting, such as demonstrating corkscrews and wine glasses emblazoned with the winery logo. Purchase of these items becomes free advertising for your winery.

Sales to the Wholesale Channel

Wine markets which are geographically removed from where the vintner can effectively operate directly should be relegated to a well-established wholesale–distributor. The cost of time and money to meet with individual retailers and restaurateurs is usually far more than the profit difference lost. As would be expected, the closer an independent wholesaler is located to a winery supplier, the more attractive that line may be to handle because of all the local marketing mix activity spilling over into the wholesaler's operating territory—resulting in a much easier task of distribution.

Brokers, if ambitious and reputable, can also be effectively employed for winery representation. The additional risk with brokers is usually in the area of limited product knowledge and consigned inventories outside of the vintner's possession.

A vintner should be fully familiar with a potential remote market before commencing business there. Demographics should be studied so that an appropriate wholesaler can be selected. For example, a wholesaler who concentrates on beer or spirits distribution in primarily blue-collar locales should not be the first choice for a premium wine line. A better bet would be a distributor of fine wine brands who concentrates on more cosmopolitan retail and restaurant accounts.

In reality, first-choice distributors nowadays already have full portfolios of wine products which enjoy high consumer demand. Aggravating this even more is the trend across the United States for distributor consolidation—for few, larger wholesale operations. On the brighter side, this has led to some larger firms setting up divisions in their sales force whereby product specialists are assigned those brands of wine which need individual attention for selected placements.

A new-entry vintner should expect to provide assistance to the total distributor effort in launching the line. This would typically include an initial "post-off" sale price in order to help make the new product more attractive. It may also include making a "blitz" of sales calls introducing both vintner and product, as well as to explain sales promotion opportunities. Further, there should be tastings conducted for various enophile groups, and an extensive participation in various media and trade relations activities.

Sales to the Retail Channel

In more adjacent environs, a vintner may want to consider the feasibility of obtaining a wholesaler's permit and operating as an in-house distributor. One attractive aspect of being an endogenous wholesaler is the higher gross revenue. Distributor margins are created from a base of about 30% over the FOB winery price. Thus, a wine selling to independent wholesalers at $50 per case would sell to retail outlets at about $65, plus taxes, shipping, and other attending costs. Another feature is that inventories can be properly warehoused and controlled without possession risks. Plus, sales personnel will have a detailed knowledge about each product.

On the other hand, the operation of day-to-day distribution can be expensive and problematic to maintain. Retailers can sometimes make unreasonable demands for post-off prices and exclusivity rights. Restaurateurs are often fickle; they may demand contributions toward the cost of reprinting wine lists and excessive quantity discounts. These examples exist among many others.

Sales to Consumers

Good marketing plans are highly sensitive to selective demographics. It is well documented that most people interested in wine generally have higher incomes, are better educated, and have an affinity for the arts. This should obviously translate in winery and product design such as to attract these types of visitors. However, these types of people represent a smaller share of the entire tourism market; marketing must be designed to reach these ideals to the exclusion of others.

Psychographics, or "life-styles" of the people who fit these demographics, are equally important. A good example of this are two people who are similarly educated, earn similar incomes, and live in the same neighborhood. One person may be single, a keen spectator sports fan, and drive a performance sports car. The other may be married, enjoying the theater and art shows, and drive a conservative sedan. The diversity of these life-styles dictates how marketing efforts should be targeted in order to optimize sales effectiveness. Consequently, it is essential to correlate market potential with visitor potential.

Selling from larger lines of wine and gift items is important because the more products and services customers can obtain from a vintner, the more committed they are to that relationship. Anticipate consumer needs and suggest how to combine various products for various purposes, such as seasonal gift-giving, upcoming family and business events, professional gatherings, and so on.

In-store signs generate impulse purchases and allows for cross-merchandising of the products available. Use quotes from celebrities, educational information, and frequent display changes in order to keep the retail center fresh for returning customers.

MARKET FEEDBACK

Gathering and processing information fed back from visitors, consumers, retailers, restaurateurs, wholesalers, and the media provides the basis on which strategic changes must be made in the marketing mix.

Market evaluation is a prime function which successful vintners strive to maintain in close parallel between the demands of the marketplace and the supply of the wine product line. Market feedback information can save precious resources and avoid recurring errors in management decision making.

Nevertheless, market evaluation is an essential marketing task which is often neglected. Build specific goals into each of the marketing elements—public relations, sales promotion, advertising, and sales. Devise a specific tracking mechanism for the marketing mix in order to measure the effectiveness of resources budgeted for each activity. For example, one goal might be to increase tasting room traffic on Sundays by 30%. Results from the marketing effort should be evaluated to ascertain how each part of the project met, exceeded, or fell below the expectation. Although some efforts must be evaluated on a short-term basis, keep in mind others are a long-term project and do not necessarily create immediate results.

Reselling

This facet of marketing incorporates many different elements from in-store promotions to direct mail campaigns. Any given combination of marketing mix activities may comprise a successful reselling program among existing customers. The single most important element is, however, a fine reputation for both the vintner and the products marketed.

Attracting new customers costs many times more than reselling existing customers. As mentioned previously, it can take seven to nine contacts with a new prospect to make a sale—far fewer than for existing or previous customers.

Resources are usually scarce in wineries, and especially so in new ones. Thus, it is important that reselling efforts be focused on promotions which promise the largest profits. Although medal-winning wines may have generated the most attention and profit during the Holiday Season, good database research may reveal that a far different wine type may be best to offer during the spring, and yet other products for summer and fall.

Utilize the direct contact opportunity with consumers visiting the winery to take their comments and suggestions. Address their needs and interests. Their responses can bring about answers, ideas, and other insights which vintners should add to the database. Formal market research can also be conducted with visitors at the winery. Develop surveys which are simple to answer, in nonjargon, and with some type of incentive for their response. Some marketers indicate that most unhappy customers will leave a place of business without complaining. Unidentified written responses can transcend this reluctance to complain or to criticize.

Keeping in Touch

Vintners should keep in contact with their customers. Develop a process and a commitment. Capitalize on contact with customers. Most consumer contacts made to vintners are for placing orders or requests for information. Ensure that their inquiry is handled properly. Phone message machines may be practical when the winery is closed, but it is a poor substitute for a friendly voice during business hours.

Mail a thank-you note and a customer satisfaction survey following a visit with a self-addressed stamped envelope. Follow up in a few months with additional information, newsletters, or even just a postcard. Try mailings to individual consumer segments suggested by database information in order to test response quality and quantity.

Develop a system by setting a schedule for reselling outreach. It may be one hour a day, one morning a week, one day a month, during lunch hours, or just before closing. January may not be the best time to attract visitors, but it could be an ideal time to plan for a summer or fall promotional month or the next winter holiday promotion.

Testimonials

Vintner credibility can dramatically increase when well-known wine authorities praise winery achievements and products. These are valuable words in advertising, brochures, or other materials. Tributes from consumers can also be employed effectively in marketing. Testimonials complement commercial advertising. Personal messages endorsing a winery and/or its products are often more believable and acceptable by most consumers.

Written permission should be obtained from every source to be quoted. Celebrities may need to delay giving permission in order to ensure their testimony does not conflict with any existing contractual agreements. Most people, public or private, who are asked to be quoted will be flattered. There are others, however, who wish to remain private and this must be honored.

Testimonials can be taken from spontaneous remarks made on a tour through the winery, acquired from surveys conducted in the tasting room, or received via mail from visitors. Letters from people who are obviously impressed with the winery are an especially good source.

Referrals

One of the best sources for new visitors is old ones. Encourage visitors to provide names and addresses of their friends who might enjoy receiving information on the winery. Remind them of how pleasant it was to have them visit the winery and that it would be great to receive similar visits from their friends.

Actively seek referrals and express appreciation to them for telling their acquaintances. At the end of the sales transaction, when they are happy, thank them for any future friends they might refer to your winery. Consider placing a small referral message on the cash register receipt, or on a small flyer which can be placed in paper bags, or even a sign over the exit door. Most people enjoy helping a nice place of business succeed. Further, satisfied consumers are likely to become ambassadors of good will for a vintner they respect.

In conclusion, a marketing plan, a marketing budget, and a marketing resource are most important in successfully trading fine wines for consumer dollars. A vineyard manager needs the proper tools and resources—post, wire, plants, and tractors—to grow grapes. A winemaker requires adequate tanks, pumps, and hoses to make wine. The very best results of these will fail unless they can be marketed. Consequently, a vintner must devises a marketing program with adequate funding, personnel, and tools to carry out a continuous, cohesive program.

APPENDIX A

SOURCES

MICROBIOLOGY LABORATORY MATERIALS AND EQUIPMENT

Fisher Scientific, IL   (800) 766-7000
Micro Filtration Systems, CA   (800) 334-7132
Millipore, MA   (800) 645-5476
Precision Laboratories, IL   (800) 323-6280
Presque Isle Wine Cellars, PA   (814) 725-1314
Scott Laboratories, CA   (707) 765-6666
Spectro Services, NY   (716) 654-9500
Vinquiry, CA   (707) 433-8869
VWR Scientific, CA   (415) 468-7150
The Wine Lab, CA   (707) 224-7903

NEW EQUIPMENT

Boordy, MD   (410) 823-4624
Budde & Westermann, NJ   (201) 744-5363

Criveller Company, Ontario, Canada   (905) 357-2930
Enotech Corporation, CA   (415) 851-2040
KLR Machines, Inc. NY   (607) 776-4193
Presque Isle Wine Cellars, PA   (814) 725-1314
Prospero Equipment Corporation, NY   (914) 769-6252
RLS Equipment Company, NJ   (609) 965-0074
Scott Laboratories, Inc. CA   (707) 765-6666
The Compleat Winemaker, CA   (714) 963-9681

NEW TANKAGE

Atlantic Welding, NJ   (609) 641-1966
Barrel Builders, Inc., CA   (707) 942-4291
Blue Grass Cooperage, KY   (502) 364-4550
Canton Cooperage Company, KY   (800) 692-9888
Demptos Napa Cooperage, CA   (707) 257-2628
Mutual Stamping Stainless Steel, CT   (203) 877-3933
Paul Mueller Company, MO   (417) 831-3000
Santa Rosa Stainless Steel, CA   (707) 544-7777
Seguin Moreau, USA, CA   (707) 252-3408
World Cooperage/Independent Stave, MO   (417) 588-4151

USED EQUIPMENT AND TANKAGE

Aaron Equipment, IL   (708) 350-2200
Brewers & Bottlers Corp, FL   (813) 621-6599
Brinks, Inc., TN   (800) 873-4684
Carmel Equipment, NJ   (201) 656-4030
Eischen Enterprises, CA   (209) 834-0013
FM Custom Food Machinery, MI   (616) 675-7050
JACO Equipment, NY   (716) 836-3755
Kelly Trailer & Container, MA   (800) 628-0497
RLS Equipment, NJ   (609) 965-0074
Rowlands Sales, PA   (717) 455-5813

GRAPES, MUST, AND JUICE

Barber and Sons, NY   (716) 326-4692
Brennan Farm Market, NY   (716) 672-2727

Fruitland Juices, Inc., Ontario, Canada   (416) 856-5700
Fulkerson's Winery and Juice Plant, NY   (607) 243-7883
Meier's Wine Cellars, Inc., OH   (513) 891-6370
Presque Isle Wine Cellars, PA   (814) 488-7492
St. Julian Wine Co., MI   (616) 657-5568
Serenity Vineyards, NY   (315) 536-6701
Vintners Edge, CA   (707) 573-0807
Walker's Fruit Basket/Press House, NY   (716) 679-1292

LABORATORY WARE AND COMPUTER SOFTWARE

Agrisystems, Inc., CO   (303) 267-0024
Compliant Data Systems, CA   (408) 624-8920
Data Consulting Associates, CA   (707) 874-3067
Fisher Scientific, CA   (408) 727-0660
Presque Isle Wine Cellars, PA   (814) 488-7492
Scott Laboratories, Inc., CA   (707) 765-6666
The Lab Mart, NJ   (908) 561-1234
Vinquiry, CA   (707) 433-8869
VWR Scientific, CA   (415) 468-7150
Wine Appreciation Guild, Ltd., CA   (415) 864-1202

MATERIALS

Budde & Westermann, NJ   (201) 744-5363
California Glass Company, CA   (510) 635-7700
Cellulo Co., NJ   (908) 272-9400
KLR Machines, Inc., NY   (607) 776-4193
Pickering Winery Supply Co., CA   (415) 474-1588
Presque Isle Wine Cellars, PA   (814) 725-1314
Scott Laboratories, Inc., CA   (707) 765-6666
The Wine Lab., CA   (707) 224-7903
Vitro Packaging, TX   (214) 960-9693
Waterloo Container, NY   (315) 539-3922

PERIODICALS

American Vineyard
441 West Paul Avenue
Clovis, CA 93612

American Wine Society Journal
3006 Latta Road
Rochester, NY 14612

Fruit Winemaking Quarterly
4330 Gunderson Court
Sebastopol, CA 95472

Practical Winery & Vineyard
15 Grande Paseo
San Rafael, CA 94903-1534

The Wine Trader
P.O. Box 1598
Carson City, NV 89702

Vineyard & Winery Management
103 3rd Street
Watkins Glen, NY 14891

Wine Business Insider
867 West Napa Street
Sonoma, CA 95476

Wine Country Classifieds
P.O. Box 389
St. Helena, CA 94574

Wine East
620 North Pine Street
Lancaster, PA 17603

Wines and Vines
1800 Lincoln Avenue
San Rafael, CA 94901-1298

APPENDIX B

Analytical Procedures

MICROSCOPY

The microscope (Fig. B–1) is a very delicate, precision instrument and should be placed only in the hands of an operator who will accept responsibility for properly using and maintaining it. Given that, the operation of this device can be rather simple and highly rewarding.

There are general operational instructions which should be observed in using any microscope:

1. Moving parts of the microscope should never be forced. They should move freely. If something binds or becomes inoperable, the microscope should be serviced by a qualified person.
2. The lenses in the objectives should never be touched with anything but appropriate cleaning materials. Other items may deposit film, oils, or even scratch the precision surfaces.
3. The objectives should never touch slides, cover slips, or the stage.
4. Specimens should be examined first with a low-power objective, increasing magnification as necessary.

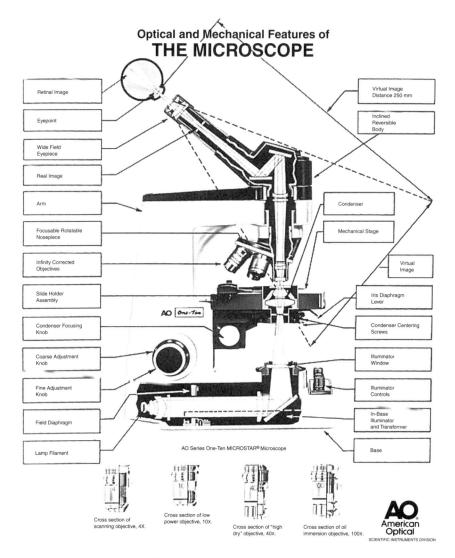

Fig. B–1 Light microscope illustration.
Courtesy: American Optical.

5. The body of the microscope should never be adjusted downward with the coarse knob while the operator is looking through the eyepiece. At eye level, the objective can be brought close to the coverslip and then adjusted for focus upward while viewing through the eyepiece. This will aid in preventing the objective from touching the slide and perhaps ruining the lens.

6. The microscope should be stored with the low-power, 10×, objective in the focusing position.
7. The microscope should be carried only by the arm, ensuring that it is maintained in the upright position.
8. Eye strain can be minimal if both eyes are kept open when viewing through the eyepiece. Squinting causes the eyes to tire quickly and, apart from the discomfort, much can be lost or overlooked in specimen observation.
9. One should become totally familiar with a new or different microscope. It may help to "dummy" the instrument without a slide in position on the stage in order to become accustomed to the feel and position of adjustment locations and operation.
10. Securing the microscope from operation should be preceded with a careful dusting and lubrication as may be instructed by the owner's manual, if available, or as may be advised by a qualified serviceman. Cover with dust cover jacket and store in a wooden storage box or cupboard.

The following procedure is provided, although if manufacturer's operating instructions are available, they should take precedence:

1. Open the aperture diaphragm of the condenser.
2. Turn on illuminator or adjust mirror to an external light source.
3. Turn coarse adjustment knob to raise nosepiece and objectives sufficiently for insertion of a glass slide.
4. Place properly prepared specimen slide and coverslip on the stage.
5. Position the low-power objective over slide and lower the body with the coarse adjustment knob until the objective is about $1/8$ in. from the slide. It is important that the objective does not actually touch the slide in order to avoid scratching the objective and causing other damage to the instrument. This operation is best accomplished by keeping the eye level with the stage.
6. Slowly elevate the body with the coarse adjustment knob while looking through the eyepiece. Once the image appears in approximate focus, stop adjustment. Adjust only upward with the adjustment knobs when viewing through the eyepiece. Adjust downward only when the eye is level with the stage.
7. Fine-tune focus by adjustment with the fine adjustment knob. Further adjustment may be necessary with the light source in order to achieve the optimum image.
8. Adjust light by moving the iris diaphragm level to the desired opening while viewing the image through the eyepiece.
9. Place the desired object or specimen in the exact center of the image by manipulation of the stage.
10. Turn coarse adjustment so as to move body upward allowing positioning of 40× objective over the slide. Be careful not to move slide on stage or the centered image will be lost.
11. With eye at stage level, lower body with coarse adjustment knob about $1/8$ in. from the coverslip.

12. Repeat Step 6.
13. Repeat Step 7.
14. Repeat Step 8.
15. Repeat Step 9.
16. Turn coarse adjustment so as to move body upward, allowing positioning of 100× oil-immersion objective over slide. Be careful not to move slide on stage or the centered image will be lost.
17. With eye at stage level, lower body until objective is approximately ¼ in. from the slide.
18. With extreme patience and care, place one drop of immersion oil on coverslip just below objective, again being sure not to move the slide.
19. With eye at stage level, lower body with coarse adjustment knob until the lens of the 100× objective comes in contact with oil. Continue to lower body very carefully until objective nearly touches the coverslip. Do not, however, allow the objective to actually come in contact with the slide.
20. Viewing through the eyepiece, slowly elevate body by turning the fine adjustment knob until a focus is achieved.
21. Make adjustments of light source and diaphragm which are necessary for the optimum image.
22. When observation is finished, raise body so that objective is about 1 in. from coverslip and return 10× objective to the focusing position.
23. The oil-immersion objective should be polished dry with dry lens paper after every use. The other lower-power objectives should never be used with the oil-immersion technique. If, however, some immersion oil should come in contact with them, they should be immediately cleaned with lens paper moistened with xylol, or commercial eyeglass cleaner and polished with dry lens paper.
24. Remove slide from stage. If any oil spills or other liquids are on the stage, they should be cleaned with a cheesecloth moistened with xylol and then dried with an untreated cheesecloth.

GRAM STAIN PROCEDURE (SEE FIG. B–2)

Apparatus and Reagents

Crystal Violet solution
(10 mL Crystal Violet, saturated alcohol solution and 40 mL ammonium oxalate 1.0% aqueous solution)
Gram's iodine solution
(1 g iodine crystals, 2 g potassium iodide, and 300 mL distilled or deionized water)
Safranin solution, 2.5 g in 100 mL 95% ethanol
Bunsen burner
Slide and coverslip

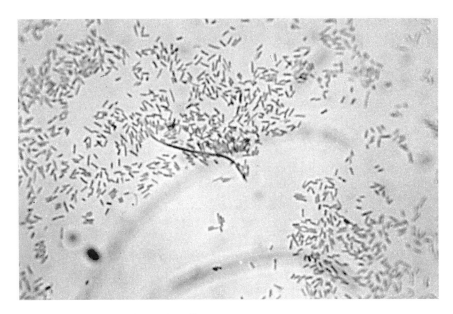

Fig. B–2 Gram stain of bacteria cells.

Distilled or deionized water
95% alcohol
Bibulous paper

Procedure

1. Place a drop of the medium or culture on the center of a clean slide, spreading the material over an area of about $1/2$ in^2.
2. Allow the sample material to dry and then fix by quickly passing the slide over the burner flame several times. Do not allow the slide to become too hot so as to avoid burning the sample.
3. Flood slide with Crystal Violet and hold for 30 s. Rinse with distilled or deionized water.
4. Cover the sample with Gram's iodine and hold for 30 s. Rinse with distilled or deionized water.
5. Decolorize in 95% alcohol for 20–30 s. Rinse with distilled or deionized water.
6. Counterstain with Safranin, allowing a reaction time of about 10 s. Rinse with distilled or deionized water.
7. Dry with bibulous paper.
8. Observe under microscope. Gram-positive organisms will stain a purple–black. Gram-negative organisms will stain pink or red.

MALOLACTIC FERMENTATION DETERMINATION BY PAPER CHROMATOGRAPHY*

The use of a paper chromatogram (Fig. B–3) in the detection of malolactic fermentation distinguishes qualitatively between malic acid and lactic acids existing in the wine sample analyzed. The absence of malic acid is an indication that this bacterial fermentation has taken place. The formation of lactic acid in itself is not valid evidence, as this may result from other microbial activity.

This analytical method determines the ratio of distance from a "baseline" to an acid "spot" called the R_f on each paper chromatogram by the use of standard acid solutions. Once these standards are made, a comparison can be made with chromatogram run with wine samples.

Apparatus and Reagents

Chromatographic grade filter paper cut into 20 × 30 cm rectangles
1.2 × 75-mm micropipets
Separatory funnel
One-gallon wide-mouth glass jars with covers
Solvent constituents
  100 mL distilled water
  100 mL n-butyl alcohol
  10.7 mL concentrated formic acid
  15 mL 1% water-soluble Bromcresol Green
Standard solutions
  0.3% tartaric acid
  0.3% citric acid
  0.3% malic acid
  0.3% lactic acid

Procedure

1. A pencil line (baseline) is drawn approximately 2.5 cm parallel to the long edge of the filter paper. The wine sample (or standard) is spotted from the micropipet on this line about 2.5 cm apart. Each spot is made four times (allowed to dry in between) at a volume of 10 μL from the micropipet.
2. A cylinder is made from the paper by stapling the short ends, without overlapping.
3. Place solvent constituents in separatory funnel and mix. After about 20 min, the lower aqueous phase is drawn off and discarded.
4. Transfer 70 mL of the upper layer into the wide-mouth jar. Stand the paper cylinder up in the jar such that the spotted edge (baseline) is in the solvent and cover the jar.

---

*Adapted from Kunkee, R.E. 1974. In "Malo-Lactic Fermentation and Winemaking," Part 7 of *Chemistry of Winemaking*. A.D. Webb, Ed. American Chemical Society, Washington, DC.

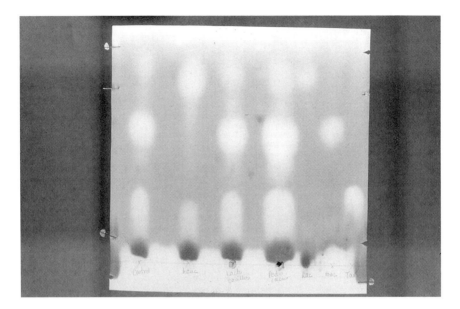

Fig. B–3 Paper chromatogram.

5. The chromatogram should develop in about 6–8 h.
6. Remove chromatogram (now yellow in color) and store in a ventilated area until dry and the formic acid has vaporized, leaving a blue–green background with yellow spots of acid having the following approximate R_f values:
   Tartaric acid   0.28
   Citric acid   0.45
   Malic acid   0.51
   Lactic acid   0.78
7. Standards and wine samples should be run simultaneously or one immediately following the other.
8. Solvent may be used repeatedly if care is taken to remove any aqueous layer which may have separated after each run.

BOTTLE STERILITY EVALUATION

Select a monitor kit such as the Millipore 55 PLUS® which is designed for measuring living or viable organisms in wine. (See Fig. B–4.) These systems allow for easy use and minimal contamination. Monitors function by filtering a sample of juice or wine, trapping any yeasts or bacteria on the surface of the filter membrane. Nutrient media is applied to that membrane and then incubated at 75°F for

Fig. B–4   Microbiological Analysis Monitor, 37mm
Courtesy: Millipore® Corporation, Bedford, MA.

7–10 days. Organisms will grow into "colonies" which can be enumerated and identified.

Carefully wash the exterior of the bottle to be evaluated in order to minimize the potential for contamination. The neck, cork, and cork puller should each be rinsed with 70% alcohol solution. Shake the bottle thoroughly just before opening and remove the cork gently, avoiding the cork puller forcing particles of cork into the wine.

Extensive directions are provided by the manufacturer for each type of monitor unit. Typically, 25–100 mL are poured from the bottle into the funnel and filtered through the membrane. This is followed by filtering sterile water (which can be easily made by boiling several minutes in a microwave and then cooling in a refrigerator) in order to flush the membrane free of any residual wine. Any wine left in the membrane may contain sufficient sulfur dioxide and/or sorbates which could inhibit viable organisms from forming colonies.

In general, yeast colonies are smooth, shiny cream-colored mounds, 2–4 mm in diameter, growing in 2–5 days. Bacteria will take 5–10 days to form tiny colonies, sometimes nearly transparent and requiring magnification to identify them.

The absence of colonies on the filter after 10 days provides confidence, not a guarantee, that the bottled wine is microbiologically stable.

The presence of a few yeast and/or bacteria colonies suggests that the bottled wine should be held for several weeks before being fully released into marketable cased goods inventory. Additional bottles should be then be taken from various case locations and each run for bottle sterility again. If there is no apparent gas formation, haziness, or other indications of fermentation in the bottle and the bottle sterility tests remains the same as before, the wine is probably stable but should undergo a sensory examination at least once per week for the next several weeks or so.

If more than 10 colonies are counted on the incubated filter, immediate action should be taken before deterioration goes any further. This may mean making the difficult decision to empty all the bottles back into a tank and taking appropriate action and treatments to surmount the situation.

Molds will not grow in the relatively high alcohol levels found in wine. The presence of mold colonies or other nonspoilage organisms indicate contamination from poor technique. Read the directions of the monitor again and resume analysis from the beginning.

WINERY SANITATION EVALUATION

The first step in any microbiological procedure is to minimize contamination sources from the laboratory itself. The technician should wear a clean labcoat laundered in hot water and hypochlorite in order to cover clothing which has been exposed to countless other organisms both airborne and waterborne. Hair should be covered and long hair done up. Hands should be washed with soap, then sprayed or rinsed with 70% ethanol solution.

Swab Test

Yeast and mold contamination can be analyzed by the use of swab test kits. If the directions supplied by the swab test manufacturer are followed carefully, there should be accurate estimation of contamination on areas such as tank interiors, valves, corks, filler tubes, and so forth. The methodology involves rubbing the area in question with the swab, transferring the swab to a buffer solution to suspend any organisms which have been picked up, then dipping a media-impregnated filter membrane into the buffer. The filter membrane is shaken to remove excess buffer, then replaced in its sterile housing and incubated, filter surface down, for 5–7 days at 75°F.

Each viable organism which lands on the filter will begin to multiply and grow to form colonies. Look for this growth after 48 h, as mold colonies may grow fuzzy and very large—obscuring the entire filter after a few days and making it impossible to evaluate other growth.

Yeasts will produce visible small shiny creamy-white mounds in 3–5 days. Molds will begin as very small rough gray, green, or black colonies—becoming larger and fuzzier with time. Most bacteria will not grow on this media, although, occasionally, certain species of *bacilli* will grow to form yeast-like colonies or appear as a slimy, flat, rapidly spreading white or nearly transparent film on the test membrane. (See Fig. B–5.)

Counting the number of colonies of yeast and/or mold provides an estimate of how many living organisms were on the surface tested. If the results of the swab culture show more than several colonies of yeast or mold, a better sanitation pro-

Fig. B–5 Colonies on swab test filter media.
Courtesy: Millipore® Corporation.

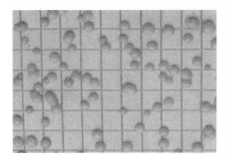
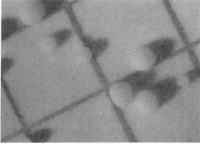
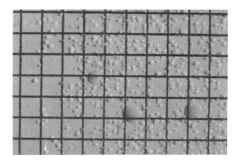

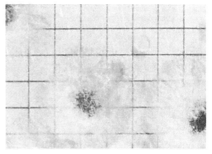

gram is necessary. Obviously, the absence of yeast or mold growth is an indication, not a guarantee, that the area is free of microbial contamination.

PH METER

A pH probe consist of a hydrogen-ion-sensitive electrode and a reference electrode. Both produce a voltage when in contact with hydrogen ions. The value of that voltage is a linear function of the pH. (See Fig. B–6.)

The following general procedure is provided. However, when manufacturer's operating instructions are available, they should take precedence in the use and care of the pH meter.

Apparatus and Reagents

pH meter with electrode
50-mL poly or glass beakers
Thermometer
Distilled water wash bottle
pH 4.0 and 7.0 buffer solution

Procedure

1. Turn pH meter on and allow several minutes of "warm up" in order to become stable.
2. Pour about 20 mL 7.0 of buffer solution into a clean beaker.

Fig. B–6 pH meter in use.

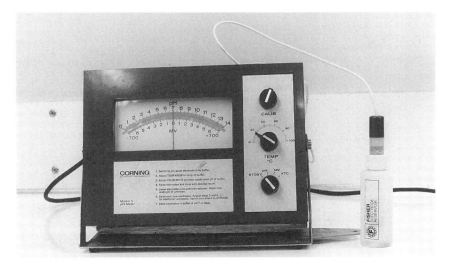

3. Immerse electrode into 7.0 buffer solution and gently swirl the beaker (be careful not to damage electrode).
4. Ensure that buffer solution and wine sample(s) are the same temperature.
5. Adjust temperature control knob of pH meter to buffer solution temperature and adjust reading of pH meter to exactly 7.0. Repeat procedure with 4.0 buffer.
6. Remove buffer solution and rinse electrode with distilled water.
7. Pour about 20 mL of wine sample into a clean beaker.
8. Immerse electrode into wine sample and gently swirl the beaker (be careful not to damage electrode).
9. Take pH reading when stabilized and record result.

BRIX–BALLING BY HYDROMETER

Brix is a measurement of dissolved solids in a sample of juice or must being analyzed. Should there be any alcohol in the sample, the test would be properly called a Balling. While dissolved solids increase the specific gravity (density) of a liquid, alcohol decreases specific gravity. Consequently, the combination of dissolved solids and alcohol is measured as viscosity (mouth feel or body) in the Balling test. Hydrometers can be used for both Brix and Balling analyses. (See Fig. B–7.)

Apparatus

Brix–Balling hydrometer in the range of wine to be analyzed
Hydrometer jar or cylinder
Thermometer

Procedure

1. If the sample has been taken from a fermenter and contains carbon dioxide gas, the gas should be removed by careful agitation, filtration, or heating.
2. Adjust sample temperature to that required as indicated on the stem of the hydrometer.
3. Pour about 50 mL of the sample to be analyzed into the hydrometer jar, swirl to rinse, and discard.
4. Pour sample into rinsed hydrometer jar up to a level of about 3 in. from the top.
5. Insert clean, dry hydrometer, carefully holding top of hydrometer stem in a pendulum effect.
6. Spin hydrometer carefully to free the instrument from the surface tension on the inside of the hydrometer jar.
7. Read stem directly at the bottom meniscus just before spinning stops and record result.

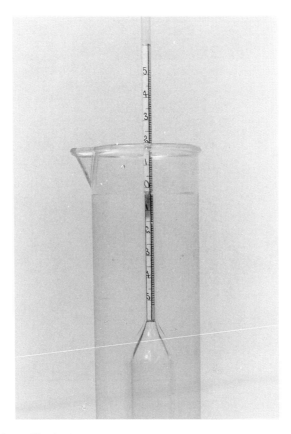

Fig. B–7 Brix–Balling hydrometer test.

BRIX BY REFRACTOMETER

The concentration of dissolved solids in juice or must (not wine) by this method is determined by measuring the refractive index of the solution. A drop or two of representative juice sample is placed on a clean refractometer prism. Exposure to a bright light source refracts incident light correlating to dissolved solids concentration which is measured upon a graduated scale on a screen inside the instrument. (See Fig. B–8.) Refractometers should not be used for Balling analysis, as samples containing alcohol will distort the refraction.

Apparatus

Refractometer
Bright light source

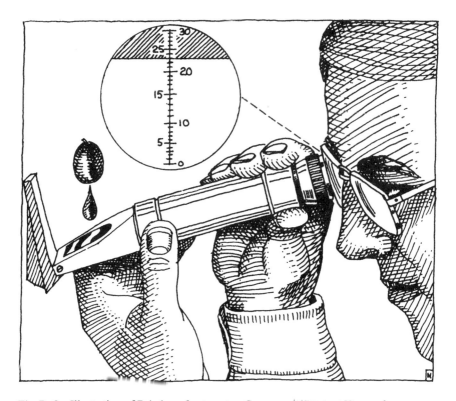

Fig. B-8 Illustration of Brix by refractometer. *Courtesy: Sebastiani Vineyards.*

Procedure

1. Adjust juice sample to room temperature or to that required by operating instructions of the instrument.
2. Open prism cover and rinse prism surface with several drops of sample. Gently wipe dry with absorbent lens paper.
3. Apply several drops of sample again and close prism cover. Point refractometer toward bright light source and hold in the same manner as a telescope. Adjust focus and read Brix at light–dark dividing line. Record result.
4. Rinse refractometer prism surface and prism cover three times with distilled or deionized water and wipe dry with absorbent lens paper. Avoid scratching prism surface.

Table B–1 lists the Brix of soluble solids at 20°C.

Table B-1 Brix (Soluble Solids) at 20°C; Sucrose (Cane Sugar) Scale

Degrees Brix	Specific Gravity	Lb Total Weight per U.S. Gallon	Lb Solids per U.S. Gallon	G Solids per Liter	Lb Water per U.S. Gallon
0.1	1.00040	8.34878	.00834	1.000	8.34044
0.2	1.00079	8.35196	.01675	2.007	8.33521
0.3	1.00118	8.35514	.02513	3.011	8.33001
0.4	1.00156	8.35828	.03351	4.015	8.32477
0.5	1.00195	8.36145	.04189	5.020	8.31956
0.6	1.00233	8.36462	.05026	6.024	8.31436
0.7	1.00272	8.36779	.05864	7.028	8.30915
0.8	1.00310	8.37096	.06702	8.031	8.30394
0.9	1.00349	8.37413	.07539	9.035	8.29874
1.0	1.00387	8.37730	.08377	10.039	8.29353
1.1	1.00425	8.38047	.09219	11.047	8.28828
1.2	1.00464	8.38372	.10060	12.055	8.28312
1.3	1.00503	8.38698	.10903	13.065	8.27795
1.4	1.00542	8.39023	.11746	14.075	8.27277
1.5	1.00581	8.39348	.12590	15.086	8.26758
1.6	1.00620	8.39674	.13435	16.099	8.26239
1.7	1.00659	8.39999	.14280	17.111	8.25719
1.8	1.00698	8.40325	.15126	18.125	8.25199
1.9	1.00737	8.40650	.15972	19.139	8.24678
2.0	1.00776	8.40976	.16820	20.155	8.24156
2.1	1.00815	8.41301	.17667	21.170	8.23634
2.2	1.00854	8.41627	.18516	22.187	8.23111
2.3	1.00893	8.41952	.19365	23.205	8.22587
2.4	1.00933	8.42286	.20215	24.223	8.22071
2.5	1.00972	8.42611	.21065	25.242	8.21546
2.6	1.01011	8.42937	.21916	26.262	8.21021
2.7	1.01051	8.43271	.22768	27.282	8.20503
2.8	1.01090	8.43596	.23621	28.305	8.19975
2.9	1.01129	8.43922	.24474	29.327	8.19448
3.0	1.01169	8.44255	.25328	30.350	8.18927
3.1	1.01208	8.44581	.26182	31.373	8.18399
3.2	1.01248	8.44915	.27037	32.398	8.17878
3.3	1.01287	8.45240	.27893	33.423	8.17347
3.4	1.01327	8.45574	.28750	34.450	8.16824
3.5	1.01366	8.45899	.29606	35.476	8.16293
3.6	1.01406	8.46233	.30464	36.504	8.15769
3.7	1.01445	8.46559	.31323	37.534	8.15236
3.8	1.01485	8.46892	.32182	38.563	8.14710
3.9	1.01524	8.47218	.33042	39.593	8.14176
4.0	1.01564	8.47552	.33902	40.624	8.13650
4.1	1.01603	8.47877	.34763	41.656	8.13114
4.2	1.01643	8.48211	.35625	42.689	8.12586
4.3	1.01683	8.48545	.36487	43.722	8.12058
4.4	1.01723	8.48878	.37351	44.757	8.11527
4.5	1.01763	8.49212	.38215	45.792	8.10997
4.6	1.01802	8.49538	.39079	46.827	8.10459
4.7	1.01842	8.49871	.39944	47.864	8.09927
4.8	1.01882	8.50205	.40810	48.902	8.09395
4.9	1.01922	8.50539	.41676	49.939	8.08863
5.0	1.01962	8.50873	.42544	50.980	8.08329
5.1	1.02002	8.51207	.43412	52.020	8.07795
5.2	1.02042	8.51540	.44280	53.060	8.07260
5.3	1.02082	8.51874	.45149	54.101	8.06725
5.4	1.02122	8.52208	.46019	55.144	8.06189
5.5	1.02163	8.52550	.46890	56.187	8.05660
5.6	1.02203	8.52884	.47762	57.232	8.05122
5.7	1.02243	8.53218	.48633	58.276	8.04585
5.8	1.02283	8.53552	.49506	59.322	8.04046
5.9	1.02323	8.53885	.50379	60.368	8.03506

Analytical Procedures

Table B–1 (*continued*)

Degrees Brix	Specific Gravity	Lb Total Weight per U.S. Gallon	Lb Solids per U.S. Gallon	G Solids per Liter	Lb Water per U.S. Gallon
6.0	1.02364	8.54228	.51254	61.417	8.02974
6.1	1.02404	8.54561	.52128	62.464	8.02433
6.2	1.02444	8.54895	.53003	63.512	8.01892
6.3	1.02485	8.55237	.53880	64.563	8.01357
6.4	1.02525	8.55571	.54757	65.614	8.00814
6.5	1.02566	8.55913	.55634	66.665	8.00279
6.6	1.02606	8.56247	.56512	67.717	7.99735
6.7	1.02646	8.56581	.57391	68.770	7.99190
6.8	1.02687	8.56923	.58271	69.825	7.98652
6.9	1.02727	8.57257	.59151	70.879	7.98106
7.0	1.02768	8.57599	.60032	71.935	7.97567
7.1	1.02808	8.57933	.60913	72.991	7.97020
7.2	1.02849	8.58275	.61796	74.049	7.96479
7.3	1.02890	8.58617	.62679	75.107	7.95938
7.4	1.02931	8.58989	.63563	76.166	7.95396
7.5	1.02972	8.59301	.64448	77.227	7.94853
7.6	1.03012	8.59635	.65332	78.286	7.94303
7.7	1.03053	8.59977	.66218	79.348	7.93759
7.8	1.03094	8.60319	.67105	80.410	7.93214
7.9	1.03135	8.60662	.67992	81.473	7.92670
8.0	1.03176	8.61004	.68880	82.538	7.92124
8.1	1.03216	8.61338	.69768	83.602	7.91570
8.2	1.03257	8.61680	.70658	84.668	7.91022
8.3	1.03298	8.62022	.71548	85.735	7.90474
8.4	1.03339	8.62364	.72439	86.802	7.89925
8.5	1.03380	8.62706	.73330	87.870	7.89376
8.6	1.03421	8.63048	.74222	88.939	7.88826
8.7	1.03462	8.63390	.75115	90.009	7.88275
8.8	1.03503	8.63733	.76009	91.080	7.87724
8.9	1.03544	8.64075	.76903	92.151	7.87172
9.0	1.03585	8.64417	.77798	93.224	7.86619
9.1	1.03626	8.64759	.78693	94.296	7.86066
9.2	1.03667	8.65101	.79589	95.370	7.85512
9.3	1.03708	8.65443	.80486	96.445	7.84957
9.4	1.03750	8.65794	.81385	97.522	7.84409
9.5	1.03791	8.66136	.82283	98.598	7.83853
9.6	1.03832	8.66478	.83182	99.675	7.83296
9.7	1.03874	8.66829	.84082	100.754	7.82747
9.8	1.03915	8.67171	.84983	101.833	7.82188
9.9	1.03956	8.67513	.85884	102.913	7.81629
10.0	1.03998	8.67863	.86786	103.994	7.81077
10.1	1.04039	8.68205	.87689	105.076	7.80516
10.2	1.04081	8.68556	.88593	106.159	7.79963
10.3	1.04122	8.68898	.89496	107.241	7.79402
10.4	1.04164	8.69249	.90402	108.327	7.78847
10.5	1.04205	8.69591	.91307	109.411	7.78284
10.6	1.04247	8.69941	.92214	110.498	7.77727
10.7	1.04288	8.70283	.93120	111.584	7.77163
10.8	1.04330	8.70634	.94028	112.672	7.76606
10.9	1.04371	8.70976	.94936	113.760	7.76040
11.0	1.04413	8.71326	.95846	114.850	7.75480
11.1	1.04454	8.71669	.96755	115.940	7.74914
11.2	1.04496	8.72019	.97666	117.031	7.74353
11.3	1.04538	8.72370	.98578	118.124	7.73792
11.4	1.04580	8.72720	.99490	119.217	7.73230
11.5	1.04622	8.73071	1.00403	120.311	7.72668
11.6	1.04663	8.73413	1.01316	121.405	7.72097
11.7	1.04705	8.73763	1.02230	122.500	7.71533
11.8	1.04747	8.74114	1.03145	123.597	7.70969

Table B–1 (continued)

Degrees Brix	Specific Gravity	Lb Total Weight per U.S. Gallon	Lb Solids per U.S. Gallon	G Solids per Liter	Lb Water per U.S. Gallon
11.9	1.04789	8.74464	1.04061	124.694	7.70403
12.0	1.04831	8.74815	1.04978	125.795	7.69837
12.1	1.04873	8.75165	1.05895	126.892	7.69270
12.2	1.04915	8.75516	1.06813	127.992	7.68703
12.3	1.04957	8.75866	1.07732	129.093	7.68134
12.4	1.04999	8.76217	1.08651	130.194	7.67566
12.5	1.05041	8.76567	1.09571	131.297	7.66996
12.6	1.05083	8.76918	1.10492	132.400	7.66426
12.7	1.05125	8.77268	1.11413	133.504	7.65855
12.8	1.05167	8.77619	1.12335	134.609	7.65284
12.9	1.05209	8.77969	1.13258	135.715	7.64711
13.0	1.05251	8.78320	1.14182	136.822	7.64138
13.1	1.05293	8.78670	1.15106	137.929	7.63564
13.2	1.05335	8.79021	1.16031	139.038	7.62990
13.3	1.05378	8.79379	1.16957	140.147	7.62422
13.4	1.05420	8.79730	1.17884	141.258	7.61846
13.5	1.05463	8.80089	1.18812	142.370	7.61277
13.6	1.05505	8.80439	1.19740	143.482	7.60699
13.7	1.05547	8.80790	1.20668	144.594	7.60122
13.8	1.05590	8.81149	1.21599	145.710	7.59550
13.9	1.05632	8.81499	1.22528	146.823	7.58971
14.0	1.05675	8.81858	1.23460	147.940	7.58398
14.1	1.05717	8.82208	1.24391	149.055	7.57817
14.2	1.05760	8.82567	1.25325	150.174	7.57242
14.3	1.05803	8.82926	1.26258	151.292	7.56668
14.4	1.05846	8.83285	1.27193	152.412	7.56092
14.5	1.05889	8.83644	1.28128	153.533	7.55516
14.6	1.05931	8.83994	1.29063	154.654	7.54931
14.7	1.05974	8.84353	1.30000	155.776	7.54353
14.8	1.06017	8.84712	1.30937	156.899	7.53775
14.9	1.06060	8.85071	1.31876	158.044	7.53195
15.0	1.06103	8.85430	1.32815	159.150	7.52615
15.1	1.06146	8.85788	1.33754	160.275	7.52034
15.2	1.06189	8.86147	1.34794	161.401	7.51453
15.3	1.06232	8.86506	1.35635	162.529	7.50871
15.4	1.06275	8.86865	1.36577	163.657	7.50288
15.5	1.06318	8.87224	1.37520	164.787	7.49704
15.6	1.06361	8.87583	1.38463	165.917	7.49120
15.7	1.06404	8.87941	1.39407	167.049	7.48534
15.8	1.06447	8.88300	1.40351	168.180	7.47949
15.9	1.06490	8.88659	1.41297	169.313	7.47362
16.0	1.06534	8.89026	1.42244	170.448	7.46782
16.1	1.06577	8.89385	1.43191	171.583	7.46194
16.2	1.06620	8.89744	1.44139	172.719	7.45605
16.3	1.06663	8.90103	1.45087	173.855	7.45016
16.4	1.06707	8.90470	1.46037	174.993	7.44433
16.5	1.06750	8.90829	1.46987	176.132	7.43842
16.6	1.06793	8.91188	1.47937	177.270	7.43251
16.7	1.06837	8.91555	1.48890	178.412	7.42665
16.8	1.06880	8.91914	1.49842	179.553	7.42072
16.9	1.06923	8.92272	1.50794	180.693	7.41478
17.0	1.06967	8.92640	1.51749	181.838	7.40891
17.1	1.07010	8.92998	1.52703	182.981	7.40295
17.2	1.07054	8.93366	1.53659	184.127	7.39707
17.3	1.07098	8.93733	1.54616	185.273	7.39117
17.4	1.07141	8.94092	1.55572	186.419	7.38520
17.5	1.07185	8.94459	1.56530	187.567	7.37929
17.6	1.07229	8.94826	1.57489	188.716	7.37337
17.7	1.07272	8.95185	1.58448	189.865	7.36737

Table B–1 (*continued*)

Degrees Brix	Specific Gravity	Lb Total Weight per U.S. Gallon	Lb Solids per U.S. Gallon	G Solids per Liter	Lb Water per U.S. Gallon
17.8	1.07316	8.95552	1.59408	191.015	7.36144
17.9	1.07360	8.95919	1.60370	192.168	7.35549
18.0	1.07404	8.96286	1.61331	193.320	7.34955
18.1	1.07448	8.96654	1.62294	194.474	7.34360
18.2	1.07492	8.97021	1.63258	195.629	7.33763
18.3	1.07536	8.97388	1.64222	196.784	7.33166
18.4	1.07580	8.97755	1.65187	197.940	7.32568
18.5	1.07624	8.98122	1.66153	199.098	7.31969
18.6	1.07668	8.98489	1.67119	200.255	7.31370
18.7	1.07712	8.98857	1.68086	201.414	7.30771
18.8	1.07756	8.99224	1.69054	202.574	7.30170
18.9	1.07800	8.99591	1.70023	203.735	7.29568
19.0	1.07844	8.99958	1.70992	204.896	7.28966
19.1	1.07888	9.00325	1.71962	206.059	7.28363
19.2	1.07932	9.00693	1.72933	207.222	7.27760
19.3	1.07977	9.01068	1.73906	208.388	7.27162
19.4	1.08021	9.01435	1.74878	209.559	7.26557
19.5	1.08066	9.01811	1.75853	210.721	7.25958
19.6	1.08110	9.02178	1.76827	211.888	7.25351
19.7	1.08154	9.02545	1.77801	213.055	7.24744
19.8	1.08199	9.02921	1.78778	214.226	7.24143
19.9	1.08243	9.03288	1.79754	215.396	7.23534
20.0	1.08288	9.03663	1.80733	216.569	7.22930
20.1	1.08332	9.04031	1.81710	217.739	7.22321
20.2	1.08377	9.04406	1.82690	218.914	7.21716
20.3	1.08421	9.04773	1.83669	220.087	7.21104
20.4	1.08466	9.05149	1.84650	221.262	7.20499
20.5	1.08510	9.05516	1.85631	222.438	7.19885
20.6	1.08555	9.05891	1.86614	223.616	7.19277
20.7	1.08599	9.06259	1.87596	224.793	7.18663
20.8	1.08644	9.06634	1.88580	225.972	7.18054
20.9	1.08688	9.07001	1.89563	227.150	7.17438
21.0	1.08733	9.07377	1.90549	228.331	7.16828
21.1	1.08777	9.07744	1.91534	229.511	7.16210
21.2	1.08822	9.08120	1.92521	230.694	7.15599
21.3	1.08867	9.08495	1.93509	231.878	7.14986
21.4	1.08912	9.08871	1.94498	233.063	7.14373
21.5	1.08957	9.09246	1.95488	234.249	7.13758
21.6	1.09002	9.09622	1.96478	235.436	7.13144
21.7	1.09047	9.09997	1.97469	236.623	7.12528
21.8	1.09092	9.10373	1.98461	237.812	7.11912
21.9	1.09137	9.10748	1.99454	239.014	7.11294
22.0	1.09182	9.11124	2.00447	240.192	7.10677
22.1	1.09227	9.11499	2.01441	241.383	7.10058
22.2	1.09272	9.11875	2.02436	242.575	7.09439
22.3	1.09317	9.12250	2.03432	243.768	7.08818
22.4	1.09362	9.12626	2.04428	244.962	7.08198
22.5	1.09408	9.13010	2.05427	246.159	7.07583
22.6	1.09453	9.13385	2.06425	247.355	7.06960
22.7	1.09498	9.13761	2.07424	248.552	7.06337
22.8	1.09543	9.14136	2.08423	249.749	7.05713
22.9	1.09588	9.14512	2.09423	250.947	7.05089
23.0	1.09634	9.14896	2.10426	252.149	7.04470
23.1	1.09679	9.15271	2.11428	253.350	7.03843
23.2	1.09725	9.15655	2.12432	254.553	7.03223
23.3	1.09770	9.16031	2.13435	255.755	7.02596
23.4	1.09816	9.16415	2.14441	256.960	7.01974
23.5	1.09862	9.16798	2.15448	258.167	7.01350
23.6	1.09907	9.17174	2.16453	259.371	7.00721

Table B–1 (*continued*)

Degrees Brix	Specific Gravity	Lb Total Weight per U.S. Gallon	Lb Solids per U.S. Gallon	G Solids per Liter	Lb Water per U.S. Gallon
23.7	1.09953	9.17558	2.17461	260.579	7.00097
23.8	1.09998	9.17933	2.18468	261.786	6.99465
23.9	1.10044	9.18317	2.19478	262.996	6.98839
24.0	1.10090	9.18701	2.20488	264.206	6.98213
24.1	1.10135	9.19085	2.21499	265.418	6.97586
24.2	1.10182	9.19469	2.22511	266.630	6.96958
24.3	1.10228	9.19853	2.23524	267.844	6.96329
24.4	1.10274	9.20237	2.24538	269.059	6.95699
24.5	1.10320	9.20620	2.25552	270.236	6.95068
24.6	1.10366	9.21004	2.26567	271.491	6.94437
24.7	1.10412	9.21388	2.27583	272.708	6.93805
24.8	1.10458	9.21772	2.28599	273.926	6.93173
24.9	1.10504	9.22156	2.29617	275.145	6.92539
25.0	1.10550	9.22540	2.30635	276.365	6.91905
25.1	1.10596	9.22924	2.31654	277.586	6.91270
25.2	1.10642	9.23307	2.32673	278.807	6.90634
25.3	1.10688	9.23691	2.33694	280.031	6.89997
25.4	1.10734	9.24075	2.34715	281.254	6.89360
25.5	1.10781	9.24467	2.35739	282.481	6.88728
25.6	1.10827	9.24851	2.36762	283.707	6.88089
25.7	1.10873	9.25235	2.37785	284.933	6.87450
25.8	1.10919	9.25619	2.38810	286.161	6.86809
25.9	1.10965	9.26003	2.39835	287.389	6.86168
26.0	1.11012	9.26395	2.40863	288.621	6.85532
26.1	1.11058	9.26779	2.41889	289.851	6.84890
26.2	1.11105	9.27171	2.42919	291.085	6.84252
26.3	1.11151	9.27555	2.43947	292.317	6.83608
26.4	1.11198	9.27947	2.44978	293.552	6.82969
26.5	1.11245	9.28340	2.46010	294.789	6.82330
26.6	1.11291	9.28723	2.47040	296.023	6.81683
26.7	1.11338	9.29116	2.48074	297.262	6.81042
26.8	1.11384	9.29499	2.49106	298.499	6.80393
26.9	1.11431	9.29892	2.50141	299.739	6.79751
27.0	1.11478	9.30284	2.51177	300.980	6.79107
27.1	1.11525	9.30676	2.52213	302.222	6.78463
27.2	1.11572	9.31068	2.53250	303.464	6.77818
27.3	1.11619	9.31461	2.54289	304.709	6.77172
27.4	1.11666	9.31853	2.55328	305.954	6.76525
27.5	1.11713	9.32245	2.56367	307.199	6.75878
27.6	1.11760	9.32637	2.57408	308.447	6.75229
27.7	1.11807	9.33029	2.58449	309.694	6.74580
27.8	1.11854	9.33422	2.59491	310.943	6.73931
27.9	1.11901	9.33814	2.60534	312.193	6.73280
28.0	1.11948	9.34206	2.61578	313.444	6.72628
28.1	1.11995	9.34598	2.62622	314.695	6.71976
28.2	1.12042	9.34990	2.63667	315.947	6.71323
28.3	1.12089	9.35383	2.64713	317.200	6.70670
28.4	1.12137	9.35783	2.65762	318.457	6.70021
28.5	1.12184	9.36175	2.66810	319.713	6.69365
28.6	1.12231	9.36568	2.67858	320.969	6.68710
28.7	1.12279	9.36968	2.68910	322.229	6.68058
28.8	1.12326	9.37360	2.69960	323.488	6.67400
28.9	1.12373	9.37753	2.71011	324.747	6.66742
29.0	1.12421	9.38153	2.72064	326.009	6.66089
29.1	1.12468	9.38545	2.73117	327.271	6.65428
29.2	1.12516	9.38946	2.74172	328.535	6.64774
29.3	1.12564	9.39347	2.75229	329.801	6.64118
29.4	1.12611	9.39739	2.76283	331.064	6.63456
29.5	1.12659	9.40139	2.77341	332.332	6.62798

328 *Analytical Procedures*

Table B–1 (*continued*)

Degrees Brix	Specific Gravity	Lb Total Weight per U.S. Gallon	Lb Solids per U.S. Gallon	G Solids per Liter	Lb Water per U.S. Gallon
29.6	1.12707	9.40540	2.78400	333.601	6.62140
29.7	1.12754	9.40932	2.79457	334.868	6.61475
29.8	1.12802	9.41333	2.80517	336.138	6.60816
29.9	1.12850	9.41733	2.81578	337.409	6.60155
30.0	1.12898	9.42134	2.82640	338.682	6.59494
30.1	1.12945	9.42526	2.83700	339.952	6.58826
30.2	1.12993	9.42927	2.84764	341.227	6.58163
30.3	1.13041	9.43327	2.85828	342.502	6.57499
30.4	1.13089	9.43728	2.86893	343.778	6.56835
30.5	1.13137	9.44128	2.87959	345.056	6.56169
30.6	1.13185	9.44529	2.89026	346.334	6.55503
30.7	1.13223	9.44929	2.90093	347.613	6.54836
30.8	1.13281	9.45330	2.91162	348.894	6.54168
30.9	1.13329	9.45731	2.92231	350.175	6.53500
31.0	1.13377	9.46131	2.93301	351.457	6.52830
31.1	1.13425	9.46532	2.94371	352.739	6.52161
31.2	1.13473	9.46932	2.95443	354.023	6.51489
31.3	1.13521	9.47333	2.96515	355.308	6.50818
31.4	1.13570	9.47742	2.97591	356.597	6.50151
31.5	1.13618	9.48142	2.98665	357.884	6.49477
31.6	1.13666	9.48543	2.99740	359.172	6.48803
31.7	1.13715	9.48952	3.00818	360.464	6.48134
31.8	1.13763	9.49352	3.01894	361.754	6.47458
31.9	1.13811	9.49753	3.02971	363.044	6.46782
32.0	1.13860	9.50162	3.04052	364.339	6.46110
32.1	1.13908	9.50562	3.05130	365.631	6.45432
32.2	1.13957	9.50971	3.06213	366.929	6.44758
32.3	1.14005	9.51372	3.07293	368.223	6.44079
32.4	1.14054	9.51781	3.08377	369.522	6.43404
32.5	1.14103	9.52190	3.09462	370.820	6.42728
32.6	1.14151	9.52590	3.10544	372.119	6.42046
32.7	1.14200	9.52999	3.11631	373.421	6.41368
32.8	1.14248	9.53400	3.12715	374.720	6.40685
32.9	1.14297	9.53808	3.13803	376.024	6.40005
33.0	1.14346	9.54217	3.14892	377.329	6.39325
33.1	1.14395	9.54626	3.15981	378.634	6.38645
33.2	1.14444	9.55035	3.17072	379.941	6.37963
33.3	1.14493	9.55444	3.18163	381.248	6.37281
33.4	1.14542	9.55853	3.19255	382.557	6.36598
33.5	1.14591	9.56262	3.20248	383.867	6.35914
33.6	1.14640	9.56671	3.21441	385.176	6.35230
33.7	1.14689	9.57080	3.22536	386.488	6.34544
33.8	1.14738	9.57489	3.23631	387.801	6.33858
33.9	1.14787	9.57898	3.24727	389.114	6.33171
34.0	1.14836	9.58306	3.25824	390.428	6.32482
34.1	1.14885	9.58715	3.26922	391.744	6.31793
34.2	1.14934	9.59124	3.28020	393.060	6.31104
34.3	1.14984	9.59541	3.29123	394.382	6.30418
34.4	1.15033	9.59550	3.30223	395.700	6.29727
34.5	1.15083	9.60368	3.31327	397.023	6.29041
34.6	1.15132	9.60777	3.32429	398.343	6.28348
34.7	1.15181	9.61185	3.33531	399.664	6.27654
34.8	1.15231	9.61603	3.34638	400.990	6.26965
34.9	1.15280	9.62012	3.35742	402.313	6.26270
35.0	1.15330	9.62429	3.36850	403.641	6.25579
35.1	1.15379	9.62838	3.37956	404.966	6.24882
35.2	1.15429	9.63255	3.39066	406.296	6.24189
35.3	1.15479	9.63672	3.40176	407.626	6.23496
35.4	1.15528	9.64081	3.41285	408.955	6.22796

Table B–1 (*continued*)

Degrees Brix	Specific Gravity	Lb Total Weight per U.S. Gallon	Lb Solids per U.S. Gallon	G Solids per Liter	Lb Water per U.S. Gallon
35.5	1.15578	9.64498	3.42397	410.287	6.22101
35.6	1.15628	9.64916	3.43510	411.621	6.21406
35.7	1.15677	9.65325	3.44621	412.952	6.20704
35.8	1.15727	9.65742	3.45736	414.289	6.20006
35.9	1.15777	9.66159	3.46851	415.625	6.19308
36.0	1.15827	9.66576	3.47967	416.962	6.18609
36.1	1.15877	9.66994	3.49085	418.302	6.17909
36.2	1.15927	9.67411	3.50203	419.641	6.17208
36.3	1.15977	9.67828	3.51322	420.982	6.16506
36.4	1.16027	9.68245	3.52441	422.323	6.15804
36.5	1.16078	9.68671	3.53565	423.670	6.15106
36.6	1.16128	9.69088	3.54686	425.013	6.14402
36.7	1.16178	9.69505	3.55808	426.358	6.13697
36.8	1.16228	9.69923	3.56932	427.704	6.12991
36.9	1.16278	9.70340	3.58055	429.050	6.12285
37.0	1.16329	9.70766	3.59183	430.402	6.11583
37.1	1.16379	9.71183	3.60309	431.751	6.10874
37.2	1.16429	9.71600	3.61435	433.100	6.10165
37.3	1.16480	9.72026	3.62566	434.456	6.09460
37.4	1.16530	9.72443	3.63694	435.807	6.08749
37.5	1.16581	9.72868	3.64826	437.164	6.08042
37.6	1.16631	9.73286	3.65956	438.518	6.07330
37.7	1.16681	9.73703	3.67086	439.872	6.06617
37.8	1.16732	9.74129	3.68221	441.232	6.05908
37.9	1.16782	9.74546	3.69353	442.588	6.05193
38.0	1.16833	9.74971	3.70489	443.950	6.04482
38.1	1.16883	9.75389	3.71623	445.308	6.03766
38.2	1.16934	9.75814	3.72761	446.672	6.03053
38.3	1.16985	9.76240	3.73900	448.037	6.02340
38.4	1.17036	9.76665	3.75039	449.402	6.01626
38.5	1.17087	9.77091	3.76180	450.769	6.00911
38.6	1.17138	9.77517	3.77322	452.137	6.00195
38.7	1.17189	9.77942	3.78464	453.506	5.99478
38.8	1.17240	9.78368	3.79607	454.875	5.98761
38.9	1.17291	9.78793	3.80750	456.245	5.98043
39.0	1.17342	9.79219	3.81895	457.617	5.97324
39.1	1.17393	9.79645	3.83041	458.990	5.96604
39.2	1.17444	9.80070	3.84187	460.364	5.95883
39.3	1.17495	9.80496	3.85335	461.739	5.95161
39.4	1.17546	9.80921	3.86483	463.115	5.94438
39.5	1.17598	9.81355	3.87635	464.495	5.93720
39.6	1.17649	9.81781	3.88785	465.873	5.92996
39.7	1.17700	9.82207	3.89936	467.253	5.92271
39.8	1.17751	9.82632	3.91188	468.633	5.91544
39.9	1.17802	9.83058	3.92240	470.013	5.90818
40.0	1.17854	9.83492	3.93397	471.340	5.90095
40.1	1.17905	9.83917	3.94551	472.783	5.89366
40.2	1.17957	9.84351	3.95709	474.170	5.88642
40.3	1.18008	9.84777	3.96865	475.555	5.87912
40.4	1.18060	9.85211	3.98025	476.945	5.87186
40.5	1.18111	9.85636	3.99183	478.333	5.86453
40.6	1.18163	9.86070	4.00344	479.724	5.85726
40.7	1.18214	9.86496	4.01504	481.114	5.84992
40.8	1.18266	9.86930	4.02667	482.508	5.84263
40.9	1.18317	9.87355	4.03828	483.899	5.83527
41.0	1.18369	9.87789	4.04993	485.295	5.82796
41.1	1.18420	9.88215	4.06156	486.689	5.82059
41.2	1.18472	9.88649	4.07323	488.087	5.81326
41.3	1.18524	9.89083	4.08491	489.487	5.80592

330 Analytical Procedures

Table B–1 (continued)

Degrees Brix	Specific Gravity	Lb Total Weight per U.S. Gallon	Lb Solids per U.S. Gallon	G Solids per Liter	Lb Water per U.S. Gallon
41.4	1.18576	9.89517	4.09660	490.887	5.79857
41.5	1.18628	9.89951	4.10830	492.289	5.79121
41.6	1.18679	9.90376	4.11996	493.687	5.78380
41.7	1.18731	9.90810	4.13168	495.091	5.77642
41.8	1.18783	9.91244	4.14340	496.495	5.76904
41.9	1.18835	9.91678	4.15513	497.901	5.76165
42.0	1.18887	9.92112	4.16687	499.308	5.75425
42.1	1.18939	9.92546	4.17862	500.716	5.74684
42.2	1.18991	9.92980	4.19038	502.125	5.73942
42.3	1.19043	9.93414	4.20214	503.534	5.73200
42.4	1.19096	9.93856	4.21395	504.949	5.72461
42.5	1.19148	9.94290	4.22573	506.361	5.71717
42.6	1.19200	9.94724	4.23752	507.774	5.70972
42.7	1.19253	9.95166	4.24936	509.192	5.70230
42.8	1.19305	9.95600	4.26117	510.607	5.69483
42.9	1.19357	9.96034	4.27299	512.024	5.68735
43.0	1.19410	9.96476	4.28485	513.445	5.67991
43.1	1.19462	9.96910	4.29668	514.863	5.67242
43.2	1.19515	9.97353	4.30856	517.711	5.66497
43.3	1.19568	9.97795	4.32045	519.132	5.65750
43.4	1.19620	9.98229	4.33231	520.559	5.64998
43.5	1.19673	9.98671	4.34422	521.986	5.64249
43.6	1.19726	9.99113	4.35613	523.411	5.63500
43.7	1.19778	9.99547	4.36802	524.842	5.62745
43.8	1.19831	9.99990	4.37996	526.273	5.61994
43.9	1.19884	10.00432	4.39190	527.705	5.61242
44.0	1.19937	10.00874	4.40385	529.126	5.60489
44.1	1.19990	10.01317	4.41581	530.571	5.59736
44.2	1.20043	10.01759	4.42777	532.000	5.58982
44.3	1.20096	10.02201	4.43975	533.442	5.58226
44.4	1.20149	10.02643	4.45173	534.880	5.57470
44.5	1.20202	10.03086	4.46373	536.318	5.56713
44.6	1.20255	10.03528	4.47573	537.758	5.55955
44.7	1.20308	10.03970	4.48775	539.198	5.55195
44.8	1.20361	10.04413	4.49977	540.640	5.54436
44.9	1.20414	10.04855	4.51180	541.722	5.53675
45.0	1.20467	10.05297	4.52384	542.803	5.52913
45.1	1.20520	10.05739	4.53588	543.525	5.52151
45.2	1.20573	10.06182	4.54794	544.971	5.51388
45.3	1.20627	10.06632	4.56004	546.420	5.50628
45.4	1.20680	10.07075	4.57212	547.868	5.49863
45.5	1.20734	10.07525	4.58424	549.320	5.49101
45.6	1.20787	10.07968	4.59633	550.769	5.48335
45.7	1.20840	10.08410	4.60843	552.219	5.47567
45.8	1.20894	10.08860	4.62058	553.675	5.46802
45.9	1.20947	10.09303	4.63270	555.127	5.46033
46.0	1.21001	10.09753	4.64486	556.584	5.45267
46.1	1.21054	10.10196	4.65700	558.039	5.44496
46.2	1.21108	10.10646	4.66918	559.499	5.43728
46.3	1.21162	10.11097	4.68138	560.960	5.42959
46.4	1.21216	10.11548	4.69358	562.422	5.42190
46.5	1.21270	10.11998	4.70579	563.885	5.41419
46.6	1.21323	10.12440	4.71797	565.345	5.40643
46.7	1.21377	10.12891	4.73020	567.028	5.39871
46.8	1.21431	10.13342	4.74244	568.277	5.39098
46.9	1.21485	10.13792	4.75468	569.744	5.38324
47.0	1.21539	10.14243	4.76694	571.213	5.37549
47.1	1.21593	10.14694	4.77921	572.683	5.36773
47.2	1.21647	10.15144	4.79148	574.153	5.35996

Table B–1 (*continued*)

Degrees Brix	Specific Gravity	Lb Total Weight per U.S. Gallon	Lb Solids per U.S. Gallon	G Solids per Liter	Lb Water per U.S. Gallon
47.3	1.21701	10.15595	4.80376	575.625	5.35219
47.4	1.21755	10.16045	4.81605	577.098	5.34440
47.5	1.21809	10.16496	4.82836	578.573	5.33660
47.6	1.21863	10.16947	4.84067	580.049	5.32880
47.7	1.21917	10.17397	4.85298	581.523	5.32099
47.8	1.21971	10.17848	4.86531	583.000	5.31317
47.9	1.22025	10.18299	4.87765	584.479	5.30534
48.0	1.22080	10.18758	4.89004	585.964	5.29754
48.1	1.22134	10.19208	4.90239	587.444	5.28969
48.2	1.22189	10.19667	4.91479	588.929	5.28188
48.3	1.22243	10.20118	4.92717	590.413	5.27401
48.4	1.22298	10.20577	4.93959	591.901	5.26618
48.5	1.22353	10.21036	4.95202	593.391	5.25834
48.6	1.22407	10.21486	4.96442	594.877	5.25044
48.7	1.22462	10.21945	4.97687	596.368	5.24258
48.8	1.22516	10.22396	4.98929	597.857	5.23467
48.9	1.22571	10.22855	5.00176	599.351	5.22679
49.0	1.22626	10.23314	5.01424	600.846	5.21890
49.1	1.22680	10.23765	5.02669	602.338	5.21096
49.2	1.22735	10.24224	5.03918	603.835	5.20406
49.3	1.22790	10.24683	5.05169	605.334	5.19514
49.4	1.22845	10.25142	5.06420	606.833	5.18722
49.5	1.22900	10.25601	5.07562	608.333	5.17929
49.6	1.22955	10.26059	5.08925	609.935	5.17134
49.7	1.23010	10.26518	5.10179	611.337	5.16339
49.8	1.23065	10.26977	5.11435	612.842	5.15542
49.9	1.23120	10.27436	5.12691	614.347	5.14745
50.0	1.23175	10.27895	5.13948	615.854	5.13957
50.1	1.23230	10.28354	5.15205	617.360	5.13149
50.2	1.23285	10.28813	5.16464	618.868	5.12349
50.3	1.23340	10.29272	5.17724	620.378	5.11548
50.4	1.23396	10.29740	5.18989	621.894	5.10751
50.5	1.23451	10.30199	5.20250	623.405	5.09949
50.6	1.23506	10.30658	5.21513	624.919	5.09145
50.7	1.23562	10.31125	5.22780	626.437	5.08345
50.8	1.23617	10.31584	5.24045	627.953	5.07539
50.9	1.23672	10.32043	5.25310	629.468	5.06733
51.0	1.23728	10.32510	5.26580	630.990	5.05930
51.1	1.23783	10.32969	5.27847	632.509	5.05122
51.2	1.23839	10.33436	5.29119	634.033	5.04317
51.3	1.23894	10.33895	5.30388	635.553	5.03507
51.4	1.23950	10.34363	5.31663	637.081	5.02700
51.5	1.24006	10.34830	5.32937	638.608	5.01893
51.6	1.24061	10.35289	5.34209	640.132	5.01080
51.7	1.24117	10.35756	5.35486	641.662	5.00270
51.8	1.24172	10.36215	5.36759	643.188	4.99456
51.9	1.24228	10.36683	5.38038	644.720	4.98645
52.0	1.24284	10.37150	5.39318	646.254	4.97832
52.1	1.24340	10.37617	5.40598	647.788	4.97019
52.2	1.24396	10.38085	5.41880	649.324	4.96205
52.3	1.24452	10.38552	5.43163	650.861	4.95389
52.4	1.24508	10.39019	5.44446	652.399	4.94573
52.5	1.24564	10.39487	5.45731	653.939	4.93756
52.6	1.24620	10.39954	5.47016	655.478	4.92938
52.7	1.24676	10.40421	5.48302	657.019	4.92119
52.8	1.24732	10.40889	5.49589	658.562	4.91300
52.9	1.24788	10.41356	5.50877	660.105	4.90479
53.0	1.24845	10.41832	5.52171	661.655	4.89661
53.1	1.24901	10.42299	5.53461	663.201	4.88838

Table B–1 (*continued*)

Degrees Brix	Specific Gravity	Lb Total Weight per U.S. Gallon	Lb Solids per U.S. Gallon	G Solids per Liter	Lb Water per U.S. Gallon
53.2	1.24957	10.42766	5.54752	664.748	4.88014
53.3	1.25014	10.43242	5.56048	666.301	4.87194
53.4	1.25070	10.43709	5.57341	667.851	4.86368
53.5	1.25127	10.44185	5.58639	669.406	4.85546
53.6	1.25183	10.44652	5.59933	670.957	4.84719
53.7	1.25239	10.45119	5.61229	672.509	4.82890
53.8	1.25296	10.45595	5.62530	674.068	4.83065
53.9	1.25352	10.46062	5.63827	675.623	4.82235
54.0	1.25409	10.46538	5.65131	677.185	4.81407
54.1	1.25465	10.47005	5.66430	678.742	4.80575
54.2	1.25522	10.47481	5.67735	680.305	4.79746
54.3	1.25579	10.47957	5.69041	681.870	4.78916
54.4	1.25636	10.48432	5.70347	683.435	4.78085
54.5	1.25693	10.48908	5.71655	685.003	4.77253
54.6	1.25749	10.49375	5.72959	686.565	4.76416
54.7	1.25806	10.49851	5.74268	688.134	4.75583
54.8	1.25863	10.50327	5.75579	689.705	4.74748
54.9	1.25920	10.50802	5.76890	691.276	4.73912
55.0	1.25977	10.51278	5.78203	692.849	4.73075
55.1	1.26034	10.51754	5.79516	694.422	4.72238
55.2	1.26091	10.52229	5.80830	695.997	4.71399
55.3	1.26148	10.52705	5.82146	697.574	4.70559
55.4	1.26205	10.53181	5.83462	699.151	4.69719
55.5	1.26263	10.53665	5.84784	700.735	4.68881
55.6	1.26320	10.54140	5.86102	702.314	4.68038
55.7	1.26377	10.54616	5.87421	703.895	4.67195
55.8	1.26434	10.55092	5.88741	705.477	4.66351
55.9	1.26491	10.55567	5.90062	707.059	4.65505
56.0	1.26549	10.56051	5.91389	708.650	4.64662
56.1	1.26606	10.56527	5.92712	710.235	4.63815
56.2	1.26664	10.57011	5.94040	711.826	4.62971
56.3	1.26721	10.57487	5.95365	713.414	4.62122
56.4	1.26779	10.57971	5.96696	715.009	4.61275
56.5	1.26836	10.58446	5.98022	716.598	4.60424
56.6	1.26894	10.58930	5.99354	718.194	4.59576
56.7	1.26951	10.59406	6.00683	719.786	4.58723
56.8	1.27009	10.59890	6.02018	721.386	4.57872
56.9	1.27066	10.60366	6.03348	722.980	4.57018
57.0	1.27124	10.60850	6.04685	724.582	4.56165
57.1	1.27182	10.61334	6.06022	726.184	4.55312
57.2	1.27240	10.61818	6.07360	727.787	4.54458
57.3	1.27298	10.62302	6.08699	729.392	4.53603
57.4	1.27356	10.62786	6.10039	730.998	4.52747
57.5	1.27414	10.63270	6.11380	732.604	4.51890
57.6	1.27472	10.63754	6.12722	734.213	4.51032
57.7	1.27530	10.64238	6.14065	735.822	4.50173
57.8	1.27588	10.64722	6.15409	737.432	4.49313
57.9	1.27646	10.65206	6.16754	739.044	4.48452
58.0	1.27704	10.65690	6.18100	740.657	4.47590
58.1	1.27762	10.66174	6.19447	742.271	4.46727
58.2	1.27820	10.66658	6.20795	743.886	4.45863
58.3	1.27878	10.67142	6.22144	745.503	4.44998
58.4	1.27937	10.67634	6.23498	747.125	4.44136
58.5	1.27995	10.68118	6.24849	748.744	4.43269
58.6	1.28053	10.68602	6.26201	750.364	4.42401
58.7	1.28112	10.69095	6.27559	751.991	4.41536
58.8	1.28170	10.69579	6.28912	753.613	4.40667
58.9	1.28228	10.70063	6.30267	755.236	4.39796
59.0	1.28287	10.70555	6.31627	756.866	4.38928

Table B–1 (*continued*)

Degrees Brix	Specific Gravity	Lb Total Weight per U.S. Gallon	Lb Solids per U.S. Gallon	G Solids per Liter	Lb Water per U.S. Gallon
59.1	1.28345	10.71039	6.32984	758.492	4.38055
59.2	1.28404	10.71531	6.34346	760.124	4.37185
59.3	1.28463	10.72024	6.35710	761.759	4.36314
59.4	1.28521	10.72508	6.37070	763.388	4.35438
59.5	1.28580	10.73000	6.38435	765.024	4.34565
59.6	1.28639	10.73492	6.39801	766.661	4.33691
59.7	1.28697	10.73976	6.41164	768.294	4.32812
59.8	1.28756	10.74469	6.42532	769.933	4.31937
59.9	1.28815	10.74961	6.43902	771.575	4.31059
60.0	1.28874	10.75454	6.45272	773.217	4.30182
60.1	1.28933	10.75946	6.46644	774.861	4.29302
60.2	1.28992	10.76437	6.48016	776.505	4.28422
60.3	1.29051	10.76931	6.49389	778.150	4.27542
60.4	1.29110	10.77423	6.50763	779.796	4.26660
60.5	1.29169	10.77915	6.52139	781.445	4.25776
60.6	1.29228	10.78408	6.53515	783.094	4.24893
60.7	1.29287	10.78900	6.54892	784.744	4.24008
60.8	1.29346	10.79392	6.56270	786.395	4.23122
60.9	1.29405	10.79885	6.57650	788.049	4.22235
61.0	1.29465	10.80385	6.59035	789.708	4.21350
61.1	1.29524	10.80878	6.60416	791.363	4.20462
61.2	1.29584	10.81378	6.61803	793.025	4.19575
61.3	1.29643	10.81871	6.63187	794.684	4.18684
61.4	1.29703	10.82372	6.64576	796.348	4.17796
61.5	1.29762	10.82864	6.65961	798.008	4.16903
61.6	1.29822	10.83365	6.67353	799.676	4.16012
61.7	1.29881	10.83857	6.68740	801.338	4.15117
61.8	1.29941	10.84358	6.70133	803.007	4.14225
61.9	1.30000	10.84850	6.71522	804.671	4.13328
62.0	1.30060	10.85351	6.72918	806.344	4.12433
62.1	1.30119	10.85843	6.74309	808.011	4.11534
62.2	1.30179	10.86344	6.74706	809.685	4.10638
62.3	1.30239	10.86844	6.77104	811.360	4.09740
62.4	1.30299	10.87345	6.78503	813.037	4.08842
62.5	1.30359	10.87846	6.79904	814.715	4.07942
62.6	1.30418	10.88338	6.81300	816.388	4.07038
62.7	1.30478	10.88839	6.82702	818.068	4.06137
62.8	1.30538	10.89340	6.84106	819.751	4.05234
62.9	1.30598	10.89840	6.85509	821.432	4.04331
63.0	1.30658	10.90341	6.86915	823.117	4.03426
63.1	1.30718	10.90842	6.88321	824.801	4.02521
63.2	1.30778	10.91342	6.89728	826.487	4.01614
63.3	1.30838	10.91843	6.91137	828.176	4.00706
63.4	1.30899	10.92352	6.92551	829.870	3.99801
63.5	1.30959	10.92853	6.93962	831.561	3.98891
63.6	1.31019	10.93354	6.95373	833.252	3.97981
63.7	1.31080	10.93863	6.96791	834.951	3.97072
63.8	1.31140	10.94363	6.98204	836.644	3.96159
63.9	1.31200	10.94864	6.99618	838.338	3.95246
64.0	1.31261	10.95373	7.01039	840.041	3.94334
64.1	1.31321	10.95874	7.02455	841.738	3.93419
64.2	1.31382	10.96383	7.03878	843.443	3.92505
64.3	1.31442	10.96883	7.05296	845.142	3.91587
64.4	1.31503	10.97393	7.06721	846.850	3.90672
64.5	1.31564	10.97902	7.08147	848.558	3.89755
64.6	1.31624	10.98402	7.09568	850.261	3.88834
64.7	1.31685	10.98911	7.10995	851.971	3.87916
64.8	1.31745	10.99412	7.12419	853.677	3.86993

Table B–1 (*continued*)

Degrees Brix	Specific Gravity	Lb Total Weight per U.S. Gallon	Lb Solids per U.S. Gallon	G Solids per Liter	Lb Water per U.S. Gallon
64.9	1.31806	10.99921	7.13849	855.391	3.86072
65.0	1.31867	11.00430	7.15280	857.106	3.85150
65.1	1.31928	11.00939	7.16711	858.820	3.84228
65.2	1.31989	11.01448	7.18144	860.538	3.83304
65.3	1.32050	11.01957	7.19578	862.256	3.82379
65.4	1.32111	11.02466	7.21013	863.975	3.81453
65.5	1.32172	11.02975	7.22449	865.696	3.80526
65.6	1.32233	11.03484	7.23886	867.418	3.79598
65.7	1.32294	11.03993	7.25323	869.140	3.78670
65.8	1.32355	11.04502	7.26762	870.864	3.77740
65.9	1.32416	11.05012	7.28203	872.591	3.76809
66.0	1.32477	11.05521	7.29644	874.318	3.75877
66.1	1.32538	11.06030	7.31086	876.046	3.74944
66.2	1.32600	11.06547	7.32534	877.781	3.74013
66.3	1.32661	11.07056	7.33978	879.511	3.73078
66.4	1.32723	11.07573	7.35428	881.249	3.72145
66.5	1.32784	11.08082	7.36875	882.983	3.71207
66.6	1.32846	11.08600	7.38328	884.724	3.70272
66.7	1.32907	11.09109	7.39776	886.459	3.69333
66.8	1.32969	11.09626	7.41230	888.201	3.68396
66.9	1.33030	11.10135	7.42680	889.939	3.67455
67.0	1.33092	11.10653	7.44138	891.686	3.66515
67.1	1.33153	11.11162	7.45590	893.426	3.65572
67.2	1.33215	11.11679	7.47048	895.173	3.64631
67.3	1.33277	11.12197	7.48509	896.923	3.63688
67.4	1.33338	11.12706	7.49964	898.667	3.62742
67.5	1.33400	11.13223	7.51426	900.419	3.61797
67.6	1.33462	11.13740	7.52888	902.171	3.60852
67.7	1.33523	11.14249	7.54347	903.919	3.59902
67.8	1.33585	11.14767	7.55812	905.674	3.58955
67.9	1.33647	11.15284	7.57278	907.431	3.58006
68.0	1.33709	11.15802	7.58745	909.189	3.57057
68.1	1.33771	11.16319	7.60213	910.948	3.56106
68.2	1.33833	11.16836	7.61682	912.708	3.55154
68.3	1.33895	11.17354	7.63153	914.471	3.54201
68.4	1.33957	11.17871	7.64624	916.234	3.53247
68.5	1.34020	11.18397	7.66102	918.005	3.52295
68.6	1.34082	11.18914	7.67575	919.770	3.51339
68.7	1.34144	11.19432	7.69050	921.537	3.50382
68.8	1.34206	11.19949	7.70525	923.305	3.49424
68.9	1.34268	11.20466	7.72001	925.073	3.48465
69.0	1.34331	11.20992	7.73484	926.850	3.47508
69.1	1.34393	11.21510	7.74963	928.623	3.46547
69.2	1.34456	11.22035	7.76448	930.402	3.45587
69.3	1.34518	11.22553	7.77929	932.177	3.44624
69.4	1.34581	11.23078	7.79416	933.959	3.43662
69.5	1.34643	11.23596	7.80899	935.736	3.42697
69.6	1.34706	11.24122	7.82389	937.521	3.41733
69.7	1.34768	11.24639	7.83873	939.299	3.40766
69.8	1.34831	11.25165	7.85365	941.087	3.39800
69.9	1.34893	11.25682	7.86852	942.869	3.38830
70.0	1.34956	11.26208	7.88346	944.659	3.37862
70.1	1.35019	11.26734	7.89841	946.451	3.36893
70.2	1.35083	11.27268	7.91342	948.249	3.35926
70.3	1.35147	11.27802	7.92845	950.050	3.34957
70.4	1.35210	11.28327	7.94342	951.844	3.33985
70.5	1.35274	11.28862	7.95848	953.649	3.33014
70.6	1.35338	11.29396	7.97354	955.453	3.32042
70.7	1.35401	11.29921	7.98854	957.251	3.31067

Table B–1 (*continued*)

Degrees Brix	Specific Gravity	Lb Total Weight per U.S. Gallon	Lb Solids per U.S. Gallon	G Solids per Liter	Lb Water per U.S. Gallon
70.8	1.35465	11.30455	8.00362	959.058	3.30093
70.9	1.35529	11.30990	8.01872	960.867	3.29118
71.0	1.35593	11.31524	8.03382	962.677	3.28142
71.1	1.35656	11.32049	8.04887	964.480	3.27162
71.2	1.35720	11.32583	8.06399	966.292	3.26184
71.3	1.35784	11.33117	8.07912	968.105	3.25205
71.4	1.35847	11.33643	8.09421	969.913	3.24222
71.5	1.35911	11.34177	8.10937	971.730	3.23240
71.6	1.35975	11.34711	8.12453	973.546	3.22258
71.7	1.36038	11.35237	8.13965	975.358	3.21272
71.8	1.36102	11.35771	8.15484	977.178	3.20287
71.9	1.36166	11.36305	8.17003	978.998	3.19302
72.0	1.36230	11.36839	8.18524	980.821	3.18315
72.1	1.36293	11.37365	8.20040	982.638	3.17325
72.2	1.36357	11.37899	8.21563	984.463	3.16336
72.3	1.36421	11.38433	8.23087	986.289	3.15346
72.4	1.36484	11.38959	8.24606	988.109	3.14353
72.5	1.36548	11.39493	8.26132	989.937	3.13361
72.6	1.36612	11.40027	8.27660	991.768	3.12367
72.7	1.36675	11.40553	8.29182	993.592	3.11371
72.8	1.36739	11.41087	8.30711	995.424	3.10376
72.9	1.36803	11.41621	8.32242	997.259	3.09379
73.0	1.36867	11.42155	8.33773	999.094	3.08382
73.1	1.36930	11.42681	8.35300	1000.923	3.07381
73.2	1.36994	11.43215	8.36833	1002.760	3.06382
73.3	1.37058	11.43749	8.38368	1004.600	3.05381
73.4	1.37121	11.44275	8.39898	1006.433	3.04377
73.5	1.37185	11.44809	8.41435	1008.275	3.03374
73.6	1.37249	11.45343	8.42972	1010.116	3.02371
73.7	1.37312	11.45869	8.44505	1011.953	3.01364
73.8	1.37376	11.46403	8.46045	1013.799	3.00358
73.9	1.37440	11.46937	8.47586	1015.645	2.99351
74.0	1.37504	11.47471	8.49129	1017.494	2.98342
74.1	1.37567	11.47997	8.50666	1019.336	2.97331
74.2	1.37631	11.48531	8.52210	1021.186	2.96321
74.3	1.37695	11.49065	8.53755	1023.038	2.95310
74.4	1.37758	11.49591	8.55296	1024.884	2.94295
74.5	1.37822	11.50125	8.56843	1026.738	2.93282
74.6	1.37886	11.50659	8.58392	1028.594	2.92267
74.7	1.37949	11.51184	8.59934	1030.442	2.91250
74.8	1.38013	11.51718	8.61485	1032.300	2.90233
74.9	1.38077	11.52253	8.63037	1034.160	2.89216
75.0	1.38142	11.52795	8.64596	1036.028	2.88199
75.1	1.38207	11.53337	8.66156	1037.897	2.87181
75.2	1.38273	11.53888	8.67724	1038.776	2.86164
75.3	1.38338	11.54431	8.69287	1041.649	2.85144
75.4	1.38404	11.54981	8.70856	1043.529	2.84125
75.5	1.38470	11.55532	8.72427	1045.412	2.83105
75.6	1.38535	11.56075	8.73993	1047.288	2.82082
75.7	1.38601	11.56625	8.75565	1049.172	2.81060
75.8	1.38666	11.57168	8.77133	1051.051	2.80035
75.9	1.38732	11.57719	8.78709	1052.939	2.79010
76.0	1.38798	11.58269	8.80284	1054.827	2.77985
76.1	1.38863	11.58812	8.81856	1056.710	2.76956
76.2	1.38929	11.59363	8.83435	1058.602	2.75928
76.3	1.38994	11.59905	8.85008	1060.487	2.74897
76.4	1.39060	11.60456	8.86588	1062.381	2.73868
76.5	1.39126	11.61006	8.88170	1064.276	2.72836
76.6	1.39191	11.61549	8.89747	1066.166	2.71802

Table B–1 (*continued*)

Degrees Brix	Specific Gravity	Lb Total Weight per U.S. Gallon	Lb Solids per U.S. Gallon	G Solids per Liter	Lb Water per U.S. Gallon
76.7	1.39257	11.62100	8.91331	1068.064	2.70769
76.8	1.39322	11.62642	8.92909	1069.955	2.69733
76.9	1.39388	11.63193	8.94495	1071.855	2.68698
77.0	1.39454	11.63744	8.96083	1073.758	2.67661
77.1	1.39519	11.64286	8.97665	1075.654	2.66621
77.2	1.39585	11.64837	8.99254	1077.558	2.65583
77.3	1.39650	11.65379	9.00838	1079.456	2.64541
77.4	1.39716	11.65930	9.02430	1081.364	2.63500
77.5	1.39782	11.66481	9.04023	1083.273	2.62458
77.6	1.39847	11.67023	9.05610	1085.174	2.61413
77.7	1.39913	11.67574	9.07205	1087.086	2.60369
77.8	1.39978	11.68116	9.08794	1088.990	2.59322
77.9	1.40044	11.68667	9.10392	1090.905	2.58275
78.0	1.40110	11.69218	9.11990	1092.819	2.57228
78.1	1.40175	11.69760	9.13583	1094.728	2.56177
78.2	1.40241	11.70311	9.15183	1096.645	2.55128
78.3	1.40306	11.70854	9.16779	1098.558	2.54075
78.4	1.40372	11.71404	9.18381	1100.478	2.53023
78.5	1.40438	11.71955	9.19985	1102.400	2.51970
78.6	1.40503	11.72498	9.21583	1104.314	2.50915
78.7	1.40569	11.73048	9.23189	1106.239	2.49859
78.8	1.40634	11.73591	9.24790	1108.157	2.48801
78.9	1.40700	11.74142	9.26398	1110.084	2.47744
79.0	1.40766	11.74692	9.28007	1112.012	2.46685
79.1	1.40831	11.75235	9.29611	1113.934	2.45624
79.2	1.40897	11.75785	9.31222	1115.865	2.44563
79.3	1.40962	11.76328	9.32828	1117.789	2.43500
79.4	1.41028	11.76879	9.34442	1119.723	2.42437
79.5	1.41094	11.77429	9.36056	1121.657	2.41373
79.6	1.41159	11.77972	9.37666	1123.586	2.40306
79.7	1.41225	11.78523	9.39283	1125.524	2.39240
79.8	1.41290	11.79065	9.40894	1127.454	2.38171
79.9	1.41356	11.79616	9.42513	1129.394	2.37103

ALCOHOL BY EBULLIOMETER

This analytical procedure should not be used for wines which have an alcohol content greater than 14% by volume, or for wines which have a Balling of more than 0.0°.

The rationale of the ebulliometer (Fig. B–9) is to compare the boiling point of water with the boiling point of a wine sample—the difference being due to the alcohol content in the wine. In that the boiling point of water ranges from about 99°C to 101°C from day to day, in response to rising and falling barometric pressure, it is necessary to recheck the boiling point of water in the ebulliometer every hour or so.

Apparatus and Reagents

Salleron–DuJardin ebulliometer with thermometer and circular slide scale
100-mL graduated cylinder

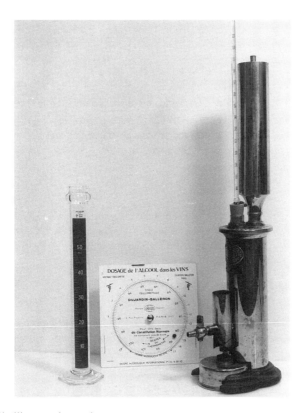

Fig. B–9 Ebulliometer in service.

Distilled or deionized water
Cold tap water

Procedure

1. Rinse all inside surfaces of the ebulliometer with distilled or deionized water. Drain valve and close.
2. Fill upper reflux condenser jacket with cold tap water.
3. Measure 50 mL of distilled or deionized water into a clean 100-mL graduated cylinder and carefully pour into lower chamber inlet.
4. Very carefully insert thermometer into lower chamber inlet, holding top of thermometer in one hand in a pendulum effect and holding rubber stopper portion in the other hand. Slowly and gently twist rubber stopper into position for a snug fit. These thermometers are very delicate and break rather easily.
5. Ignite ethanol burner and carefully position under lower chamber in the proper position.

6. Observe thermometer mercury rising until it stops and holds for 15–20 s at the same temperature. Remove ethanol burner and close carefully to extinguish flame.
7. Remove thermometer carefully in reverse manner to Step 4 above. Hold in vertical position until the mercury drops from the capillary. Dry with towel carefully and place upright in a safe place.
8. Adjust circular slide scale to indicate the boiling point temperature of the water.
9. Empty the ebulliometer carefully and rinse inner surfaces with a few milliliters of the wine sample to be analyzed. Drain the instrument and fill upper reflux condenser with cold tap water. Ensure that no water goes down the inner tube.
10. Rinse the 100-mL graduate with a few milliliters of the wine sample to be analyzed. Empty and refill with 50 mL of the wine.
11. Repeat Steps 4–7.
12. Compare reading of the thermometer to corresponding alcohol percentage on the circular slide scale. For example, with water boiling at 99.8°C and a wine boiling at 91.1°C, the alcohol would be 12.1% by volume.

EXTRACT BY NOMOGRAPH

Extract is virtually the same as Brix—except extract infers that the alcohol of the wine has been removed and replaced with water. As this is a rather tedious process, extract can be determined by carefully using a special graph called a nomograph (See Fig. B–10.)

Use of a nomograph requires the prior knowledge of alcohol content and Balling of the wine to be analyzed. The procedure is simply to find the extract by placing a straightedge, such as a ruler, so as to intersect the known alcohol and Balling levels.

For example, using the following nomograph, an alcohol content of 12.0% by volume and a Balling of $-2.0°$ results in an extract of 2.0 g/100 mL of soluble solids.

TOTAL ACIDITY

This analysis measures the unknown concentration of all wine acids by neutralizing those acids with a known concentration of a basic (alkaline) reagent. The method is shown in Fig. B–11.

Apparatus and Reagents

10-mL volumetric pipette
250-mL wide-mouth Erlenmeyer flask
25- or 50-mL (in 0.1-mL gradations) buret with stopcock

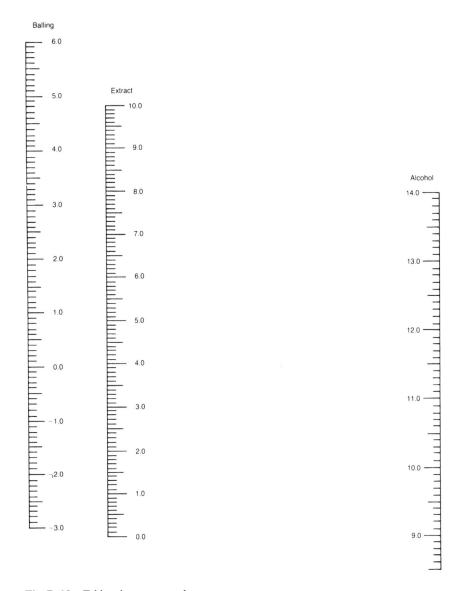

Fig. B–10 Table wine nomograph.
Courtesy: J.M. Vahl, Am. J. Enol. Vitic., Volume 30, No. 3, 1979. 30 (3) 1979.

340   *Analytical Procedures*

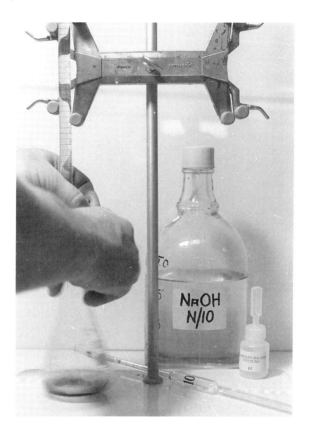

Fig. B–11 Total acidity titration.

$1/10$ N sodium hydroxide reagent
1% phenolphthalein indicator solution
Distilled or deionized water

Procedure

1. If the sample has been taken from a freshly pressed lot of juice or must, there may be too many suspended solids for the pipette to allow flow through the narrow capillary at the tip. In this case, the juice sample should be carefully centrifuged or filtered through neutral paper. Adjust sample temperature to 68°F or to whatever temperature is indicated on the 10-mL volumetric pipette.
2. Pipette the 10-milliliter sample into clean 250 mL wide-mouth Erlenmeyer flask and add 5 drops of 1% phenolphthalein indicator into sample.
3. Fill burette with $1/10$ N sodium hydroxide reagent and slowly drain down to the 0.0 starting point. Ensure that there is no air remaining in the outlet capillary.

4. Titrate carefully drop by drop until the slight pink color from the indicator turns pink (green for red wine) and holds for at least 15 s. Note burette reading for milliliters of $1/10$ N sodium hydroxide used.
5. Grams/100 mL total acidity is found by using the Total Acidity Table (Table B–2). For example, when 9.0 mL of $1/10$ N sodium hydroxide is used in titration the resulting T.A. is 0.675 g/100 mL.

Table B–2   Total Acidity Table (g/100 mL Expressed as Tartaric Acid)

1/10 N NaOH	g/100 ml T.A.	1/10 N NaOH	g/100 ml T.A.	1/10 N NaOH	g/100 ml T.A.
5.1	.383	9.1	.683	13.1	.983
5.2	.390	9.2	.690	13.2	.990
5.3	.398	9.3	.698	13.3	.998
5.4	.405	9.4	.705	13.4	1.005
5.5	.413	9.5	.713	13.5	1.013
5.6	.420	9.6	.720	13.6	1.020
5.7	.428	9.7	.728	13.7	1.028
5.8	.435	9.8	.735	13.8	1.035
5.9	.443	9.9	.743	13.9	1.043
6.0	.450	10.0	.750	14.0	1.050
6.1	.458	10.1	.758	14.1	1.058
6.2	.465	10.2	.765	14.2	1.065
6.3	.473	10.3	.773	14.3	1.073
6.4	.480	10.4	.780	14.4	1.080
6.5	.488	10.5	.788	14.5	1.088
6.6	.495	10.6	.795	14.6	1.096
6.7	.503	10.7	.803	14.7	1.103
6.8	.510	10.8	.810	14.8	1.110
6.9	.518	10.9	.818	14.9	1.118
7.0	.525	11.0	.825	15.0	1.125
7.1	.533	11.1	.833	15.1	1.133
7.2	.540	11.2	.840	15.2	1.140
7.3	.548	11.3	.848	15.3	1.148
7.4	.555	11.4	.855	15.4	1.155
7.5	.563	11.5	.863	15.5	1.163
7.6	.570	11.6	.870	15.6	1.170
7.7	.578	11.7	.878	15.7	1.178
7.8	.585	11.8	.885	15.8	1.185
7.9	.593	11.9	.893	15.9	1.193
8.0	.600	12.0	.900	16.0	1.200
8.1	.608	12.1	.908	16.1	1.208
8.2	.615	12.2	.915	16.2	1.215
8.3	.623	12.3	.923	16.3	1.223
8.4	.630	12.4	.930	16.4	1.230
8.5	.638	12.5	.938	16.5	1.238
8.6	.645	12.6	.945	16.6	1.245
8.7	.653	12.7	.953	16.7	1.253
8.8	.660	12.8	.960	16.8	1.260
8.9	.668	12.9	.968	16.9	1.268
9.0	.675	13.0	.975	17.0	1.275

VOLATILE ACIDITY BY CASH STILL

The rationale for this method of determining volatile acidity (V.A.) is to separate the volatile, or distillable, acids from the fixed acids in the Cash still (See Fig. B–12.) The distilled acids are then titrated separately in the same manner as for a total acidity determination.

Apparatus and Reagents

Cash volatile acid apparatus complete with condenser, support clamps, and rods
$1/4$-in.-inside diameter Tygon® tubing
10-mL volumetric pipette
250-mL Erlenmeyer flask with rubber stopper to fit

Fig. B–12 Cash volatile acid still in service.

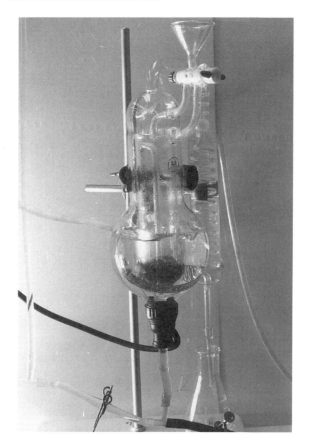

25 or 50 mL (in 0.1-mL gradations) burette with stopcock
$1/10$ N sodium hydroxide reagent
1% phenolphthalein indicator
Distilled or deionized water
Continuous cold tap water connected into condenser and out into drain

Procedure

1. Turn stopcock on Cash still so that passage runs from funnel to inner tube.
2. Adjust sample temperature to 68°F or to whatever temperature is indicated on 10-ml volumetric pipette.
3. Fill lower (outer) pot with distilled water up to a level which is about 1 in. above heating element.
4. Pipette sample into funnel and drain into inner tube of Cash still. Wash funnel into inner tube with 10–15 ml of distilled water.
5. Turn stopcock so that passage goes to outer tube. Place a clean 250-ml Erlenmeyer flask under condenser outlet. (It is recommended to place about 1 in. of Tygon® tubing on the condenser outlet in order to avoid chipping of both the outlet and the flask.)
6. Commence cold tap water running through condenser at a moderate rate and turn on electricity to heating coil in the lower pot.
7. Small bubbles should form upon the heating coil which should be allowed to boil off, keeping the stopcock turned to the outer tube position. When "spitting" commences, turn stopcock to the horizontal position closing off both tubes.
8. Distill over approximately 100 mL of distillate. Be sure that distillate runs out of condenser at a temperature not exceeding 80°F. Make several trial runs in order to determine ideal cold water flow rate. Increase condenser water flow as needed, but not so fast to break the condenser from excessive pressure.
9. After distillate is collected, turn off electric power to heating coil and turn stopcock to inner tube passage. Remove the 250-mL Erlenmeyer flask carefully and close with rubber stopper.
10. Open outlet tube to drain and add approximately 50 mL distilled water into funnel. The Cash apparatus should "cough" as it cleans itself by the vacuum created by water draining out of the lower pot. Repeat twice more, spacing the "coughing" about equally at the beginning, middle, and end of outer tube drainage. Turn off condenser water and allow apparatus to drain.
11. Remove stopper from distillate flask, add 5 drops phenolphthalein indicator solution. Titrate very carefully with $1/10$ N sodium hydroxide solution from burette in the same manner as for the total acidity analysis above. Note burette reading for milliters of $1/10$ N sodium hydroxide used.
12. Grams/100 mL volatile acidity is found by using the Volatile Acidity Table (Table B–3). For example, when 0.5 mL of $1/10$ N sodium hydroxide is used in titration the resulting V.A. is 0.030 g/100 mL.

Table B–3 Volatile Acidity Table (g/100 mL Expressed as Acetic Acid) (For use when distilling 10 mL sample in Cash V.A. apparatus and titrating using N/10 NaOH)

(Ml) NaOH	G/100 ml VA
.1	.006
.2	.012
.3	.018
.4	.024
.5	.030
.6	.036
.7	.042
.8	.048
.9	.054
1.0	.060
1.1	.066
1.2	.072
1.3	.078
1.4	.084
1.5	.090
1.6	.096
1.7	.102
1.8	.108
1.9	.114
2.0	.120
2.1	.126
2.2	.132
2.3	.138
2.4	.144
2.5	.150
2.6	.156
2.7	.162
2.8	.168
2.9	.174
3.0	.180

FREE SULFUR DIOXIDE

Free sulfur dioxide, as the term suggests, is dynamic and can quickly change in concentration. Conditions which contribute to rapidly diminishing free sulfur dioxide in wine are agitation and warmer temperatures. Consequently, wine samples taken for free sulfur dioxide determination should be analyzed as soon as possible.

There are inexpensive kits available for instant free SO_2 analysis. These can serve as quick checks but should not be used for finite determinations. The most accurate and dependable method for analyzing free SO_2 is by a modified Monier–Williams procedure. This requires a rather complex glassware apparatus and extensive process. The most common procedure for free SO_2 analysis in a small winery is by the Ripper method (Fig. B–13). Although this is simplistic, results should be given a confidence interval of at least 10% for white wines and twice that for reds.

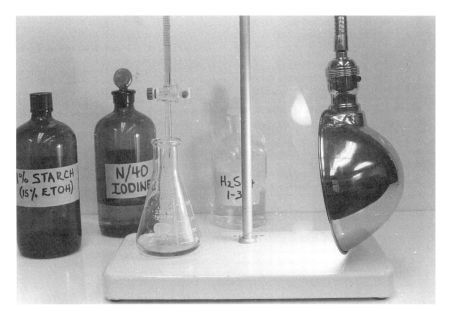

Fig. B–13 Free sulfur dioxide titration by Ripper method.

Apparatus and Reagents

20-mL volumetric pipette
250-mL wide-mouth Erlenmeyer flask with fitted stopper
25- or 50-mL (in .1 mL gradations) burette with stopcock
$1/40$ N iodine reagent (stored in a refrigerator)
1% starch indicator solution
Bicarbonate of soda
25% sulfuric acid
25% sodium hydroxide
Distilled or deionized water

Procedure

1. Adjust sample temperature to 68°F, or to whatever temperature indicated on 20-ml volumetric pipette.
2. Pipette the sample into a clean 250-ml wide-mouth Erlenmeyer flask.
3. Add 5 mL of 25% sulfuric acid, then a pinch of bicarbonate of soda and 5 mL of 1% starch indicator solution.
4. Fill burette with 1/40 N iodine and drain down to the 0.0 mark. Ensure that no air remains in outlet capillary.

Analytical Procedures

5. Titrate carefully with 1/40 N iodine from burette until color changes to light blue (blue–green in reds) and holds for at least 15 s. Note burette reading for milliliters of $1/40$ N iodine used. (Note: Addition of up to 20 mL of distilled or deionized water and/or an appropriately situated light may help with some very dark or turbid samples.)
6. Milligrams per liter (parts per million) of free sulfur dioxide is found by using the Sulfur Dioxide Table (Table B–4). For example, when 1.4 mL of 1/40 N iodine reagent is used in titration, the resulting free SO_2 is 56 mg/L (ppm).

Table B–4 Sulfur Dioxide Table (20 mL Sample $\times$ 1/40 N Iodine $\times$ 1% Starch Indicator)

Ml Iodine	Ppm SO_2	Ml Iodine	Ppm SO_2
.1	4	4.1	164
.2	8	4.2	168
.3	12	4.3	172
.4	16	4.4	176
.5	20	4.5	180
.6	24	4.6	184
.7	28	4.7	188
.8	32	4.8	192
.9	36	4.9	196
1.0	40	5.0	200
1.1	44	5.1	204
1.2	48	5.2	208
1.3	52	5.3	212
1.4	56	5.4	216
1.5	60	5.5	220
1.6	64	5.6	224
1.7	68	5.7	228
1.8	72	5.8	232
1.9	76	5.9	236
2.0	80	6.0	240
2.1	84	6.1	244
2.2	88	6.2	248
2.3	92	6.3	252
2.4	96	6.4	256
2.5	100	6.5	260
2.6	104	6.6	264
2.7	108	6.7	268
2.8	112	6.8	272
2.9	116	6.9	276
3.0	120	7.0	280
3.1	124	7.1	284
3.2	128	7.2	288
3.3	132	7.3	292
3.4	136	7.4	296
3.5	140	7.5	300
3.6	144	7.6	304
3.7	148	7.7	308
3.8	152	7.8	312
3.9	156	7.9	316
4.0	160	8.0	320

Table B–5 Approximate Effective Sulfur Dioxide Conversion Table

ppm molecular SO_2	ppm @ pH 3.10	ppm @ pH 3.20	ppm @ pH 3.30	ppm @ pH 3.40	ppm @ pH 3.50	ppm @ pH 3.60
0.5	10	13	16	20	25	31
0.8	16	21	26	32	30	50
1.1	22	29	36	44	55	69
1.4	28	37	46	56	70	88
1.7	34	45	56	68	85	108
2.0	40	53	66	80	100	126

It should be fully understood that the pH level in wine has a major influence on the effectiveness of free sulfur dioxide as a preservative. The active portion of free SO_2 is referred to as "molecular sulfur dioxide" which is correlated to the pH level in every wine. Consequently, the results from every free SO_2 determination should be evaluated on the basis of the Approximate Effective Sulfur Dioxide Table (Table B–5).

For example, if the molecular SO_2 desired is 1.4 ppm, then sufficient sulfite would need to be added to reach a free sulfur dioxide level of 37 ppm at a pH of 3.20. To reach the same 1.4 molecular level at a pH of 3.40 would require sufficient sulfite to reach a free sulfur dioxide level of 56 ppm.

TOTAL SULFUR DIOXIDE

Each wine has some portion of the total amount of sulfur dioxide existing in combination with other wine components. This is called "bound" sulfur dioxide. The combination of bound and free SO_2 is called "total SO_2" and is limited to a maximum of 350 ppm by ATF regulations in 27 CFR, Part 4, Section 4.22.

One of the most popular methods for determining total sulfur dioxide is by the following modified Ripper method.

Apparatus and Reagents

Same as listed for Free Sulfur Dioxide.

Procedure

1. Adjust sample temperature to 68°F, or to whatever temperature is indicated on 20-ml volumetric pipette.
2. Pipette the sample into clean 250-ml wide-mouth Erlenmeyer flask and add 5 mL of 25% sodium hydroxide. Seal with rubber stopper. Swirl for a few seconds and set aside for about 15 min.

3. Remove stopper, add 5 mL of 25% sulfuric acid, a pinch of bicarbonate of soda, and 5 drops of 1% starch solution to sample in flask.
4. Proceed with titration and determination in the same manner as for Free Sulfur Dioxide.

SENSORY EVALUATION

The single most important element of analysis in quality control is in sensory examination (see Figs. B–14 and B–15) and criticism. There is no substitute for building learning and experience in wine judging, which is achieved, of course, through learning proper procedure and continued practice.

Fig. B–14 Human sensory organs.

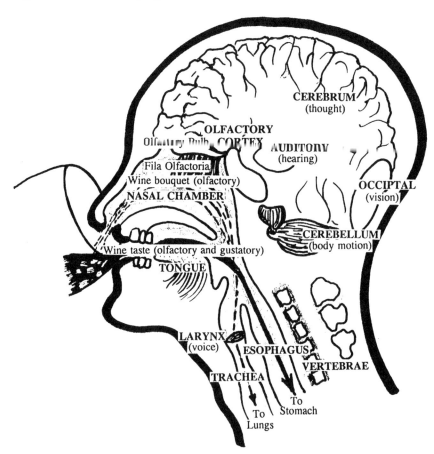

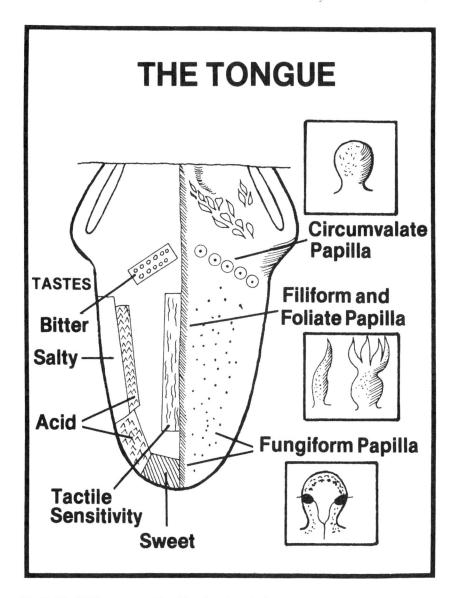

Fig. B–15 Different types of papillae found on the human tongue.
Adapted from J. Puisais and R.L. Chabanon. 1974. Initiation into the Art of Wine Tasting, Interpublish, Madison, WI.

Beginning winemakers should sample a large number of positive and negative wine constituents and components in water and/or water–ethanol solutions. This will help in learning to identify these compounds without the masking effects in more complex wine solutions. The ability to commit these to sensory memory will help develop the critical eye, nose, and palate which are essential to long-term success in commercial wine production.

Apparatus and Conditions

Several dozen identical tulip-shape wine glasses
Solid white tabletop or countertop
Mild cheese
Salt-free crackers
Noncarbonated bottled water
Hot water for cleaning
Deionized water for rinsing
Drain racks for drying
Cool room temperature
Noiseless and odorless room maintained at cool temperature and moderate humidity with adjustable incandescent light
Sufficient pencils and score sheets

Procedure

Visual Mode

The visual mode (Fig. B–16) in wine sensory evaluation is divided into two major categories: clarity and color—and, in the case of sparkling wines, effervescence too.

The best procedure in evaluating clarity is to hold the wine glass by the stem or base so that a constant light source behind the glass can filter through the wine. Wine with perfect clarity has no trace of suspended particles or lint. The four generally accepted echelons of wine clarity are as follows:

Brilliant
Clear
Hazy
Cloudy

To replicate a hazy condition, fill an 8-oz. wine glass half full with water and add 1 drop of whole milk—2 or 3 more drops will render a cloudy mixture.

The first concern for color is its *hue,* or an identity of specific color value such as straw–gold or ruby red. This is followed by making a judgment on color *inten-*

Fig. B–16 The visual mode.
Photo: David King, Purdue University Agricultural Communications.

sity—the quantity of hue present in the wine. Color evaluations are best made upon a stark white background such as plain paper or a table napkin. By looking downward through the glass, one can perceive variances in both hue and intensity as stationary light passes through different depths of the wine. Obviously, the wine will have the same color hue throughout, but the intensity will be denser in the center of the glass where the wine is deeper. Conversely, intensity will be lighter around the edge of the glass where the wine is more shallow. The dilution of intensity in the shallow portion of the glass can reveal nuances of color, such as tones of tawny brown in older red wines, which are hidden by the heavier density in the center of the glass.

Some judges prefer to make this examination by holding the glass at various angles and looking from side to side through the bowl of the glass. Whichever approach is adopted, a judge should remain consistent.

Following are some examples of typical color judgments:

Wine Type	Color Hue	Color Intensity
Dry Vermouth	Pale celery	Extremely light
Sauvignon Blanc	Pale straw	Very light
Chardonnay	Straw	Light
Sauternes	Golden	Moderate

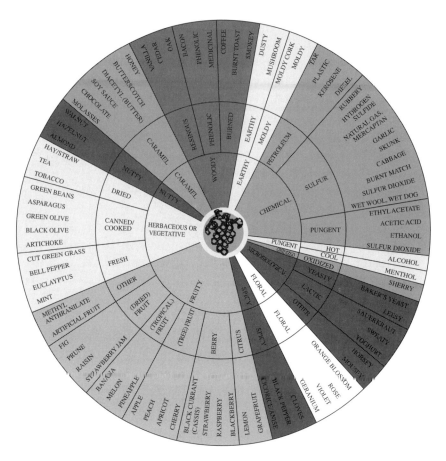

Fig. B–17 The aroma wheel.

COPYRIGHT 1990 A.C. Noble
For additional information about the aroma wheel, please contact A.C. Noble.
Department of Viticulture and Enology, University of California, Davis, CA 95616,
acnoble @ucdavis.edu or visit www.wineserver.edu

Using the Wine Aroma Wheel

For each wine variety, which has a distinctive flavor, specific descriptors can be used. Below are listed terms for aromas most often encountered in varietal wines. The terms which are not underlined come from the grape; those which are underlined come from winemaking practices such as malolactic fermentation (buttery) or oak aging (vanilla).

Blush	Pink	Light
Anjou Rosé	Rosé	Moderate
Pinot Noir	Crimson red	Light
Merlot	Scarlet red	Moderate
Cabernet Sauvignon	Garnet red	Dark
Port	Ruby red	Dense
Tawny Port	Amber–red	Moderate
Oloroso Sherry	Amber	Dark

Olfactory Mode

The olfactory mode is the single most important element in wine judging. It involves the most sensitive human organ, the nose, and is generally the primary factor in deciding on wine quality. For all these reasons, the entire scope of the olfactory mode comprises the most points on every good cardinal scale of wine judging.

The first concern is that of *bouquet*—often referred to as the "nose" by wine judges. Wine bouquet is perhaps best compared to a bouquet of flowers. Each flower in the bouquet expresses its own characteristic odor which contributes to a complex array of smells overall. In the case of wine, that part of the bouquet which is grape flavor is called the *aroma*. The aroma, combined with odors resulting from the winemaking process, such as malolactic fermentation and barrel aging, results in a complex array of smells collectively referred to as bouquet. A typical Chardonnay wine, for example, may have a slight ripe olive and herbal aroma. The nose may also detect a "buttery" flavor resulting from a special malolactic bacterial fermentation and a "vanilla" flavor due to oak barrel aging. The combined olive–herbal aroma and buttery–vanilla flavors comprise the overall bouquet of this Chardonnay. As mentioned earlier, a properly trained human olfactory and temporal system can identify, classify, store, and recall hundreds of different flavors. Some of these are delicious experiences; others are unpleasant odors resulting from winemaking problems. An experienced wine judge will be familiar with both.

◄ **Fig. B–17** *(continued).*

CABERNET SAUVIGNON, MERLOT, MALBEC, CABERNET FRANC: Berry, Vegetative (Bell pepper, Asparagus, Olives), Mint, Black Pepper, *Vanilla, Buttery, Soy.*
PINOT NOIR: Berry, Berry Jam (Strawberry), *Vanilla, Buttery.*
ZINFANDEL: Berry, Black Pepper, Raisin, *Soy, Buttery, Vanilla.*
CHARDONNAY: Fruity (Apple, Peach, Citrus, Pineapple), Spicy, *Cloves, Vanilla, Buttery.*
SAUVIGNON BLANC: Floral, Fruity (Citrus, Peach, Apricot), Vegetative (Bell Pepper, Asparagus), *Vanilla, Buttery.*
WHITE RIESLING: Floral, Fruity (Citrus, Peach, Apricot, Pineapple), Honey.
GEWÜRZTRAMINER: Floral, Fruity (Citrus, Grapefruit, Peach), Honey, Spicy.
CHENIN BLANC: Floral, Fruity (Peach, Citrus).

One system of wine flavor terminology is organized in the form of an "aroma wheel" (Fig. B–17). It employs a three-tier system of primary, secondary, and tertiary levels of flavor definition. Although not yet perfect, it does establish a basis on which people can now communicate effectively in discussing the many varied olfactory impressions which exist in the vast world of wine.

In many white grape varieties we find *primary floral aromas*. Among these are acacia, grapefruit, honeysuckle, linden, peach, pear, rose, peony, and violets. These flavor values are most often found in grapes which have been harvested slightly before maturity, prior to when *primary fruit aromas have* formed during full maturation.

Many of the primary floral and fruit aromas exist in the form of higher terpene alcohols such as citronellol, linalol, and geraniol. These are commonly found in Johannisberg Riesling, Gewürztraminer, Vidal Blanc, Vignoles, and most of the Muscat varieties.

Other aromas are identified in the form of esters. Among these are the isoamyl acetate associated with bananas, ethyl propionate associated with apples, and methyl anthranilate found in the native American *Vitis labrusca* cultivars such as Niagara and Concord.

Primary fruit aromas in red grapes are often identified as cherry, black currant or cassis, strawberry, raspberry, and plum, which are often used as flavor descriptives in evaluating wines made from Chambourcin, Gamay, and Pinot Noir. Many of these are found in ester forms. Strawberry flavor is attributed principally to several butyrate compounds. Raspberry is identified with ethyl caproate.

Another flavor group is called *primary vegetal aromas*. Good examples of these are anise, green pepper, tobacco leaf, and mint. This group is often structured in the form of carbonyl compounds, such as the methoxy-isobutyl-pyrazine which is the principal aroma component associated with bell peppers. These compounds can be very complex such as the aroma associated with Cabernet Sauvignon, 2-methoxy-3-isobutylpyrazine. Vegetal aromas can also exist as phenols, such as cinnamic acid found in grass and tobacco. This character is most often desired in very heavy red wines such as those traditionally made from Cabernet Sauvignon, Chancellor, and Merlot.

Yet another group, *primary wood aromas,* is exemplified by briar, cedar, hazelnut, resin, oak, and eucalyptus. These generally exist in the form of phenolic compounds such as vanillin which is the aroma component associated with oak. Wood aromas can also be carbonyl compounds such benzaldehyde—the aroma associated with bitter almond.

The first phase of olfactory judgment commences with the examination of bouquet and should be made while the wine is in a glass on the countertop or tabletop. The judge will be smelling only the vapors which the wine gives off at rest. The nose is inserted in the glass and several deep sniffs are taken and a mental note made of each reaction. Sometimes, closing the eyes may help concentration (See Fig. B–18.) These primary judgment notes are then recorded.

Fig. B–18 The olfactory mode.
Photo: David King, Purdue University Agricultural Communications.

After a few seconds rest, another examination of bouquet should be made immediately after the wine has been fully swirled around the inside walls of the glass. Grasp the glass stem and gently rotate the bowl. This significantly increases the surface area of wine from which flavor esters can evaporate into the chamber for the nose to receive. Again the nose in inserted in the glass and several deep sniffs are taken and a mental note made of each reaction. These secondary judgment notes are also recorded.

The second phase of olfactory judgment takes place in the mouth, with the taste buds employed as evaluators of wine flavor constituents. The mouth should be freshly rinsed clean with water and then a small sip of wine taken from the glass and rolled around the mouth. The flavors may, or may not, be the same as identified in the nose. The palate may identify new flavors altogether, or the same flavors with different intensity. These primary judgment notes are recorded and the wine spit out into the bucket.

After a few seconds rest, take a small nibble of cheese and unsalted cracker; the mouth is rinsed clean again for making the second flavor judgment. This time a sip of wine is drawn in the mouth along with a little air—almost like whistling in reverse. Roll the wine around in the mouth, drawing in air again to stimulate taste buds and accentuate shy flavors. Record secondary judgment impressions and spit into the bucket as before.

356   *Analytical Procedures*

Fig. B–19   The gustatory mode.
Photo: David King, Purdue University Agricultural Communications.

Gustatory Mode

The *gustatory mode* (Fig. B–19) evaluates wine on the palate, principally the "feel" of the wine. Conventional wisdom often relates this to four human gustatory senses—*acidity, sweetness, bitterness,* and *saltiness.* These senses are supplemented by *tactile,* or mouth feel, sensations of touch and temperature on the palate.

There is comparatively little gustatory sensation in the nose. High alcohol content, carbon dioxide bubbles, and some measures of acidity and phenolic compounds are detectable in the nose and have flavor. Consequently, experienced judges usually make olfactory and gustatory evaluations simultaneously while the wine is on the palate. For beginners, it helps to make gustatory judgments more deliberately until an expertise is developed.

The upper surface of the tongue is constructed of several thousand various types of *papillae,* which are expanded skin protuberances—each papilla containing about 250 taste buds. Each bud delivers taste information directly to the brain via special transmitter nerves.

Total acidity is the sum total of acids present in wine. These consist primarily of *tartaric, malic, lactic,* and *acetic* acids. Typically total acidity in wine is measured between 0.5% and 1.0% of total wine composition. Tartaric acid has very little perceptible flavor. Malic is a bit green apple-like in character, whereas lactic has a buttery–cheesy flavor, and acetic acid is vinegary.

Acidity is perceived mostly on the sides of the tongue by *foliate papilla* cells, in various degrees of tartness—from which the term *tartaric* acid is derived. The following is a typical range of acidity expressed in wine judging:

Insipid
Bland
Balanced
Tart
Harsh

Sweetness can mask total acidity on the palate. A beginning wine enthusiast should devote some time studying the interaction of these two gustatory influences. One suggestion is to pour three glasses of the same wine. In one, dissolve a pinch of tartaric acid and in another dissolve a pinch of sugar. Compare all three to ascertain whether or not differences can be detected. If not, dissolve another pinch and try it again. Continue until the additions are readily obvious on the palate. In a fourth glass, blend together equal parts of the acid and sugar-treated wines. Compare the resulting blend to both the high-acid wine and the high-sugar wine. This exercise can help illustrate how acidity and sweetness serve to neutralize each other on the palate—and also to gather meaning from the term "sugar–acid balance." The following is a typical range of sweetness expressed in wine judging:

Bone dry
Dry
Semidry
Sweet
Cloying

Glucose, or common corn sugar, is the least sweet of the common sugars. Sucrose, which is cane sugar, is a bit sweeter on the palate, but fructose, or natural fruit sugar, has the greatest gustatory sensation of sweetness.

The *bitterness* in wines is usually attributed to phenolic compounds extracted mostly from grape seeds. A good example of this gustatory sense can be experienced by biting into a few grape seeds and chewing them for a minute or so before spitting them out.

A related gustatory term is *astringency*—also resulting from phenolic compounds extracted principally from grape skins and stems. Wine judges will often use the term *tannin* in relating to puckery astringency—a rather leathery, aspirin-like sensation on the palate.

Tannins are a special group of phenols which serve to slow the wine-aging process by inhibiting oxidation potential. Wine aging is, thus, wine oxidation.

Vintners desiring long-lived red wines purposely conduct fermentations with extended periods of contact with grape skins (and sometimes a portion of stems, too) in order to extract greater tannin content. As would be expected, these young red wines will have unpleasant astringent tannins. Over years of aging, however, these tannins will eventually give way to the inevitable oxidation–aging process and will develop an attractive delicate tea-like flavor. During that time, all of the other flavor constituency in the wine will also have matured into what has hopefully made the wait worthwhile.

The following is a typical range of bitterness expressed in wine judging:

Smooth
Astringent
Coarse
Tannic
Harsh

The tactile phase of gustatory examination has to do with the manner in which a wine "lays" on the palate—more often referred to as mouth feel by food scientists.

A wine judge will consider wine *body* by whether it feels light and thin—as opposed to full and heavy. Heavy-bodiedness results from dissolved solids in the wine, such as sugars, color pigments, and higher alcohols. Intuitively, sweet wines have more body than drys, but it is the relative degree of body for each individual wine type which is the judgment criterion. A little glycerine mixed to a light dry white wine can create the effect of heavier body.

Higher concentrations of ethyl alcohol in wine can contribute to a slight burning sensation in the nose and on the palate. Mix enough unflavored vodka with water to make a solution of about 13% alcohol—a common level in table wines. If the vodka is 80 proof, it is about 40% alcohol by volume. Mixing two parts of water with one part of vodka results in a little over 13% alcohol by volume.

Smell and taste the plain water side-by-side with the ethyl alcohol solution at room temperature. The ethanol mixture should have no flavor, but there should be a gustatory sensation—perhaps a slight burning in the nostrils and on the tongue. Then chill both the water and the ethanol solution to refrigerator temperature and repeat the evaluation. The result should significantly mute the gustatory effect of the alcohol—a term called "closed." Then heat both solutions up to luke warm temperature and try the experiment again. This should have a profound burning effect—a term called "hot" or "strong" by wine judges.

Aftertaste relates to how long both the olfactory and gustatory effects remain on the palate after the wine has been spit out. The common judgment terminology for aftertaste is simply "short," or "lingering," or "long." As a rule of thumb, lengthy aftertaste is a virtue. Some very high-quality wines are, however, also delicate and shy—precluding any measure of lengthy aftertaste.

Purdue University
WINE EVALUATION CHART

NAME: _____ PLACE: _____ DATE: _____

WINE	PRICE	APPEARANCE 3 MAX	AROMA/ BOUQUET 6 MAX	TASTE 6 MAX	AFTERTASTE 3 MAX	OVERALL 2 MAX	TOTAL 20 MAX
1							
2							
3							
4							
5							
6							
7							
8							
9							

COMMENTS:

Fig. B–20 Modified 20-point cardinal scale.

APPEARANCE (the visual mode - what one is seeing in a wine)
The color and clarity of a wine can be an indication of it's character - of potential qualities and flaws. Darker colors (both reds and whites) generally indicate heavier, more full-bodied wines. Rosé wines should be pink, perhaps just a bit of orange tint to older selections. All wines should be transparent - younger whites brilliant, while older selections may exhibit a very slight haziness, and very old 'reds often "throw" a noticeable sediment.

- 3 - Excellent — brilliant with outstanding characteristic color
- 2 - Good — clear with characteristic color
- 1 - Poor — excessive haziness and/or uncharacteristic color
- 0 - Objectionable — cloudiness and/or very poor color

AROMA AND BOUQUET (the olfactory mode - what one is smelling in a wine)
This second step in wine evaluation is directed towards aroma (fruit flavor) and bouquet (fruit flavor plus added odors from vinification). Take several large whiffs. Some wines may be herbaceous, while others are fruity - experience serving to distinguish which should be which. In every case the result should be balanced (not either neutral or overpowering) and, most importantly, pleasant. This mode is often called the "nose" in organoleptic testing.

- 6 - Extraordinary — outstanding in character with exceptional balance in bouquet constituency
- 5 - Excellent — characteristic and well-balanced
- 4 - Good — distinguishable and adequately balanced
- 3 - Fair — somewhat neutral and/or slightly unbalanced
- 2 - Poor — undistinguishable and detectable "off" odors
- 1 - Unacceptable — obvious "off" odors and unbalanced
- 0 - Objectionable — offensive odors and very unbalanced

TASTE (the gustation mode - what one is tasting in a wine)
The human tongue can detect variances in acidity, bitterness, and sweetness, levels in wine - typical values for each wine type are learned with experience. Other portions of the mouth are anatomically connected to the olfactory lobe and are, therefore, sensitive to specific flavors and essences associated with the "nose" of each wine - again judged authoritatively through continued practice. Take a small sip and "wash" it around the mouth so that all surfaces are given a chance to experience the wine - "whistling" in some air to activate taste bud activity. Generally, after ten to fifteen seconds of examination, the wine is spit out.

- 6 - Extraordinary — outstanding in character with exceptional balance of acidity and sweetness
- 5 - Excellent — characteristic and well-balanced
- 4 - Good — distinguishable and adequately balanced
- 3 - Fair — somewhat neutral and/or slightly unbalanced
- 2 - Poor — undistinguishable and detectable "off" flavors and/or unbalanced
- 1 - Unacceptable — obvious "off" flavors and unbalanced
- 0 - Objectionable — offensive flavors and very unbalanced

AFTERTASTE (the taste and flavor values that linger following completion of the gustatory mode)
Take a second, perhaps smaller, sip of the wine and swallow - endeavoring to judge how long the taste and flavor constituents remain detectable in the mouth. Lighter wines may linger only momentarily, while heavier wines can last for perhaps a dozen seconds or more.

- 3 - Excellent — outstandingly pleasant aftertaste, lasting beyond normal duration of time
- 2 - Good — pleasant aftertaste, lasting a normal duration of time
- 1 - Poor — little or no distinguishable aftertaste
- 0 - Objectionable — unpleasant aftertaste

OVERALL IMPRESSION (the trueness of type and price/value judgement one determines in a wine)
The total quality level achieved by a wine in comparison to other wines of the same variety or type, and in regard to the monetary price, is addressed in this mode. This not a "fudge factor" used indiscriminately to give or take points in response to subjective notions.

- 2 - Excellent — outstanding example of the variety or type and/or exceptional value for the money
- 1 - Good — representative of the variety or type and/or an acceptable value for the money
- 0 - Poor — uncharacteristic of the variety or type and/or an unacceptable value for the money

SCORES
- 18 - 20 — truly great wines, unusually superior attributes orchestrated in a faultless character
- 15 - 17 — excellent wines, perhaps faulted only by a shyness of one or another attributes
- 12 - 14 — good wines, with perhaps only one or two detectable faults
- 9 - 11 — poor wines, with at least two obvious faults
- 0 - 8 — objectionable wines, having many faults

Fig. B-21 Modified 20-point cardinal scale rationale.

Proceed with judging the gustatory mode in precisely the same manner as for tasting the olfactory mode. After a few seconds rest take a small bite of cheese and unsalted cracker. Rinse out the mouth with water and take a sip of wine. Roll it around the mouth to obtain the full effects of acidity, sweetness, and bitterness. Spit out the wine and record all impressions. While recording, make a judgment as to the length of the aftertaste.

Remember that this entire discussion has to do with wine judging and not wine drinking. When wine is swallowed, there is an additional set of olfactory and gustatory impressions made in the back of the mouth and in the pharyngeal passage of the throat. For obvious reasons, wine judges cannot swallow every wine presented them in a whole day of wine competition.

Overall Impression

The *overall impression* is used to rate the combined visual, olfactory, and gustatory impressions of each wine.

More often than not, judgment of overall impression is relegated to a rather "fudge-factor" role. A given wine might possibly score only moderately when scrutinized mode by mode, whereas all the modes may fit together very nicely—in which case a judge would be generous in scoring overall impression. The opposite holds, as well.

Cardinal Scales

There are many cardinal (numerical) scales for the sensory scoring of wines. One of the most commonly used is the Modified Davis 20-Point Scale (Fig. B–20 and B–21) modeled after the original devised by the University of California, Davis, years ago.

Appendix C

Charts, Tables, and Conversion Tables

Table C-1 Tons of Grapes per Acre

(Acre = 202 ft × 202 ft = 20 rows at 200 ft spaced 10 ft apart)

Lb Grapes per Vine	Vines at 5 ft (800 Vines)	Vines at 6 ft (667 Vines)	Vines at 7 ft (571 Vines)	Vines at 8 ft (500 Vines)	Vines at 15 ft (267 Vines)	Vines at 20 ft (200 Vines)
2	.80	.67	.57	.50	.27	.20
4	1.60	1.33	1.14	1.00	.54	.40
6	2.40	2.00	1.71	1.50	.81	.60
8	3.20	2.67	2.28	2.00	1.08	.80
10	4.00	3.34	2.85	2.50	1.35	1.00
12	4.80	4.00	3.42	3.00	1.62	1.20
14	5.60	4.67	3.99	3.50	1.89	1.40
16	6.40	5.34	4.56	4.00	2.16	1.60
18	7.20	6.00	5.13	4.50	2.43	1.80
20	8.00	6.67	5.70	5.00	2.70	2.00
22	8.80	7.34	6.27	5.50	2.97	2.20
24	9.60	8.00	6.84	6.00	3.24	2.40
26	10.40	8.67	7.41	6.50	3.51	2.60
28	11.20	9.34	7.98	7.00	3.78	2.80
30	12.00	10.01	8.55	7.50	4.05	3.00
32	12.80	10.67	9.12	8.00	4.32	3.20
34	13.60	11.34	9.69	8.50	4.59	3.40
36	14.40	12.01	10.26	9.00	4.86	3.60
38	15.20	12.67	10.83	9.50	5.13	3.80
40	16.00	13.34	11.40	10.00	5.40	4.00
42	16.80	14.07	11.97	10.50	5.67	4.20
44	17.60	14.74	12.54	11.00	5.94	4.40
46	18.40	15.41	13.11	11.50	6.21	4.60
48	19.20	16.08	13.68	12.00	6.48	4.80
50	20.00	16.75	14.25	12.50	6.75	5.00
60	24.00	20.04	17.10	15.00	8.10	6.00
70	28.00	23.38	19.95	17.50	9.45	7.00
80	32.00	26.72	22.80	20.00	10.80	8.00
90	36.00	30.06	25.65	22.50	12.15	9.00
100	40.00	33.40	28.50	25.00	13.50	10.00
110	44.00	36.74	31.35	27.50	14.85	11.00
120	48.00	40.08	34.20	30.00	16.20	12.00
130	52.00	43.42	37.05	32.50	17.55	13.00
140	56.00	46.76	39.90	35.00	18.90	14.00
150	60.00	50.10	42.75	37.50	20.25	15.00

Table C-1 (*continued*)

Lb Grapes per Vine	Vines at 5 ft (800 Vines)	(Acre = 202 ft × 202 ft = 20 rows at 200 ft spaced 10 ft apart)				
		Vines at 6 ft (667 Vines)	Vines at 7 ft (571 Vines)	Vines at 8 ft (500 Vines)	Vines at 15 ft (267 Vines)	Vines at 20 ft (200 Vines)
160		53.44	45.60	40.00	21.60	16.00
170		56.78	48.45	42.50	22.95	17.00
180		60.12	51.30	45.00	24.30	18.00
190			54.15	47.50	25.65	19.00
200			57.00	50.00	27.00	20.00
210			59.85	52.50	28.35	21.00
220				55.00	29.70	22.00
230				57.50	31.05	23.00
240				60.00	32.40	24.00
250					33.75	25.00
260					35.10	26.00
270					36.45	27.00
280					37.80	28.00
290					39.15	29.00
300					40.50	30.00

Table C-2 Traditional Bulk Wine Vessels

Vessel	Origin	Capacity in U.S. Gal.	Liters
Aroba	Spain	3–5	12–18
Aum	Germany	42.27	160.00
Baril	Lisbon	4.42	16.74
Baril	Malaga	7.93	30.00
Barile	Rome	15.41	58.34
Barrel	U.S.	31.50	119.24
Barril	Madeira	4.08	15.44
Barrique	Algeria	58.12	220.00
Barrique	Beaujolais	57.07	216.00
Barrique	Bordeaux	59.45	225.00
Barrique	Champagne	52.84	200.00
Barrique	Côte d'Or	60.24	228.00
Barrique	Côtes du Rhone	59.45	225.00
Barrique	Macon	56.80	215.00
Barrique	Yonne	65.52	248.00
Bocoy	Spain	162.00	613.17
Botte (small)	Sardinia	11.77	44.54
Botte (large)	Sardinia	132.10	500.00
Brente	Switzerland	13.21	50.00
Butt	Spain	132.00	499.62
Butte (Buttig)	Hungary	3.59	13.6
Demi-queue	Burgundy	60.24	228.00
Double-Aum	Germany	84.54	320.00
Dreiling	Vienna	358.78	1358.00
Eimer	Switzerland	9.91	37.50
Eimer	Vienna	14.95	56.58
Eimer	Württemberg	77.65	293.92
Fuder	Mosel	253.63	960.00
Fuder	Germany (Prussian)	217.81	824.42
Fuder	Württemberg	465.94	1763.57
Gallon	England	1.20	5.54
Gallon	U.S.	1.00	3.79
Halb-Fuder	Mosel	126.82	480.00
Halb-Stuck	Rhine and Palatinate	169.09	640.00
Hogshead	England	66.00	249.81
Octavilla	England and Spain	16.50	62.45
Ohm	Alsace	13.21	50.00
Ohm	Baden	39.63	150.00
Ohm	Bavaria	33.82	128.00
Ohm	Saar	38.04	144.00
Ohm	Switzerland	10.57	40.00
Oka	Balkans	.34	1.28
Oxhoft	Hamburg	59.71	226.00
Piece	Burgundy	60.24	228.00
Piece	Champagne	52.84	200.00
Piece	Saumur	58.12	220.00
Piece	Vouvray	66.05	250.00
Pipe	England	126.02	477.00
Pipe	Lisbon	132.63	502.00
Pipe	Madeira	109.91	416.00
Pipe	Oporto	141.35	535.00
Pipe	Tarragona	134.74	510.00
Pipe	Valencia	19.92	75.39
Puncheon	England	84.00	317.94
Queue	Burgundy	120.48	456.00
Stuck	Rhineland	317.04	1200.00
Tierce	England	42.00	158.97
Tonneau	Bordeaux	237.78	900.00
Tun	England	252.05	954.00
Vedro	Russia	3.27	12.39

Table C–3 Alcohol Measurements

Specific Gravity[2] 20°C/4°C	Percent by Volume at 20°C	Percent by Weight	G per 100 ml
0.99823	0.00	0.00	0.00
0.99675	1.00	0.79	0.79
0.99528	2.00	1.59	1.58
0.99384	3.00	2.38	2.37
0.99243	4.00	3.18	3.16
0.99106	5.00	3.98	3.95
0.98973	6.00	4.78	4.74
0.98845	7.00	5.59	5.53
0.98718	8.00	6.40	6.32
0.98596	9.00	7.20	7.10
0.98476	10.00	8.02	7.89
0.98416	11.00	8.83	8.68
0.98296	11.50	9.23	9.08
0.98238	12.00	9.64	9.47
0.98180	12.50	10.05	9.87
0.98122	13.00	10.40	10.26
0.98066	13.50	10.86	10.66
0.98009	14.00	11.28	11.05
0.97953	14.50	11.68	11.44
0.97897	15.00	12.09	11.84
0.97841	15.50	12.50	12.23
0.97786	16.00	12.92	12.63
0.97732	16.50	13.33	13.02
0.97678	17.00	13.74	13.42
0.97624	17.50	14.15	13.81
0.97570	18.00	14.56	14.21
0.97517	18.50	14.97	14.60
0.97464	19.00	15.39	15.00
0.97412	19.50	15.80	15.39
0.97359	20.00	16.21	15.79
0.97306	20.50	16.63	16.18
0.97252	21.00	17.04	16.58
0.97199	21.50	17.46	16.97
0.97145	22.00	17.88	17.37
0.97091	22.50	18.29	17.76
0.97036	23.00	18.71	18.16
0.96980	23.50	19.13	18.55
0.96925	24.00	19.55	18.94
0.96869	24.50	19.96	19.34
0.96812	25.00	20.38	19.73
0.96755	25.50	20.80	20.13
0.96699	26.00	21.22	20.52
0.96641	26.50	21.64	20.92
0.96583	27.00	22.07	21.31
0.96525	27.50	22.49	21.71
0.96465	28.00	22.91	22.10
0.96406	28.50	23.33	22.50
0.96346	29.00	23.76	22.89
0.96285	29.50	24.18	23.29
0.96224	30.00	24.61	23.68
0.96163	30.50	25.04	24.08
0.96100	31.00	25.46	24.47
0.96036	31.50	25.89	24.86
0.95972	32.00	26.32	25.26

Table C–3 (*continued*)

Specific Gravity[2] $\frac{20°C}{4°C}$	Percent by Volume at 20°C	Percent by Weight	G per 100 ml
0.95906	32.50	26.75	25.64
0.95839	33.00	27.18	26.05
0.95771	33.50	27.61	26.44
0.95703	34.00	28.04	26.84
0.95634	34.50	28.48	27.23
0.95563	35.00	28.91	27.63
0.95492	35.50	29.34	28.02
0.95419	36.00	29.78	28.42
0.95346	36.50	30.22	28.81
0.95272	37.00	30.66	29.21
0.95196	37.50	31.09	29.60
0.95120	38.00	31.53	29.99
0.95043	38.50	31.97	30.39
0.94964	39.00	32.42	30.79
0.94885	39.50	32.86	31.18
0.94805	40.00	33.30	31.57
0.94724	40.50	33.75	31.97
0.94643	41.00	34.19	32.36
0.94560	41.50	34.64	32.76
0.94477	42.00	35.09	33.15
0.94393	42.50	35.54	33.55
0.94308	43.00	35.99	33.94
0.94222	43.50	36.44	34.34
0.94135	44.00	36.89	34.73
0.94046	44.50	37.35	35.13
0.93957	45.00	37.80	35.52
0.93867	45.50	38.26	35.92
0.93776	46.00	38.72	36.31
0.93684	46.50	39.18	36.70
0.93591	47.00	39.64	37.10
0.93498	47.50	40.10	37.49
0.93404	48.00	40.56	37.89
0.93308	48.50	41.03	38.29
0.93213	49.00	41.49	38.68
0.93116	49.50	41.96	39.07
0.93017	50.00	42.43	39.47

[1] Calculated by U.S. Bureau of Standards.
[2] This table is valid for specific gravity determinations at temperatures between an upper limit of 20°C and a lower limit of 4°C.

Table C-4 Calulation of Partially Filled Horizontal Tanks (Cylindrical, Straight-Sided Tanks Only; Where Total Capacity of Tank Is Known)

Procedure:

(1) Determine inside diameter of tank.
(2) Determine depth of wine in tank.
(3) Compute percent depth of wine (line 2 less line 1).
(4) From table below find percent total volume that corresponds to this percent depth
(5) Multiply percent found by total volume = amount of wine in tank

Percent Depth Wine	Percent Volume	Percent Depth Wine	Percent Volume	Percent Depth Wine	Percent Volume
99	99.6	64	67.8	29	23.8
98	99.2	63	66.6	28	22.7
97	98.8	62	65.4	27	21.6
96	98.4	61	64.2	26	20.5
95	98.0	60	63.0	25	19.4
94	97.4	59	61.6	24	18.3
93	96.8	58	60.2	23	17.2
92	96.2	57	58.8	22	16.1
91	95.6	56	57.4	21	15.0
90	95.0	55	56.0	20	14.0
89	94.2	54	54.8	19	13.0
88	93.4	53	53.6	18	12.0
87	92.6	52	52.4	17	11.0
86	91.8	51	51.2	16	10.0
85	91.0	50	50.0	15	9.0
84	90.0	49	48.8	14	8.2
83	89.0	48	47.6	13	7.4
82	88.0	47	46.4	12	6.6
81	87.0	46	45.2	11	5.8
80	86.0	45	44.0	10	5.0
79	85.0	44	42.6	9	4.4
78	83.9	43	41.2	8	3.8
77	82.8	42	39.8	7	3.2
76	81.7	41	38.4	6	2.6
75	80.5	40	37.0	5	2.0
74	79.5	39	35.8	4	1.6
73	78.4	38	34.6	3	1.2
72	77.3	37	33.4	2	.8
71	76.2	36	32.2	1	.4
70	75.0	35	31.0	0	.0
69	73.8	34	29.8		
68	72.6	33	28.6		
67	71.4	32	27.4		
66	70.2	31	26.2		
65	69.0	30	25.0		

Table C–5 Wine Tank Capacities (for Cylindrical Straight-Sided Tanks Only Measured at the Waist[1])

Calculation of total capacity in U.S. gal.:
Diam. (in.) × diam. (in.) × height (in.) × 0.0034 = U.S. gal capacity of tank.
Example:
72 in. diam. × 72 in. diam × 96 in. height (or length) × 0.0034 = 1692.1 U.S. gal. capacity of tank.

Calculation of U.S. gal. per in.:
Total U.S. gal. capacity divided by total number of in. = U.S. gal. per in.
Example:
1692.1 U.S. gal. capacity divided by 96 in. = 17.626 U.S. gal. per in.

Calculation of total capacity in liters:
Total U.S. gal. capacity × 3.785 = liters capacity of tank.
Example 1692.1 U.S. gal. capacity × 3.785 = 6404.6 liters capacity of tank.

Calculation of liters per cm:
U.S. gal. per in. × 1.49016 = liters per cm.
Example:
17.626 gal. per in. × 1.49016 = 26.266 liters per cm.

More Examples:

Diameter In. I.D.	Length In. I.D.	Capacity Gal. U.S.	Gal. per In. U.S.	Capacity Liters	Liters per Cm
36	36	158.6	4.41	600.3	6.44
36	48	211.5	4.41	800.5	6.44
36	60	264.4	4.41	1000.8	6.44
48	48	376.0	7.83	1423.2	11.67
48	60	470.0	7.83	1779.0	11.67
48	72	564.0	7.83	2134.7	11.67
60	60	734.4	12.24	2779.7	18.24
60	72	881.3	12.24	3335.7	18.24
60	84	1028.2	12.24	3891.7	18.24
60	96	1175.0	12.24	4447.4	18.24
72	72	1269.0	17.63	4803.2	26.27
72	84	1480.6	17.63	5604.1	26.27
72	96	1692.1	17.63	6404.6	26.27
72	108	1903.6	17.63	7205.1	26.27
72	120	2115.1	17.63	8005.7	26.27
84	84	2015.2	23.99	7627.5	35.75
84	96	2303.1	23.99	8717.2	35.75
84	108	2591.0	23.99	9806.9	35.75
84	120	2878.8	23.99	10,896.3	35.75
84	132	3166.7	23.99	11,986.0	35.75
84	144	3454.6	23.99	13,075.7	35.75
96	96	3008.1	31.33	11,385.7	46.69
96	108	3384.1	31.33	12,810.0	46.69
96	120	3760.1	31.33	14,232.0	46.69
96	132	4136.1	31.33	15,655.1	46.69
96	144	4512.2	31.33	17,078.7	46.69

[1] Whether the tank is horizontal or vertical, the diameter must be measured at the center of the tank between the two parallel, flat ends.

Table C-6 Vat Capacity

(Per 1 Ft of Depth—Inside Measure)

Diameter (Ft)	(In.)	Area (Ft)	U.S. Gal.
1	0	.7854	5.8735
1	2	1.0690	7.9944
1	4	1.3962	10.4413
1	6	1.7671	13.2150
1	8	2.1816	16.3148
1	10	2.6398	19.7414
2	0	3.1416	23.4940
2	2	3.6869	27.5720
2	4	4.2760	32.6976
2	6	4.9087	36.7092
2	8	5.5850	41.7668
2	10	6.3049	47.1505
3	0	7.0686	52.8618
3	2	7.8757	58.8976
3	4	8.7265	65.2602
3	6	9.6211	73.1504
3	8	10.5591	78.9652
3	10	11.5409	86.3074
4	0	12.5664	93.9754
4	2	13.6353	101.9701
4	4	14.7479	110.2907
4	6	15.9043	118.9386
4	8	17.1041	127.9112
4	10	18.3476	137.2105
5	0	19.6350	146.8384
5	2	20.9656	156.7891
5	4	22.3400	167.0674
5	6	23.7583	177.6740
5	8	25.2199	188.6045
5	10	26.7251	199.8610
6	0	28.2744	211.4472
6	6	33.1831	248.1564
7	0	38.4846	287.8230
7	6	44.1787	330.3859
8	0	50.2656	375.9062
8	6	56.7451	424.3625
9	0	63.6174	475.7563
9	6	70.8823	530.0861
10	0	78.5400	587.3534
10	6	86.5903	647.5568
11	0	95.0034	710.6977
11	6	103.8691	776.7746
12	0	113.0976	848.1890

Table C–7   Example of Tank Chart for Straight-Sided Horizontal Tank[1]

Tank No. S-41		Gal. per In.		Wet In.	
In.	Gal.	In.	Gal.	In.	Gal.
1	10	37	1855	73	3896
2	23	38	1920	74	3928
3	41	39	1987	75	3955
4	66	40	2052	76	3980
5	91	41	2118	77	3998
6	120	42	2183	78	4008
7	153	43	2248		
8	189	44	2314		
9	228	45	2379		
10	271	46	2445		
11	317	47	2508		
1 ft	362	4 ft	2573		
13	409	49	2638		
14	459	50	2703		
15	512	51	2765		
16	568	52	2827		
17	622	53	2888		
18	677	54	2949		
19	733	55	3010		
20	791	56	3070		
21	842	57	3130		
22	903	58	3188		
23	962	59	3245		
2 ft	1023	5 ft	3300		
25	1084	61	3356		
26	1147	62	3411		
27	1209	63	3463		
28	1271	64	3515		
29	1336	65	3565		
30	1399	66	3612		
31	1464	67	3658		
32	1528	68	3703		
33	1593	69	3745		
34	1659	70	3786		
35	1724	71	3825		
3 ft	1789	6 ft	3862		

[1] Tank has an inside diameter of 122.41 in. at the waist (39 in.).

Table C-8 Example of Tank Chart for Straight-Sided Vertical Tank[1]

	Tank No. S-56 Liters per Cm			Measured at Center of Tank Wet Cm	
Cm	Liters	Cm	Liters	Cm	Liters
0	0	59	1040.64	118	2081.28
1	17.64	60	1058.28	119	2098.92
2	35.28	61	1075.92	120	2116.56
3	52.91	62	1093.56	121	2134.20
4	70.55	63	1111.19	122	2151.84
5	88.19	64	1128.83	123	2169.47
6	105.83	65	1146.47	124	2187.11
7	123.47	66	1164.11	125	2204.75
8	141.10	67	1181.75	126	2222.39
9	158.74	68	1199.38	127	2240.03
10	176.38	69	1217.02	128	2257.66
11	194.02	70	1234.66	129	2275.30
12	211.66	71	1252.30	130	2292.94
13	229.29	72	1269.94	131	2310.58
14	246.93	73	1287.57	132	2328.22
15	264.57	74	1305.21	133	2345.85
16	282.21	75	1322.85	134	2363.49
17	299.85	76	1340.49	135	2381.13
18	317.48	77	1358.13	136	2398.77
19	335.12	78	1375.76	137	2416.41
20	352.76	79	1393.40	138	2434.04
21	370.40	80	1411.04	139	2451.68
22	388.04	81	1428.68	140	2469.32
23	405.67	82	1446.32	141	2486.96
24	423.31	83	1463.95	142	2504.60
25	440.95	84	1481.59	143	2522.23
26	458.59	85	1499.23	144	2539.87
27	476.23	86	1516.87	145	2557.51
28	493.86	87	1534.51	146	2575.15
29	511.50	88	1552.14	147	2592.79
30	529.14	89	1569.78	148	2610.42
31	546.78	90	1587.42	149	2628.06
32	564.42	91	1605.06	150	2645.70
33	582.05	92	1622.70	151	2663.34
34	599.69	93	1640.33	152	2680.98
35	617.33	94	1657.97	153	2698.61
36	634.97	95	1675.61	154	2716.25
37	652.61	96	1693.25	155	2733.89
38	670.24	97	1710.89	156	2751.53
39	687.88	98	1728.52	157	2769.17
40	705.52	99	1746.16	158	2786.80
41	723.16	100	1763.80	159	2804.44
42	740.80	101	1781.44	160	2822.08
43	758.43	102	1799.08	161	2839.72
44	776.07	103	1816.71	162	2857.36
45	793.71	104	1834.35	163	2874.99
46	811.35	105	1851.99	164	2892.63
47	828.99	106	1869.63	165	2910.27
48	846.62	107	1887.27	166	2927.91
49	864.26	108	1904.90	167	2945.55
50	881.90	109	1922.54	168	2963.18
51	899.54	110	1940.18	169	2980.82
52	917.18	111	1957.82	170	2998.46
53	934.81	112	1975.46	171	3016.10
54	952.45	113	1993.09	172	3033.74
55	970.09	114	2010.73	173	3051.37
56	987.73	115	2028.37	174	3069.01
57	1005.37	116	2046.01	175	3086.65
58	1023.00	117	2063.65	176	3104.29

[1] Tank has an inside diameter of 149.86 cm. at the waist (137 cm.).

Table C–8 (*continued*)

	Tank No. S-56 Liters per Cm			Measured at Center of Tank Wet Cm	
Cm	Liters	Cm	Liters	Cm	Liters
177	3121.93	210	3703.98	242	4268.40
178	3139.56	211	3721.62	243	4286.03
179	3157.20	212	3739.26	244	4303.67
180	3174.84	213	3756.89	245	4321.31
181	3192.48	214	3774.53	246	4338.95
182	3210.12	215	3792.17	247	4356.59
183	3227.75	216	3809.81	248	4374.22
184	3245.39	217	3827.45	249	4391.86
185	3263.03	218	3845.08	250	4409.50
186	3280.67	219	3862.72	251	4427.14
187	3298.31	220	3880.36	252	4444.78
188	3315.94	221	3898.00	253	4462.41
189	3333.58	222	3915.64	254	4480.05
190	3351.22	223	3933.27	255	4497.69
191	3368.86	224	3950.91	256	4515.33
192	3386.50	225	3968.56	257	4532.97
193	3404.13	226	3986.19	258	4550.60
194	3421.77	227	4003.83	259	4568.24
195	3439.41	228	4021.46	260	4585.88
196	3457.05	229	4039.10	261	4603.52
197	3474.69	230	4056.74	262	4621.16
198	3492.32	231	4074.38	263	4638.79
199	3509.96	232	4092.02	264	4656.43
200	3527.60	233	4109.65	265	4674.07
201	3545.24	234	4127.29	266	4691.71
202	3562.88	235	4144.93	267	4709.35
203	3580.51	236	4162.57	268	4726.98
204	3598.15	237	4180.21	269	4744.62
205	3615.80	238	4197.84	270	4762.26
206	3633.43	239	4215.48	271	4779.90
207	3651.07	240	4233.12	272	4797.54
208	3668.70	241	4250.76	273	4815.70
209	3686.34				

[1] Tank has an inside diameter of 149.86 cm. at the waist (137 cm.).

Table C–9 Periodic Table of the Elements

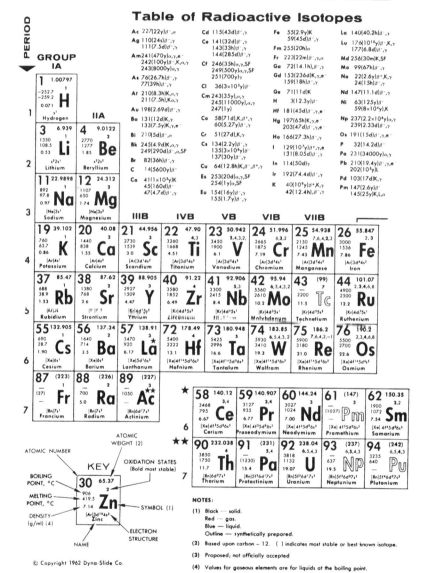

Table C–9 (continued)

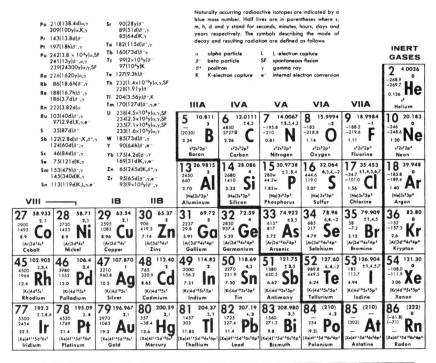

Courtesy of E.H. Sargent & Co

Table C–10 Periodic Properties of the Elements

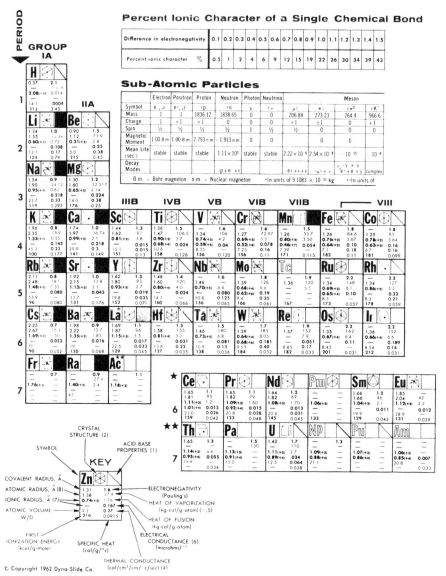

OF THE ELEMENTS

Table C-10 (continued)

[Periodic table chart showing element properties. Due to the complexity and density of the tabular data, a faithful reproduction is provided below in approximate form.]

1.6	1.7	1.8	1.9	2.0	2.1	2.2	2.3	2.4	2.5	2.6	2.7	2.8	2.9	3.0	3.1	3.2
47	51	55	59	63	67	70	74	76	79	82	84	86	88	89	91	92

Hyperon

	K^0	Λ^0	Σ	Σ	Ξ
	974.4	2181.4	2327.7	2343.2	2584
	0	±1	1		1
	0	½	½-integral	½-integral	½-integral
10^{-10}	10^{-7}	2.8×10^{-10}	5×10^{-11}	10^{-10}	10^{-10}
	$-p + \pi^-$	$-p + \pi^0$			$-\Lambda^0 + \pi^-$
complex	$-n + \pi^0$	$-n + \pi^-$		$-n + \pi$	

4.80286×10^{-10} esu. Exists as an antiparticle not listed.

INERT GASES

	IIIA	IVA	VA	VIA	VIIA	He	0.93
						—	0.171
						—	—
						—	0.005
	B	C	N	O	F	31.8	0003
						567	1.25
						Ne	1.31
	0.82	0.77	0.75	0.73	0.72	—	—
	2.0	2.5	3.0	3.5	4.0	—	4.1
	0.98	0.914	0.92	0.666	0.815	—	0.755
	75	171.7	0.086	—	0.753	—	0.080
	0.20(+3)	2.60(-4)	1.71(-3)	1.40(-3)	1.36(-1)	16.8	0.061
	5.3	0.0007	0.11(+5)	0.053	0.07(+7)	0.0001	
	—	0.15(+4)	—	0.09(+6)	—		
	10			—	—		
	4.6	5.3	17.3	14.0	16.5	21.5	
	191	0.309	260	0.165	336	0.247	314	0.218	402	0.18	497	
	Al	**Si**	**P**	**S**	**Cl**	**Ar**	1.74
	1.18	1.5	1.11	1.8	1.06	2.1	1.02	2.5	0.99	3.0	—	—
	1.43	67.9	1.32	140.61	1.28	297	1.27	3.01	—	2.44	—	1.56
	0.50(+3)	2.55	2.71(-1)	11.1	2.12(-3)	0.15	1.84(-2)	0.34	1.81(-1)	0.77	—	0.281
	—	0.382	0.41(+4)	0.10	0.34(+5)	10.17	0.29(+6)	—	0.26(+7)	—	24.2	00004
	—	0.50	—	0.20	—	—	15.5	0007	—	—	363	0.125
	10.0	—	12.1	—	17.0	—	—	—	18.7	00002		
	138	0.215	188	0.162	254	0.177	239	0.175	300	0.116		
IB	IIB										

Ni	Cu	Zn	Ga	Ge	As	Se	Br	Kr
—	1.8	1.38	1.9	1.31	1.6	1.26	1.6	1.22	1.8	1.19	2.0	1.16	2.4	1.14	2.8	1.89	—
1.24	91.0	1.28	72.8	1.38	27.4	1.41	—	1.37	68	1.39	7.75	1.40	3.34	1.58	—	—	2.16
0.78(+2)	4.21	0.96(+1)	3.11	0.74(+2)	1.76	1.48(+1)	1.34	0.93(+4)	7.6	2.22(-3)	6.62	1.98(-2)	1.25	1.95(-1)	1.26	—	1.39
0.62(+3)	0.145	0.69(+1)	0.593	—	—	—	0.167	0.53(+4)	0.022	0.47(+5)	0.029	0.42(+4)	0.08	0.39(+7)	10.18	32.2	00002
6.6	0.22	7.1	0.94	9.2	0.27	11.8	0.058	13.6	0.14	13.1	—	16.5	00001	—	0.070	323	—
176	0.105	178	0.092	216	0.0915	138	0.079	187	0.073	231	0.082	225	0.084	273			

Pd	Ag	Cd	In	Sn	Sb	Te	I	Xe
—	2.2	1.53	1.9	1.48	1.7	1.44	1.7	1.41	1.8	1.38	1.9	1.35	2.1	1.33	2.5	2.09	—
1.37	90	1.44	60.7	1.54	23.9	1.66	53.7	1.62	70	1.59	46.6	1.60	11.9	—	52	—	3.02
0.50(+3)	4.2	—	—	1.26(+1)	0.56	0.97(+3)	1.46	0.81(+4)	0.111	0.71(+4)	0.088	2.45(-2)	4.74	2.16(-1)	1.87	—	0.55
—	0.093	—	0.616	—	—	1.32(+3)	0.78	1.12(+2)	1.72	0.56(+6)	10.6	0.50(+7)	10.11	—	—	—	—
8.9	0.17	10.3	0.98	13.1	0.22	15.7	0.057	16.3	0.16	18.4	0.05	20.5	0.014	25.7	0.01	42.9	0.001
192	0.058	175	0.056	207	0.055	133	0.057	169	0.054	199	0.049	208	0.047	241	0.052	280	—

Pt	Au	Hg	Tl	Pb	Bi	Po	At	Rn
—	2.2	1.50	2.4	1.49	1.9	1.48	1.8	1.47	1.8	1.46	1.9	—	2.0	—	2.2	2.14	—
1.38	122	1.44	81.8	1.57	13.9	1.71	38.8	1.75	42.4	1.70	42.7	1.76	29	—	8	—	3.92
0.52(+2)	5.2	1.37(+1)	2.13	1.10(+2)	0.36	1.40(+1)	1.02	0.95(+4)	0.055	0.84(+4)	0.046	0.74(+4)	0.009	—	—	—	0.69
—	0.095	—	0.42	—	—	0.99(+3)	1.46	1.20(+2)	1.77	1.20(+2)	2.6	—	0.02	—	—	—	—
—	—	—	—	0.146	0.02	0.011	0.093	18.7	0.083	21.1	0.02	22.7	—	—	—	—	—
9.10	0.17	10.2	0.71	14.8	0.02	17.2	0.093										
207	0.032	213	0.031	241	0.031	141	0.031	171	0.031	185	0.034					248	—

Gd	Tb	Dy	Ho	Er	Tm	Yb	Lu
1.61	1.1	1.59	1.2	1.59	—	1.58	1.2	1.57	1.2	1.56	1.2	1.70	1.1	1.56	1.2
1.79	72	1.77	70	1.77	67	1.76	67	1.75	67	1.74	59	1.92	18	1.74	90
1.02(+3)	—	1.00(+3)	—	0.99(+3)	4.1	0.97(+3)	4.1	0.96(+3)	4.1	0.95(+3)	4.4	1.13(+3)	1.8	0.93(+3)	4.6
—	0.007	—	0.009	—	0.011	—	0.011	—	0.012	—	0.011	0.94(+3)	0.035	—	0.015
19.9	0.021	19.2	—	19.0	0.024	18.7	—	18.4	0.023	18.1	—	24.8	—	17.8	0.037
142	0.071	155	0.044	157	0.041	—	0.039	—	0.040	—	0.038	143	0.015	175	0.037

Cm	Bk	Cf	Es	Fm	Md	No	Lw

See note 3

NOTES: • Element sublimes

(1) For representative oxides (higher valence) of group. Oxide is acidic if color is red, basic if color is blue and amphoteric if both colors are shown. Intensity of color indicates relative strength.

(2) ⊡ Cubic, face centered; ⊠ cubic, body centered; ◇ diamond; □ cubic; ⬡ hexagonal; ◯ rhombohedral; ▯ tetragonal; ▭ orthorhombic; ⬠ monoclinic.

(3) Proposed; not officially accepted. (4) At room temperature. (5) At boiling point.

(6) From 0° to 20°C. (7) Ionic (crystal) radii for coordination number 6.

(8) Metallic radii for coordination number of 12.

Courtesy of E.H. Sargent & Co.

C-11 Basic Conversions: Units of Area

1 square inch	645.2 square millimeters
1 square inch	6.452 square centimeters
1 square inch	.000645 square meters
1 square inch	.006944 square feet
1 square inch	.000772 square yards
1 square foot	92,903 square millimeters
1 square foot	929.034 square centimeters
1 square foot	.092903 square meters
1 square foot	144 square inches
1 square foot	0.11111 square yards
1 square yard	8361.31 square centimeters
1 square yard	.836131 square meters
1 square yard	1296 square inches
1 square yard	9 square feet
1 square rod	30.25 square feet
1 acre	40.47 ares
1 acre	0.4047 hectares
1 acre	4480 square yards
1 square mile	2.58998 kilometers
1 square mile	640 acres
1 square centimeter	100 square millimeters
1 square centimeter	0.1550 square inches
1 square centimeter	.001076 square feet
1 square centimeter	.00012 square yards
1 square centimeter	.00010 square meters
1 square meter	10000 square centimeters
1 square meter	1550 square inches
1 square meter	10.7639 square feet
1 square meter	1.19598 square yards
1 square kilometer	247.1 acres
1 square kilometer	0.3861 square miles

C-12 Area Conversion Formulas

To Convert	into	Multiply by
square inches	square centimeters	6.451600
square feet	square meters	0.092903
square yards	square meters	0.836127
acres	square meters	4046.856422
square miles	square kilometers	2.589980
square centimeters	square inches	0.155000
square meters	square feet	10.763910
square meters	square yards	1.195990
square kilometers	square miles	0.386102

Table C–13 Square Inches into Square Centimeters

Sq. In	0	1	2	3	4	5	6	7	8	9
					Sq. Cm					
0	—	6.45	12.90	19.36	25.81	32.26	38.71	45.16	51.61	58.06
10	64.52	70.97	77.42	83.87	90.32	96.77	103.23	109.68	116.13	122.58
20	129.03	135.48	141.94	148.39	154.84	161.29	167.74	174.19	180.65	187.10
30	193.55	200.00	206.45	212.90	219.35	225.81	232.26	238.71	245.16	251.61
40	258.06	264.52	270.97	277.42	283.87	290.32	296.77	303.23	309.68	316.13
50	322.58	329.03	335.48	341.94	348.39	354.84	361.29	367.74	374.19	380.64
60	387.10	393.55	400.00	406.45	412.90	419.35	425.81	432.26	438.71	455.16
70	451.61	458.06	464.52	470.97	477.42	483.87	490.32	496.77	503.23	509.68
80	516.13	522.58	529.03	535.48	541.93	548.39	554.84	561.29	567.74	574.19
90	580.64	587.10	593.55	600.00	606.45	612.90	619.35	625.81	632.26	638.71
100	645.16	651.61	658.06	664.52	670.97	677.42	683.87	690.32	696.77	703.22

Table C–14 Square Centimeters into Square Inches

Sq. cm	0	1	2	3	4	5	6	7	8	9
					Sq. In					
0	—	0.155	0.310	0.465	0.620	0.775	0.930	1.085	1.240	1.395
10	1.550	1.705	1.860	2.015	2.170	2.325	2.480	2.635	2.790	2.945
20	3.100	3.255	3.410	3.565	3.720	3.875	4.030	4.185	4.340	4.495
30	4.650	4.805	4.960	5.115	5.270	5.425	5.580	5.735	5.890	6.045
40	6.200	6.355	6.510	6.665	6.820	6.975	7.130	7.285	7.440	7.595
50	7.750	7.905	8.060	8.215	8.370	8.525	8.680	8.835	8.990	9.145
60	9.300	9.455	9.610	9.765	9.920	10.075	10.230	10.385	10.540	10.695
70	10.850	11.005	11.160	11.315	11.470	11.625	11.780	11.935	12.090	12.245
80	12.400	12.555	12.710	12.865	13.020	13.175	13.330	13.485	13.640	13.795
90	13.950	14.105	14.260	14.415	14.570	14.725	14.880	15.035	15.190	15.345
100	15.500	15.655	15.810	15.965	16.120	16.275	16.430	16.585	16.740	16.895

Table C-15 Square Feet into Square Meters

Sq. Ft	0	1	2	3	4	5	6	7	8	9
					Sq. M					
0	—	0.0929	0.1858	0.2787	0.3716	0.4645	0.5574	0.6503	0.7432	0.8361
10	0.9290	1.0219	1.1148	1.2077	1.3006	1.3936	1.4865	1.5794	1.6723	1.7652
20	1.8581	1.9510	2.0439	2.1368	2.2297	2.3226	2.4155	2.5084	2.6013	2.6942
30	2.7871	2.8800	2.9729	3.0658	3.1587	3.2516	3.3445	3.4374	3.5303	3.6232
40	3.7161	3.8090	3.9019	3.9948	4.0877	4.1806	4.2735	4.3664	4.4594	4.5523
50	4.6452	4.7381	4.8310	4.9239	5.0168	5.1097	5.2026	5.2955	5.3884	5.4813
60	5.5742	5.6671	5.7600	5.8529	5.9458	6.0387	6.1316	6.2245	6.3174	6.4103
70	6.5032	6.5961	6.6890	6.7819	6.8748	6.9677	7.0606	7.1535	7.2464	7.3393
80	7.4322	7.5252	7.6181	7.7110	7.8039	7.8968	7.9897	8.0826	8.1755	8.2684
90	8.3613	8.4542	8.5471	8.6400	8.7329	8.8258	8.9187	9.0116	9.1045	9.1974
100	9.2903	9.3832	9.4761	9.5690	9.6619	9.7548	9.8477	9.9406	10.0335	10.1264

Table C-16 Square Meters into Square Feet

Sq. M	0	1	2	3	4	5	6	7	8	9
					Sq. Ft					
0	—	10.76	21.53	32.29	43.06	53.82	64.58	75.35	86.11	96.88
10	107.64	118.40	129.17	139.93	150.70	161.46	172.22	182.99	193.75	204.51
20	215.28	226.04	236.81	247.57	258.33	269.10	279.86	290.63	301.39	312.15
30	322.92	333.68	344.45	355.21	365.97	376.74	387.50	398.27	409.03	419.79
40	430.56	441.32	452.08	462.85	473.61	484.38	495.14	505.90	516.67	527.43
50	538.20	548.96	559.72	570.49	581.25	592.02	602.78	613.54	624.31	635.07
60	645.84	656.60	667.36	678.13	688.89	699.65	710.42	721.18	731.95	742.71
70	753.47	764.24	775.00	785.77	796.53	807.29	818.06	828.82	839.59	850.35
80	861.11	871.88	882.64	893.41	904.17	914.93	925.70	936.46	947.22	957.99
90	968.75	979.52	990.28	1001.04	1011.81	1022.57	1033.34	1044.10	1054.86	1065.63
100	1076.39	1087.15	1097.92	1108.68	1119.45	1130.21	1140.97	1151.74	1162.50	1173.27

Table C-17 Acres to Hectares

Acres	0	10	20	30	40	50	60	70	80	90
					Hectares					
0	—	4.047	8.094	12.141	16.187	20.234	24.281	28.328	32.375	36.422
100	40.469	44.515	48.562	52.609	56.656	60.703	64.750	68.797	72.843	76.890
200	80.937	84.984	89.031	93.078	97.125	101.171	105.218	109.265	113.312	117.359
300	121.406	125.453	129.499	133.546	137.593	141.640	145.687	149.734	153.781	157.827
400	161.874	165.921	169.968	174.015	178.062	182.109	186.155	190.202	194.249	198.296
500	202.343	206.390	210.437	214.483	218.530	222.577	226.624	230.671	234.718	238.765
600	242.811	246.858	250.905	254.952	258.999	263.046	267.093	271.139	275.186	279.233
700	283.280	287.327	291.374	295.421	299.467	303.514	307.561	311.608	315.655	319.702
800	323.749	327.795	331.842	335.889	339.936	343.983	348.030	352.077	356.123	360.170
900	364.217	368.264	372.311	376.358	380.405	384.451	388.498	392.545	396.592	400.639
1000	404.686	—	—	—	—	—	—	—	—	—

Table C-18 Hectares to Acres

Hectares	0	1	2	3	4	5	6	7	8	9
					Acres					
0	—	24.71	49.42	74.13	98.84	123.55	148.26	172.97	197.68	222.40
100	247.11	271.82	296.53	321.24	345.95	370.66	395.37	420.08	444.79	469.50
200	484.21	518.92	543.63	568.34	593.05	617.76	642.47	667.19	691.90	716.61
300	741.32	766.03	790.74	815.45	840.16	864.87	889.58	914.29	939.00	963.71
400	988.42	1013.13	1037.84	1062.55	1087.26	1111.97	1136.68	1161.40	1186.11	1210.82
500	1235.53	1260.24	1284.95	1309.66	1334.37	1359.08	1383.79	1408.50	1433.21	1457.92
600	1482.63	1507.34	1532.05	1556.76	1581.47	1606.18	1630.90	1655.61	1680.32	1705.03
700	1729.74	1754.45	1779.16	1803.87	1828.58	1853.29	1878.00	1902.71	1927.42	1952.13
800	1976.84	2001.55	2026.26	2050.97	2075.69	2100.40	2125.11	2149.82	2174.53	2199.24
900	2223.95	2248.66	2273.37	2298.08	2322.79	2347.50	2372.21	2396.92	2421.63	2446.34
1000	2471.05	—	—	—	—	—	—	—	—	—

Table C–19 Basic Conversions: Units of Capacity and Volume

1 cubic inch	16.38706 cubic centimeters
1 cubic inch	.01638706 liters
1 cubic inch	0.5541 fluid ounces
1 cubic inch	.017316 U.S. quarts
1 cubic inch	.004329 U.S. gallons
1 cubic foot	0.02832 cubic meters
1 cubic foot	1728 cubic inches
1 cubic foot	28.317 liters
1 cubic foot	.03704 cubic yards
1 cubic foot	7.48052 U.S. gallons
1 cubic yard	764,559 cubic centimeters
1 cubic yard	46,656 cubic inches
1 cubic yard	27 cubic feet
1 cubic yard	0.7646 cubic meters
1 fluid ounce	8 fluid drams
1 fluid ounce	.03125 U.S. quarts
1 fluid ounce	.007812 U.S. gallons
1 fluid ounce	.29573 liters
1 U.S. quart	256 fluid drams
1 U.S. quart	32 fluid ounces
1 U.S. quart	.94633 liters
1 U.S. gallon	3.78533 liters
1 U.S. gallon	268.8 cubic inches
1 cubic centimeter	0.061 cubic inches
1 milliliter	0.03381 fluid ounces
1 milliliter	0.001057 U.S. quarts
1 milliliter	0.000264 U.S. gallons
1 cubic decimeter	1 liter
1 liter	61.023 cubic inches
1 liter	0.0353 cubic feet
1 liter	1.0567 U.S. quarts
1 liter	0.2642 U.S. gallons
1 liter	33.8147 fluid ounces

Table C–20 Capacity and Volume Conversion Formulas

To Convert	into	Multiply by
U.S. pints	liters	0.4732
U.S. quarts	liters	0.9463
U.S. gallons	liters	3.78533
U.S. gallons	cubic meters	0.00455
cubic inches	cubic centimeters	16.387064
cubic feet	liters	28.316847
cubic yards	cubic meters	0.764555
liters	U.S. pints	2.1134
liters	U.S. quarts	1.0567
liters	U.S. gallons	0.2642

Table C–21 Dry Measure

1 pint	0.5506 liters
1 quart	1.1012 liters
1 peck	8.8096 liters
1 bushel	35.2383 liters
1 liter	1.8162 pints
1 liter	.9081 quarts
1 liter	.1135 pecks
1 liter	.0284 bushels

Table C–22 Liquid Measure

1 dash	6 drops—1⅓ milliliters
1 teaspoon	⅛ fluid ounces—4 milliliters
1 tablespoon	½ fluid ounces—15 milliliters
1 pony	1 fluid ounce—28⅓ milliliters
1 jigger	1½ fluid ounces—42½ milliliters
1 miniature	3.4 fluid ounces—100 milliliters
1 nip (split)	6.3 fluid ounces—187 milliliters
1 half-bottle	12.7 fluid ounces—375 milliliters
1 pint	16 fluid ounces—378½ milliliters
1 bottle	25.4 fluid ounces—750 milliliters
1 fifth	25.6 fluid ounces—757 milliliters
1 quart	32 fluid ounces—946⅓ milliliters
1 liter	33.8 fluid ounces—1000 milliliters
1½ liters	50.7 fluid ounces—1500 milliliters
3 liters	101.4 fluid ounces—3000 milliliters
1 gallon	128.0 fluid ounces—3785 milliliters

Table C-23 U.S. Gallons into Liters

Gal.	0	1	2	3	4	5	6	7	8	9
0	—	3.785	7.571	11.356	15.141	18.927	22.712	26.497	30.283	34.068
10	37.853	41.639	45.424	49.209	52.995	56.780	50.565	64.351	68.136	71.921
20	75.707	79.492	83.277	87.063	90.848	94.633	98.419	102.204	105.989	109.775
30	113.560	117.345	121.131	124.916	128.701	132.487	136.272	140.057	143.843	147.628
40	151.413	155.199	158.984	162.769	166.555	170.340	174.125	177.911	181.696	185.481
50	189.267	193.052	196.837	200.622	204.408	208.193	211.978	215.764	219.549	223.334
60	227.120	230.905	234.690	238.476	242.261	246.046	249.832	253.617	257.402	261.188
70	264.973	268.758	272.544	276.329	280.114	283.900	287.685	291.470	295.256	299.041
80	302.826	306.612	310.397	314.182	317.968	321.753	325.538	329.324	333.109	336.894
90	340.680	344.465	348.250	352.036	355.821	359.606	363.392	367.177	370.962	374.748
100	378.533	382.318	386.104	389.889	393.674	397.460	401.245	405.030	408.816	412.601

Table C-24 Liters into U.S. Gallons

Liters	0	1	2	3	4	5	6	7	8	9
0	—	.265	.528	.793	1.057	1.321	1.585	1.849	2.114	2.378
10	2.642	2.906	3.170	3.435	3.699	3.963	4.227	4.491	4.756	5.020
20	5.284	5.548	5.812	6.077	6.341	6.605	6.869	7.133	7.398	7.662
30	7.926	8.190	8.454	8.719	8.983	9.247	9.511	9.775	10.040	10.304
40	10.568	10.832	11.096	11.361	11.625	11.889	12.153	12.417	12.682	12.946
50	13.210	13.474	13.738	14.003	14.267	14.531	14.795	15.059	15.324	15.588
60	15.852	16.116	16.380	16.645	16.909	17.173	17.437	17.701	17.966	18.230
70	18.494	18.758	19.022	19.287	19.551	19.815	20.079	20.343	20.608	20.872
80	21.136	21.400	21.664	21.929	22.193	22.457	22.721	22.985	23.250	23.514
90	23.778	24.042	24.306	24.571	24.835	25.099	25.363	25.627	25.892	26.156
100	26.420	26.684	26.948	27.213	27.477	27.741	28.005	28.269	28.534	28.798

Table C-25 Cubic Inches into Cubic Centimeters

Cu. In.	0	1	2	3	4	5	6	7	8	9
					Cu. Cm					
0	—	16.39	32.77	49.16	65.55	81.94	98.32	114.71	131.10	147.48
10	163.87	180.26	196.64	213.03	229.42	245.81	262.19	278.58	294.97	311.35
20	327.74	344.13	360.51	376.90	393.29	409.68	426.06	442.45	458.84	475.22
30	491.61	508.00	524.38	540.77	557.16	573.55	589.93	606.32	622.71	639.09
40	655.48	671.87	688.25	704.64	721.03	737.42	753.80	770.19	786.58	802.96
50	819.35	835.74	852.12	868.51	884.90	901.29	917.67	934.06	950.45	966.83
60	983.22	999.61	1016.0	1032.4	1048.8	1065.2	1081.5	1097.9	1143.3	1130.7
70	1147.1	1163.5	1179.9	1196.3	1212.6	1229.0	1245.4	1261.8	1278.2	1294.6
80	1311.0	1327.4	1343.7	1360.1	1376.5	1392.9	1409.3	1425.7	1442.1	1458.5
90	1474.8	1491.2	1507.6	1524.0	1540.4	1556.8	1573.2	1589.6	1605.9	1622.3
100	1638.7	1655.1	1671.5	1687.9	1704.3	1720.7	1737.1	1753.4	1769.8	1786.2

Table C-26 Cubic Centimeters into Cubic Inches

Cu. Cm	0	1	2	3	4	5	6	7	8	9
					Cu. In					
0	—	0.0610	0.1221	0.1831	0.2441	0.3051	0.3661	0.4272	0.4882	0.5492
10	0.6102	0.6713	0.7323	0.7933	0.8543	0.9154	0.9764	1.0374	1.0984	1.1595
20	1.2205	1.2815	1.3425	1.4036	1.4646	1.5256	1.5866	1.6477	1.7087	1.7697
30	1.8307	1.8917	1.9528	2.0138	2.0748	2.1358	2.1969	2.2579	2.3189	2.3799
40	2.4410	2.5020	2.5630	2.6240	2.6850	2.7461	2.8071	2.8681	2.9291	2.9902
50	3.0512	3.1122	3.1732	3.2343	3.2953	3.3563	3.4173	3.4784	3.5394	3.6004
60	3.6614	3.7225	3.7835	3.8445	3.9055	3.9665	4.0276	4.0886	4.1496	4.2106
70	4.2717	4.3327	4.3937	4.4548	4.5158	4.5768	4.6378	4.6988	4.7599	4.8209
80	4.8819	4.9429	5.0040	5.0650	5.1260	5.1870	5.2480	5.3091	5.3701	5.4311
90	5.4921	5.5532	5.6142	5.6752	5.7362	5.7973	5.8583	5.9193	5.9803	6.0414
100	6.1024	6.1634	6.2244	6.2854	6.3465	6.4075	6.4685	6.5295	6.5906	6.6516

Table C–27 Cubic Feet into Liters

Cu. Ft	0	1	2	3	4	5	6	7	8	9
					Liters					
0	—	28.32	56.63	84.95	113.26	141.58	169.90	198.21	226.53	254.84
10	283.16	311.48	339.79	368.11	396.42	424.74	453.06	481.37	509.69	538.01
20	566.32	594.64	622.95	651.27	679.59	707.90	736.22	764.53	792.85	821.17
30	849.48	877.80	906.11	934.43	962.75	991.06	1019.4	1047.7	1076.0	1104.3
40	1132.6	1161.0	1189.3	1217.6	1245.9	1274.2	1302.5	1330.9	1359.2	1387.5
50	1415.8	1444.1	1472.4	1500.8	1529.1	1557.4	1585.7	1614.0	1642.3	1670.7
60	1699.0	1727.3	1755.6	1783.9	1812.2	1840.5	1868.9	1897.2	1925.5	1953.8
70	1982.1	2010.4	2038.8	2067.1	2095.4	2123.7	2152.0	2180.3	2208.7	2237.0
80	2265.3	2293.6	2321.9	2350.2	2378.6	2406.9	2435.2	2463.5	2491.8	2520.1
90	2548.4	2576.8	2605.1	2633.4	2661.7	2690.0	2718.3	2746.7	2775.0	2803.3
100	2831.6	2859.9	2888.2	2916.6	2944.9	2973.2	3001.5	3029.8	3058.1	3086.5

Table C–28 Liters into Cubic Feet

Liters	0	1	2	3	4	5	6	7	8	9
					Cu. Ft					
0		0.0353	0.0706	0.1060	0.1413	0.1766	0.2119	0.2472	0.2825	0.3178
10	0.3532	0.3885	0.4238	0.4591	0.4944	0.5297	0.5651	0.6004	0.6357	0.6710
20	0.7063	0.7416	0.7769	0.8123	0.8476	0.8829	0.9182	0.9535	0.9888	1.0242
30	1.0595	1.0948	1.1301	1.1654	1.2007	1.2361	1.2714	1.3067	1.3420	1.3773
40	1.4126	1.4479	1.4833	1.5186	1.5539	1.5892	1.6245	1.6598	1.6952	1.7305
50	1.7658	1.8011	1.8364	1.8717	1.9071	1.9424	1.9777	2.0130	2.0483	2.0836
60	2.1189	2.1543	2.1896	2.2249	2.2602	2.2955	2.3308	2.3662	2.4015	2.4368
70	2.4721	2.5074	2.5427	2.5780	2.6134	2.6487	2.6840	2.7193	2.7546	2.7899
80	2.8253	2.8606	2.8959	2.9312	2.9665	3.0018	3.0372	3.0725	3.1078	3.1431
90	3.1784	3.2137	3.2490	3.2844	3.3197	3.3550	3.3903	3.4256	3.4609	3.4963
100	3.5315	3.5669	3.6021	3.6375	3.6728	3.7081	3.7434	3.7787	3.8140	3.8493

Table C–29 Cubic Feet into Cubic Meters

Cu. Ft	0	1	2	3	4	5	6	7	8	9
					Cu. M					
0	—	0.0283	0.0566	0.0850	0.1133	0.1416	0.1689	0.1982	0.2265	0.2549
10	0.2832	0.3115	0.3398	0.3681	0.3964	0.4248	0.4531	0.4814	0.5097	0.5380
20	0.5663	0.5947	0.6230	0.6513	0.6796	0.7079	0.7352	0.7646	0.7929	0.8212
30	0.8495	0.8778	0.9061	0.9345	0.9628	0.9911	1.0194	1.0477	1.6760	1.1044
40	1.1327	1.1610	1.1893	1.2176	1.2459	1.2743	1.3026	1.3309	1.3592	1.3875
50	1.4158	1.4442	1.4725	1.5008	1.5291	1.5574	1.5857	1.6141	1.6424	1.6707
60	1.6990	1.7273	1.7556	1.7840	1.8123	1.8406	1.8689	1.8972	1.9256	1.9539
70	1.9822	2.0105	2.0388	2.0671	2.0955	2.1238	2.1521	2.1804	2.2087	2.2370
80	2.2654	2.2937	2.3220	2.3503	2.3786	2.4069	2.4353	2.4636	2.4919	2.5202
90	2.5485	2.5768	2.6052	2.6335	2.6618	2.6901	2.7184	2.7467	2.7751	2.8034
100	2.8317	2.8600	2.8883	2.9166	2.9450	2.9733	3.0016	3.0299	3.0582	3.0865

Table C–30 Cubic Meters into Cubic Feet

Cu. M	0	1	2	3	4	5	6	7	8	9
					Cu. Ft					
0	—	35.3	70.6	105.9	141.3	176.6	211.9	247.2	282.5	317.8
10	353.1	388.5	423.8	459.1	494.4	529.7	565.0	600.3	635.7	671.0
20	706.3	741.6	776.9	812.2	847.6	882.9	918.2	953.5	988.8	1024.1
30	1059.4	1094.8	1130.1	1165.4	1200.7	1236.0	1271.3	1306.6	1342.0	1377.3
40	1412.6	1447.9	1483.2	1518.5	1553.9	1589.2	1624.5	1659.8	1695.1	1730.4
50	1765.7	1801.1	1836.4	1871.7	1907.0	1942.3	1977.6	2012.9	2048.3	2083.6
60	2118.9	2154.2	2189.5	2224.8	2260.1	2295.5	2330.8	2366.1	2401.4	2436.7
70	2472.0	2507.3	2542.7	2578.0	2613.3	2648.6	2683.9	2719.2	2754.5	2789.9
80	2825.2	2860.5	2895.8	2931.1	2966.4	3001.8	3037.1	3072.4	3107.7	3143.0
90	3178.3	3213.6	3249.0	3284.3	3319.6	3354.9	3390.2	3425.5	3460.8	3496.2
100	3531.5	3566.8	3602.1	3637.4	3672.7	3708.0	3743.4	3778.7	3814.0	3849.3

Table C–31 Basic Conversions: Units of Length

1 inch	25.4 millimeters
1 inch	2.54 centimeters
1 inch	0.0254 meters
1 inch	0.08333 feet
1 inch	0.02777 yards
1 foot	304.8 millimeters
1 foot	30.48 centimeters
1 foot	0.3048 meters
1 foot	12 inches
1 foot	0.3333 yards
1 yard	91.44 centimeters
1 yard	0.9144 meters
1 yard	36 inches
1 yard	3 feet
1 mile	1609.34 meters
1 mile	63,360 inches
1 mile	5,280 feet
1 mile	1760 yards
1 rod	16.5 feet
1 furlong	40 rods
8 furlongs	1 mile
1 league	3 miles
1 knot	6,085 feet
1 knot	1.1526 miles
1 millimeter	.03937 inches
1 millimeter	1000 micrometers (μms)
1 centimeter	0.3937 inches
1 centimeter	0.03281 feet
1 centimeter	0.01094 yards
1 centimeter	0.01 meters
1 meter	39.37 inches
1 meter	3.2808 feet
1 meter	1.0936 yards
1 meter	100 centimeters
1 kilometer	0.62137 miles

Table C–32 Length Conversion Formulas

To Convert	into	Multiply by
inches	millimeters	25.4
inches	centimeters	2.54
feet	meters	0.304800
yards	meters	0.914400
miles	kilometers	1.609344
centimeters	inches	0.393701
meters	feet	3.280840
meters	yards	1.093613
kilometers	miles	0.6213712

Table C–33 Millimeters into Inches

Mm	In.	Mm	In.
1	.03937	19	.74803
2	.07874	20	.78740
3	.11811	21	.82677
4	.15743	22	.86614
5	.19685	23	.90551
6	.23622	24	.94488
7	.27559	25	.98425
8	.31496	25.4	1.0
9	.35433	38.1	1.5
10	.39370	50.0	1.968
11	.43307	50.8	2.0
12	.47244	75.0	2.953
13	.51181	76.2	3.0
14	.55118	100.0	3.937
15	.59055	101.6	4.0
16	.62992	127.0	5.0
17	.66929	152.4	6.0
18	.70866		

Table C–34 Length Equivalents

Fraction In.	Decimal In.	Mm
1/64	.015625	0.397
1/32	.03125	0.794
3/64	.046875	1.191
1/16	.0625	1.588
5/64	.078125	1.984
3/32	.09375	2.381
1/8	.125	3.175
3/16	.1875	4.763
1/4	.25	6.35
5/16	.3125	7.938
3/8	.375	9.525
3/16	.4375	11.113
1/2	.50	12.70
9/16	.5625	14.288
5/8	.625	15.875
11/16	.6875	17.463
3/4	.75	19.05
13/16	.8125	20.638
7/8	.875	22.225
15/16	.9375	23.813
1	1.0	25.40

Table C–35 Hundredths of a Millimeter into Inches

Mm	In.	Mm	In.	Mm	In.	Mm	In.	Mm	In.
0.01	0.0004	0.21	0.0083	0.41	0.0161	0.61	0.0240	0.81	0.0319
0.02	0.0008	0.22	0.0087	0.42	0.0165	0.62	0.0244	0.82	0.0323
0.03	0.0012	0.23	0.0091	0.43	0.0169	0.63	0.0248	0.83	0.0327
0.04	0.0016	0.24	0.0094	0.44	0.0172	0.64	0.0252	0.84	0.0331
0.05	0.0020	0.25	0.0098	0.45	0.0177	0.65	0.0256	0.85	0.0335
0.06	0.0024	0.26	0.0102	0.46	0.0181	0.66	0.0260	0.86	0.0339
0.07	0.0028	0.27	0.0106	0.47	0.0185	0.67	0.0264	0.87	0.0343
0.08	0.0032	0.28	0.0110	0.48	0.0189	0.68	0.0268	0.88	0.0346
0.09	0.0035	0.29	0.0114	0.49	0.0193	0.69	0.0272	0.89	0.0350
0.10	0.0039	0.30	0.0118	0.50	0.0197	0.70	0.0276	0.90	0.0354
0.11	0.0043	0.31	0.0122	0.51	0.0201	0.71	0.0280	0.91	0.0358
0.12	0.0047	0.32	0.0126	0.52	0.0205	0.72	0.0283	0.92	0.0362
0.13	0.0051	0.33	0.0130	0.53	0.0209	0.73	0.0287	0.93	0.0366
0.14	0.0055	0.34	0.0134	0.54	0.0213	0.74	0.0291	0.94	0.0370
0.15	0.0059	0.35	0.0138	0.55	0.0217	0.75	0.0295	0.95	0.0374
0.16	0.0063	0.36	0.0142	0.56	0.0220	0.76	0.0299	0.96	0.0378
0.17	0.0067	0.37	0.0146	0.57	0.0224	0.77	0.0303	0.97	0.0382
0.18	0.0071	0.38	0.0150	0.58	0.0228	0.78	0.0307	0.98	0.0386
0.19	0.0075	0.39	0.0154	0.59	0.0232	0.79	0.0311	0.99	0.0390
0.20	0.0079	0.40	0.0157	0.60	0.0236	0.80	0.0315	1.00	0.0394

Table C–36 Inches into Millimeters

In.	Mm	In.	Mm	In.	Mm
1/64	0.3969	51/64	20.2406	2-5/32	54.7688
1/32	0.7938	13/16	20.5375	2-3/16	55.5625
3/64	1.1906	53/64	21.0344	2-9/32	56.3563
1/16	1.5875	27/32	21.4313	2-1/4	57.1500
5/64	1.9844	55/64	21.8281	2-9/32	57.9438
3/32	2.3813	7/8	22.2250	2-5/16	58.7375
7/64	2.7781	57/64	22.6219	2-11/32	59.5313
1/8	3.1750	29/32	23.0188	2-3/8	60.3250
9/64	3.5719	59/64	23.4156	2-13/32	61.1188
5/32	3.9688	15/16	23.8125	2-7/16	61.9125
11/64	4.3656	61/64	24.2094	2-15/32	62.7063
3/16	4.7625	31/32	24.6063	2-1/2	63.5000
13/64	5.1594	63/64	25.0031	2-17/32	64.2938
7/32	5.5563	1	25.4000	2-9/16	65.0875
15/64	5.9531	1-1/32	26.1938	2-19/32	65.8813
1/4	6.3500	1-1/16	26.9875	2-5/8	66.6750
17/64	6.7469	1-3/32	27.7813	2-21/32	67.4688
9/32	7.1438	1-1/6	28.5750	2-11/16	68.2625
19/64	7.5406	1-5/32	29.3688	2-23/32	69.0563
5/16	7.9375	1-3/16	30.1625	2-3/4	69.8500
21/64	8.3344	1-7/32	30.9563	2-25/32	70.6438
11/32	8.7313	1-1/4	31.7500	2-13/16	71.4375
23/64	9.1281	1-9/32	32.5438	2-27/32	72.2313
3/8	9.5250	1-5/16	33.3375	2-7/8	73.0250
25/64	9.9219	1-11/32	34.1313	2-29/32	73.8188
13/32	10.3188	1-3/8	34.9250	2-15/16	74.6125
27/64	10.7156	1-13/32	35.7188	2-31/32	75.4063
7/16	11.1125	1-7/16	36.5125	3	76.2000
29/64	11.5094	1-15/32	37.3063	3-1/32	76.9938
15/32	11.9063	1-1/2	38.1000	3-1/16	77.7875
31/64	12.3031	1-17/32	38.8938	3-3/32	78.5813
1/2	12.7000	1-9/16	39.6875	3-1/8	79.3750
33/64	13.0969	1-19/32	40.4813	3-5/32	80.1688
17/32	13.4938	1-5/8	41.2750	3-3/16	80.9625
35/64	13.8906	1-21/32	42.0688	3-7/32	81.7563
9/16	14.2875	1-11/16	42.8625	3-1/4	82.5500
37/64	14.6844	1-23/32	43.6563	3-9/32	83.3438
19/32	15.0813	1-3/4	44.4500	3-5/16	84.1375
39/64	15.4781	1-25/32	45.2438	3-11/32	84.9313
5/8	15.8750	1-13/16	46.0375	3-3/8	85.7250
41/64	16.2719	1-27/32	46.8313	3-13/32	86.5188
21/32	16.6688	1-7/8	47.6250	3-7/16	87.3125
43/64	17.0656	1-29/32	48.4188	3-15/32	88.1063
11/16	17.4625	1-15/16	49.2125	3-1/2	88.9000
45/64	17.8594	1-31/32	50.0063	3-17/32	89.6938
23/32	18.2563	2	50.8000	3-9/16	90.4875
47/64	18.6531	2-1/32	51.5938	3-19/32	91.2813
3/4	19.0550	2-1/16	52.3875	3-5/8	92.0750
49/64	19.4469	2-3/32	53.1813	3-21/32	92.8688
25/32	19.8438	2-1/8	53.9750	3-11/16	93.6625

Table C–36 (continued)

In.	Mm	In.	Mm	In.	Mm
3-23/32	94.4563	5-9/32	134.144	7-11/16	195.262
3-3/4	95.2500	5-5/16	134.938	7-3/4	196.850
3-25/32	96.0438	5-11/32	135.731	7-13/16	198.438
3-13/16	96.8375	5-3/8	136.525	7-7/8	200.025
3-27/32	97.6313	5-13/32	137.319	7-15/16	201.612
3-7/8	98.4250	5-7/16	138.112	8	203.200
3-29/32	99.2188	5-15/32	138.906	8-1/16	204.783
3-15/16	100.012	5-1/2	139.700	8-1/8	206.375
3-31/32	100.806	5-17/32	140.494	8-3/16	207.962
4	101.600	5-9/16	141.288	8-1/4	209.550
4-1/32	102.394	5-19/32	142.081	8-5/16	211.138
4-1/16	103.188	5-5/8	142.875	8-3/8	212.725
4-3/32	103.981	5-21/32	143.669	8-7/16	214.312
4-1/8	104.775	5-11/16	144.462	8-1/2	215.900
4-5/32	105.569	5-23/32	145.256	8-9/16	217.488
4-3/16	106.362	5-3/4	146.050	8-6/8	219.075
4-7/32	107.156	5-25/32	146.844	8-11/16	220.662
4-1/4	107.950	5-13/16	147.638	8-3/4	222.250
4-9/32	108.744	5-27/32	148.431	8-13/16	223.838
4-5/16	109.538	5-7/8	149.225	8-7/8	225.425
4-11/32	110.331	5-29/32	150.019	8-15/16	227.012
4-3/8	111.125	5-15/16	150.812	9	228.600
4-13/32	111.919	5-31/32	151.606	9-1/16	230.188
4-7/16	112.712	6	152.400	9-1/8	231.775
4-15/32	113.506	6-1/16	153.988	9-3/16	233.362
4-1/2	114.300	6-1/8	155.575	9-1/4	234.950
4-17/32	115.094	6-3/16	157.162	9-5/16	236.538
4-9/16	115.888	6-1/4	158.750	9-3/8	238.125
4-19/32	116.681	6-5/16	160.338	9-7/16	239.712
4-5/8	117.475	6-3/8	161.925	9-1/2	241.300
4-21/32	118.269	6-7/16	163.512	9-9/16	242.888
4-11/16	119.062	6-1/2	165.100	9-5/8	244.475
4-23/32	119.856	6-9/16	166.688	9-11/16	246.062
4-3/4	120.650	6-5/8	168.275	9-3/4	247.650
4-25/32	121.444	6-11/16	169.862	9-13/16	249.238
4-13/18	122.238	6-3/4	171.450	9-7/8	250.825
4-27/32	123.031	6-13/16	173.038	9-15/16	252.412
4-7/8	123.825	6-7/8	174.625	10	254.000
4-29/32	124.619	6-15/16	176.212	10-1/16	255.588
4-15/16	125.412	7	177.800	10-1/8	257.175
4-31/32	126.206	7-1/16	179.388	10-3/16	258.762
5	127.000	7-1/8	180.975	10-1/4	260.350
5-1/32	127.794	7-3/16	182.562	10-5/16	261.938
5-1/16	128.588	7-1/4	184.150	10-3/8	263.525
5-3/32	129.381	7-5/16	185.738	10-7/16	265.112
5-1/8	130.175	7-3/8	187.325	10-1/2	266.700
5-5/32	130.969	7-7/16	188.912	10-9/16	268.288
5-3/16	131.762	7-1/2	190.500	10-5/8	269.875
5-7/32	132.556	7-9/16	192.088	10-11/16	271.462
5-1/4	133.350	7-5/8	193.675	10-3/4	273.050

Table C–36 (*continued*)

In.	Mm	In.	Mm	In.	Mm
10-13/16	274.638	13	330.200	28	711.200
10-7/8	276.225	14	355.600	29	736.600
10-15/16	277.812	15	381.000	30	762.000
11	279.400	16	406.400	31	787.400
11-1/16	280.988	17	431.800	32	812.800
11-1/8	282.575	18	457.200	33	833.200
11-3/16	284.162	19	482.600	34	863.600
11-1/4	285.750	20	508.000	35	889.000
11-5/16	287.338	21	533.400	36	914.400
11-3/8	288.925	22	558.800	37	939.800
11-7/16	290.512	23	584.200	38	965.200
11-1/2	292.100	24	609.600	39	990.600
11-9/16	293.688	25	635.000	40	1016.00
11-5/8	295.275	26	660.400	41	1041.40
11-11/16	296.862	27	685.800	42	1066.80
11-3/4	298.450				
11-13/16	300.038				
11-7/8	301.625				
11-15/16	303.212				
12	304.800				

Table C–37 Millimeters into Inches

Mm	In.	Mm	In.	Mm	In.	Mm	In.
1	0.0394	51	2.0079	101	3.9764	151	5.9449
2	0.0787	52	2.0472	102	4.0158	152	5.9843
3	0.1181	53	2.0866	103	4.0551	153	6.0236
4	0.1575	54	2.1260	104	4.0945	154	6.0630
5	0.1969	55	2.1654	105	4.1339	155	6.1024
6	0.2362	56	2.2047	106	4.1732	156	6.1417
7	0.2756	57	2.2441	107	4.2126	157	6.1811
8	0.3150	58	2.2835	108	4.2520	158	6.2205
9	0.3543	59	2.3228	109	4.2913	159	6.2599
10	0.3937	60	2.3622	110	4.3307	160	6.2992
11	0.4331	61	2.4016	111	4.3701	161	6.3386
12	0.4724	62	2.4409	112	4.4095	162	6.3780
13	0.5118	63	2.4803	113	4.4488	163	6.4173
14	0.5512	64	2.5197	114	4.4882	164	6.4567
15	0.5906	65	2.5591	115	4.5276	165	6.4961
16	0.6299	66	2.5984	116	4.5669	166	6.5354
17	0.6693	67	2.6378	117	4.6063	167	6.5748
18	0.7087	68	2.6772	118	4.6457	168	6.6142
19	0.7480	69	2.7165	119	4.6850	169	6.6535
20	0.7874	70	2.7559	120	4.7244	170	6.6929
21	0.8268	71	2.7953	121	4.7638	171	6.7323
22	0.8661	72	2.8347	122	4.8032	172	6.7717
23	0.9055	73	2.8740	123	4.8425	173	6.8110
24	0.9449	74	2.9134	124	4.8819	174	6.8504
25	0.9843	75	2.9528	125	4.9213	175	6.8898
26	1.0236	76	2.9921	126	4.9606	176	6.9291
27	1.0630	77	3.0315	127	5.0000	177	6.9685
28	1.1024	78	3.0709	128	5.0394	178	7.0079
29	1.1417	79	3.1102	129	5.0787	179	7.0472
30	1.1811	80	3.1496	130	5.1181	180	7.0866
31	1.2205	81	3.1890	131	5.1575	181	7.1260
32	1.2598	82	3.2284	132	5.1969	182	7.1654
33	1.2992	83	3.2677	133	5.2362	183	7.2047
34	1.3386	84	3.3071	134	5.2756	184	7.2441
35	1.3780	85	3.3465	135	5.3150	185	7.2835
36	1.4173	86	3.3858	136	5.3543	186	7.3228
37	1.4567	87	3.4252	137	5.3937	187	7.3622
38	1.4961	88	3.4646	138	5.4331	188	7.4016
39	1.5354	89	3.5039	139	5.4724	189	7.4409
40	1.5748	90	3.5433	140	5.5118	190	7.4803
41	1.6142	91	3.5827	141	5.5512	191	7.5197
42	1.6535	92	3.6221	142	5.5906	192	7.5591
43	1.6929	93	3.6614	143	5.6299	193	7.5984
44	1.7323	94	3.7008	144	5.6693	194	7.6378
45	1.7717	95	3.7402	145	5.7087	195	7.6772
46	1.8110	96	3.7795	146	5.7480	196	7.7165
47	1.8504	97	3.8189	147	5.7874	197	7.7559
48	1.8898	98	3.8583	148	5.8268	198	7.7953
49	1.9291	99	3.8976	149	5.8661	199	7.8347
50	1.9685	100	3.9370	150	5.9055	200	7.8740

Table C–37 (continued)

Mm	In.	Mm	In.	Mm	In.	Mm	In.
201	7.9134	251	9.8819	301	11.8504	351	13.8189
202	7.9528	252	9.9213	302	11.8898	352	13.8583
203	7.9921	253	9.9606	303	11.9291	353	13.8976
204	8.0315	254	10.0000	304	11.9686	354	13.9370
205	8.0709	255	10.0393	305	12.0079	355	13.9764
206	8.1102	256	10.0787	306	12.0472	356	14.0157
207	8.1496	257	10.1181	307	12.0866	357	14.0551
208	8.1890	258	10.1575	308	12.1260	358	14.0945
209	8.2284	259	10.1969	309	12.1654	359	14.1339
210	8.2677	260	10.2362	310	12.2047	360	14.1732
211	8.3071	261	10.2756	311	12.2441	361	14.2126
212	8.3465	262	10.3150	312	12.2835	362	14.2520
213	8.3858	263	10.3543	313	12.3228	363	14.2913
214	8.4252	264	10.3937	314	12.3622	364	14.3307
215	8.4646	265	10.4331	315	12.4016	365	14.3701
216	8.5039	266	10.4724	316	12.4409	366	14.4094
217	8.5433	267	10.5118	317	12.4803	367	14.4488
218	8.5827	268	10.5512	318	12.5197	368	14.4882
219	8.6221	269	10.5906	319	12.5591	369	14.5276
220	8.6614	270	10.6299	320	12.5984	370	14.5669
221	8.7008	271	10.6693	321	12.6378	371	14.6063
222	8.7402	272	10.7087	322	12.6772	372	14.6457
223	8.7795	273	10.7480	323	12.7165	373	14.6850
224	8.8189	274	10.7874	324	12.7559	374	14.7244
225	8.8583	275	10.8268	325	12.7953	375	14.7638
226	8.8976	276	10.8661	326	12.8346	376	14.8031
227	8.9370	277	10.9055	327	12.8740	377	14.8425
228	8.9764	278	10.9449	328	12.9134	378	14.8819
229	9.0158	279	10.9843	329	12.9528	379	14.9213
230	9.0551	280	11.0236	330	12.9921	380	14.9606
231	9.0945	281	11.0630	331	13.0315	381	15.0000
232	9.1339	282	11.1024	332	13.0709	382	15.0394
233	9.1732	283	11.1417	333	13.1102	383	15.0787
234	9.2126	284	11.1811	334	13.1496	384	15.1181
235	9.2520	285	11.2205	335	13.1890	385	15.1575
236	9.2913	286	11.2598	336	13.2283	386	15.1969
237	9.3307	287	11.2992	337	13.2677	387	15.2362
238	9.3701	288	11.3386	338	13.3071	388	15.2756
239	9.4095	289	11.3780	339	13.3465	389	15.3150
240	9.4488	290	11.4173	340	13.3858	390	15.3543
241	9.4882	291	11.4567	341	13.4252	391	15.3937
242	9.5276	292	11.4961	342	13.4646	392	15.4331
243	9.5669	293	11.5354	343	13.5039	393	15.4724
244	9.6063	294	11.5748	344	13.5433	394	15.5118
245	9.6457	295	11.6142	345	13.5827	395	15.5512
246	9.6850	296	11.6535	346	13.6220	396	15.5906
247	9.7244	297	11.6929	347	13.6614	397	15.6299
248	9.7638	298	11.7323	348	13.7008	398	15.6693
249	9.8031	299	11.7717	349	13.7402	399	15.7087
250	9.8425	300	11.8110	350	13.7795	400	15.7480

Table C–37 (*continued*)

Mm	In.	Mm	In.	Mm	In.	Mm	In.
401	15.7874	451	17.7559	501	19.7244	551	21.6929
402	15.8268	452	17.7963	502	19.7638	552	21.7323
403	15.8661	453	17.8346	503	19.8031	553	21.7717
404	15.9055	454	17.8740	504	19.8425	554	21.8110
405	15.9449	455	17.9134	505	19.8819	555	21.8504
406	15.9832	456	17.9528	506	19.9213	556	21.8898
407	16.0236	457	17.9921	507	19.9606	557	21.9291
408	16.0630	458	18.0315	508	20.0000	558	21.9685
409	16.1024	459	18.0709	509	20.0394	559	22.0079
410	16.1417	460	18.1102	510	20.0787	560	22.0472
411	16.1811	461	18.1496	511	20.1181	561	22.0866
412	16.2205	462	18.1890	512	20.1575	562	22.1260
413	16.2598	463	18.2283	513	20.1969	563	22.1654
414	16.2992	464	18.2677	514	20.2362	564	22.2047
415	16.3386	465	18.3071	515	20.2756	565	22.2441
416	16.3780	466	18.3465	516	20.3150	566	22.2835
417	16.4173	467	18.3358	517	20.3543	567	22.3228
418	16.4567	468	18.4252	518	20.3937	568	22.3622
419	16.4961	469	18.4646	519	20.4331	569	22.4016
420	16.5354	470	18.5039	520	20.4724	570	22.4409
421	16.5748	471	18.5433	521	20.5118	571	22.4803
422	16.6142	472	18.5827	522	20.5512	572	22.5197
423	16.6535	473	18.6220	523	20.5906	573	22.5591
424	16.6929	474	18.6614	524	20.6299	574	22.5984
425	16.7323	475	18.7008	525	20.6693	575	22.6378
426	16.7716	476	18.7402	526	20.7087	576	22.6772
427	16.8110	477	18.7795	527	20.7480	577	22.7165
428	16.8504	478	18.8189	528	20.7874	578	22.7559
429	16.8898	479	18.8583	529	20.8268	579	22.7953
430	16.9291	480	18.8976	530	20.8661	580	22.8346
431	16.9685	481	18.9370	531	20.9055	581	22.8740
432	17.0079	482	18.9764	532	20.9449	582	22.9134
433	17.0472	483	19.0157	533	20.9843	583	22.9528
434	17.0866	484	19.0551	534	21.0236	584	22.9921
435	17.1260	485	19.0945	535	21.0630	585	23.0315
436	17.1654	486	19.1339	536	21.1024	586	23.0709
437	17.2047	487	19.1732	537	21.1417	587	23.1102
438	17.2441	488	19.2126	538	21.1811	588	23.1496
439	17.2835	489	19.2520	539	21.2205	589	23.1890
440	17.3228	490	19.2813	540	21.2598	590	23.2283
441	17.3622	491	19.3307	541	21.2992	591	23.2677
442	17.4016	492	19.3701	542	21.3386	592	23.3071
443	17.4409	493	19.4094	543	21.3780	593	23.3465
444	17.4803	494	19.4488	544	21.4173	594	23.3858
445	17.5197	495	19.4882	545	21.4567	595	23.4252
446	17.5591	496	19.5276	546	21.4961	596	23.4646
447	17.5984	497	19.5669	547	21.5354	597	23.5039
448	17.6378	498	19.6063	548	21.5743	598	23.5433
449	17.6772	499	19.6457	549	21.6142	599	23.5827
450	17.7165	500	19.6850	550	21.6535	600	23.6220

Table C–37 (*continued*)

Mm	In.	Mm	In.	Mm	In.	Mm	In.
601	23.6614	651	25.6299	701	27.5984	751	29.5669
602	23.7008	652	25.6693	702	27.6378	752	29.6063
603	23.7402	653	25.7087	703	27.6772	753	29.6457
604	23.7795	654	25.7480	704	27.7165	754	29.6850
605	23.8189	655	25.7874	705	27.7559	755	29.7244
606	23.8583	656	25.8268	706	27.7953	756	29.7638
607	23.8976	657	25.8661	707	27.8346	757	29.8031
608	23.9370	658	25.9055	708	27.8740	758	29.8425
609	23.9764	659	25.9449	709	27.9134	759	29.8819
610	24.0157	660	25.9843	710	27.9528	760	29.9213
611	24.0551	661	26.0236	711	27.9921	761	29.9606
612	24.0945	662	26.0630	712	28.0315	762	30.0000
613	24.1339	663	26.1024	713	28.0709	763	30.0394
614	24.1732	664	26.1417	714	28.1102	764	30.0787
615	24.2126	665	26.1811	715	28.1496	765	30.1181
616	24.2520	666	26.2205	716	28.1890	766	30.1575
617	24.2913	667	26.2598	717	28.2283	767	30.1969
618	24.3307	668	26.2992	718	28.2677	768	30.2362
619	24.3701	669	26.3386	719	28.3071	769	30.2756
620	24.4094	670	26.3780	720	28.3465	770	30.3150
621	24.4488	671	26.4173	721	28.3858	771	30.3543
622	24.4882	672	26.4567	722	28.4252	772	30.3937
623	24.5276	673	26.4961	723	28.4646	773	30.4331
624	24.5669	674	26.5354	724	28.5039	774	30.4724
625	24.6063	675	26.5748	725	28.5433	775	30.5118
626	24.6457	676	26.6142	726	28.5827	776	30.5512
627	24.6850	677	26.6535	727	28.6220	777	30.5906
628	24.7244	678	26.6929	728	28.6614	778	30.6299
629	24.7638	679	26.7323	729	28.7008	779	30.6693
630	24.8031	680	26.7717	730	28.7042	780	30.7087
631	24.8425	681	26.8110	731	28.7795	781	30.7480
632	24.8819	682	26.8504	732	28.8189	782	30.7874
633	24.9213	683	26.8898	733	28.8583	783	30.8268
634	24.9606	684	26.9291	734	28.8976	784	30.8661
635	25.0000	685	26.9685	735	28.9370	785	30.9055
636	25.0394	686	27.0079	736	28.9764	786	30.9449
637	25.0787	687	27.0472	737	29.0157	787	30.9843
638	25.1181	688	27.0866	738	29.0551	788	31.0236
639	25.1575	689	27.1260	739	29.0945	789	31.0630
640	25.1969	690	27.1654	740	29.1339	790	31.1024
641	25.2362	691	27.2047	741	29.1732	791	31.1417
642	25.2756	692	27.2441	742	29.2126	792	31.1811
643	25.3150	693	27.2835	743	29.2520	793	31.2205
644	25.3543	694	27.3228	744	29.2913	794	31.2598
645	25.3937	695	27.3622	745	29.3307	795	31.2992
646	25.4331	696	27.4016	746	29.3701	796	31.3386
647	25.4724	697	27.4409	747	29.4094	797	31.3780
648	25.5118	698	27.4803	748	29.4488	798	31.4173
649	25.5512	699	27.5197	749	29.4882	799	31.4567
650	25.5906	700	27.5591	750	29.5276	800	31.4961

Table C–37 (continued)

Mm	In.	Mm	In.	Mm	In.	Mm	In.
801	31.5354	851	33.5039	901	35.4724	951	37.4409
802	31.5748	852	33.5433	902	35.5118	952	37.4803
803	31.6142	853	33.5827	903	35.5512	953	37.5197
804	31.6535	854	33.6220	904	35.5906	954	37.5591
805	31.6929	855	33.6614	905	35.6299	955	37.5984
806	31.7323	856	33.7008	906	35.6693	956	37.6378
807	31.7717	857	33.7402	907	35.7087	957	37.6772
808	31.8110	858	33.7795	908	35.7480	958	37.7165
809	31.8504	859	33.8189	909	35.7874	959	37.7559
810	31.8898	860	33.8583	910	35.8268	960	37.7953
811	31.9291	861	33.8976	911	35.8661	961	37.8346
812	31.9685	862	33.9370	912	35.9055	962	37.8740
813	32.0079	863	33.9764	913	35.9449	963	37.9134
814	32.0472	864	34.0157	914	35.9843	964	37.9528
815	32.0866	865	34.0551	915	36.0236	965	37.9921
816	32.1260	866	34.0945	916	36.0630	966	38.0315
817	32.1654	867	34.1339	917	36.1024	967	38.0709
818	32.2047	868	34.1732	918	36.1417	968	38.1102
819	32.2441	869	34.2126	919	36.1811	969	38.1496
820	32.2835	870	34.2520	920	36.2205	970	38.1890
821	32.3228	871	34.2913	921	36.2598	971	38.2283
822	32.3622	872	34.3307	922	36.2992	972	38.2677
823	32.4016	873	34.3701	923	36.3386	973	38.3071
824	32.4409	874	34.4094	924	36.3780	974	38.3465
825	32.4803	875	34.4488	925	36.4173	975	38.3858
826	32.5197	876	34.4882	926	36.4567	976	38.4252
827	32.5591	877	34.5276	927	36.4961	977	38.4646
828	32.5984	878	34.5670	928	36.5354	978	38.5039
829	32.6378	879	34.6063	929	36.5748	979	38.5433
830	32.6772	880	34.6457	930	36.6142	980	38.5827
831	32.7165	881	34.6850	931	36.6535	981	38.6220
832	32.7559	882	34.7244	932	36.6929	982	38.6614
833	32.7953	883	34.7638	933	36.7323	983	38.7008
834	32.8346	884	34.8031	934	36.7717	984	38.7402
835	32.8740	885	34.8425	935	36.8110	985	38.7795
836	32.9134	886	34.8819	936	36.8504	986	38.8189
837	32.9528	887	34.9213	937	36.8898	987	38.8583
838	32.9921	888	34.9606	938	36.9291	988	38.8976
839	33.0315	889	35.0000	939	36.9685	989	38.9370
840	33.0709	890	35.0394	940	37.0079	990	38.9764
841	33.1102	891	35.0787	941	37.0472	991	39.0157
842	33.1496	892	35.1181	942	37.0866	992	39.0551
843	33.1890	893	35.1575	943	37.1260	993	39.0945
844	33.2283	894	35.1969	944	37.1654	994	39.1339
845	33.2677	895	35.2362	945	37.2047	995	39.1732
846	33.3071	896	35.2756	946	37.2441	996	39.2126
847	33.3465	897	35.3150	947	37.2835	997	39.2520
848	33.3858	898	35.3543	948	37.3228	998	39.2913
849	33.4252	899	35.3937	949	37.3622	999	39.3307
850	33.4646	900	35.4331	950	37.4016	1000	39.3701

Table C–38 Decimals of an Inch into Millimeters

In.	Mm	In.	Mm	In.	Mm	In.	Mm
0.001	0.025	0.200	5.08	0.470	11.94	0.740	18.80
0.002	0.051	0.210	5.33	0.480	12.19	0.750	19.05
0.003	0.076	0.220	5.59	0.490	12.45	0.760	19.30
0.004	0.102	0.230	5.84	0.500	12.70	0.770	19.56
0.005	0.127	0.240	6.10	0.510	12.95	0.780	19.81
0.006	0.152	0.250	6.35	0.520	13.21	0.790	20.07
0.007	0.178	0.260	6.60	0.530	13.46	0.800	20.32
0.008	0.203	0.270	6.86	0.540	13.72	0.810	20.57
0.009	0.229	0.280	7.11	0.550	13.97	0.820	20.83
0.010	0.254	0.290	7.37	0.560	14.22	0.830	21.08
0.020	0.508	0.300	7.62	0.570	14.48	0.840	21.34
0.030	0.762	0.310	7.87	0.580	14.73	0.850	21.59
0.040	1.016	0.320	8.13	0.590	14.99	0.860	21.84
0.050	1.270	0.330	8.38	0.600	15.24	0.870	22.10
0.060	1.524	0.340	8.65	0.610	15.49	0.880	22.35
0.070	1.778	0.350	8.89	0.620	15.75	0.890	22.61
0.080	2.032	0.360	9.14	0.630	16.00	0.900	22.86
0.090	2.286	0.370	9.40	0.640	16.26	0.910	23.11
0.100	2.540	0.380	9.65	0.650	16.51	0.920	23.37
0.110	2.794	0.390	9.91	0.660	16.76	0.930	23.62
0.120	3.048	0.400	10.16	0.670	17.02	0.940	23.88
0.130	3.302	0.410	10.41	0.680	17.27	0.950	24.13
0.140	3.56	0.420	10.67	0.690	17.53	0.960	24.38
0.150	3.81	0.430	10.92	0.700	17.78	0.970	24.64
0.160	4.06	0.440	11.18	0.710	18.03	0.980	24.89
0.170	4.32	0.450	11.43	0.720	18.29	0.990	25.15
0.180	4.57	0.460	11.68	0.730	18.54	1.000	25.40
0.190	4.83						

Table C-39 Inches into Centimeters

In.	0 Cm	1 Cm	2 Cm	3 Cm	4 Cm	5 Cm	6 Cm	7 Cm	8 Cm	9 Cm
0	—	2.54	5.08	7.62	10.16	12.70	15.24	17.78	20.32	22.86
10	25.40	27.94	30.48	33.02	35.56	38.10	40.64	43.18	45.72	48.26
20	50.80	53.34	55.88	58.42	60.96	63.50	66.04	68.58	71.12	73.66
30	76.20	78.74	81.28	83.82	86.36	88.90	91.44	93.98	96.52	99.06
40	101.60	104.14	106.68	109.22	111.76	114.30	116.84	119.38	121.92	124.46
50	127.00	129.54	132.08	134.62	137.16	139.70	142.24	144.78	147.32	149.86
60	152.40	154.94	157.48	160.02	162.56	165.10	167.64	170.18	172.72	175.26
70	177.80	180.34	182.88	185.42	187.96	190.50	193.04	195.58	198.12	200.66
80	203.20	205.74	208.28	210.82	213.36	215.90	218.44	220.98	223.52	226.06
90	228.60	231.14	233.68	236.22	238.76	241.30	243.84	246.38	248.92	251.46
100	254.00	256.54	259.08	261.62	264.16	266.70	269.24	271.78	274.32	276.86

Table C-40 Centimeters into Inches

Cm	0 In.	1 In.	2 In.	3 In.	4 In.	5 In.	6 In.	7 In.	8 In.	9 In.
0	—	0.394	0.787	1.181	1.575	1.969	2.362	2.756	3.150	3.543
10	3.937	4.331	4.724	5.118	5.512	5.906	6.299	6.693	7.087	7.480
20	7.874	8.268	8.661	9.055	9.449	9.843	10.236	10.630	11.024	11.417
30	11.811	12.205	12.598	12.992	13.386	13.780	14.173	14.567	14.961	15.354
40	15.748	16.142	16.535	16.929	17.323	17.717	18.110	18.504	18.898	19.291
50	19.685	20.079	20.472	20.866	21.260	21.654	22.047	22.441	22.835	23.228
60	23.622	24.016	24.409	24.803	25.197	25.591	25.984	26.378	26.772	27.164
70	27.559	27.953	28.346	28.740	29.134	29.528	29.921	30.315	30.709	31.102
80	31.496	31.890	32.283	32.677	33.071	33.465	33.858	34.252	34.646	35.039
90	35.433	35.827	36.220	36.614	37.008	37.402	37.795	38.189	38.583	38.976
100	39.370	39.764	40.157	40.551	40.945	41.339	41.732	42.126	42.520	42.913

Table C-41 Feet into Meters

Ft	0	1	2	3	4	5	6	7	8	9
	M	M	M	M	M	M	M	M	M	M
0	—	0.305	0.610	0.914	1.219	1.524	1.829	2.134	2.438	2.743
10	3.048	3.353	3.658	3.962	4.267	4.572	4.877	5.182	5.486	5.791
20	6.096	6.401	6.706	7.010	7.315	7.620	7.925	8.230	8.534	8.839
30	9.144	9.449	9.754	10.058	10.363	10.668	10.973	11.278	11.582	11.887
40	12.192	12.497	12.802	13.106	13.411	13.716	14.021	14.326	14.630	14.935
50	15.240	15.545	15.850	16.154	16.459	16.764	17.069	17.374	17.678	17.983
60	18.288	18.593	18.898	19.202	19.507	19.812	20.177	20.422	20.726	21.031
70	21.336	21.641	21.946	22.250	22.555	22.860	23.165	23.470	23.774	24.079
80	24.384	24.689	24.994	25.298	25.603	25.908	26.213	26.518	26.822	27.127
90	27.432	27.737	28.042	28.346	28.651	28.956	29.261	29.566	29.870	30.175
100	30.480	30.785	31.090	31.394	31.699	32.004	32.309	32.614	32.918	33.223

Table C-42 Meters into Feet

M	0	1	2	3	4	5	6	7	8	9
	Ft	Ft	Ft	Ft	Ft	Ft	Ft	Ft	Ft	Ft
0	—	3.281	6.562	9.842	13.123	16.404	19.685	22.966	26.247	29.528
10	32.808	36.089	39.370	42.661	45.932	49.212	52.493	55.774	59.055	52.336
20	65.617	68.897	72.178	75.459	78.740	82.021	85.302	88.582	91.863	95.144
30	98.425	101.71	104.99	108.27	111.55	114.83	118.11	121.39	124.67	127.95
40	131.23	134.51	137.79	141.08	144.36	147.64	150.92	154.20	157.48	160.76
50	164.04	167.32	170.60	173.88	177.16	180.45	183.73	187.01	190.29	193.57
60	196.85	200.13	203.41	206.69	209.97	213.25	216.53	219.82	223.10	226.38
70	229.66	232.94	236.22	239.50	242.78	246.06	249.34	252.62	255.90	259.19
80	262.47	265.75	269.03	272.31	275.59	278.87	282.15	285.43	288.71	291.99
90	295.27	298.56	301.84	305.12	308.40	311.68	314.96	318.24	321.52	324.80
100	328.08	331.36	334.64	337.93	341.21	344.49	347.77	351.05	354.33	357.61

Table C–43 Miles into Kilometers

Miles	0	1	2	3	4	5	6	7	8	9
	Km	Km	Km	Km	Km	Km	Km	Km	Km	Km
0	—	1.609	3.219	4.828	6.437	8.047	9.656	11.265	12.875	14.484
10	16.093	17.703	19.312	20.922	22.531	24.140	25.750	27.359	28.968	30.578
20	32.187	33.796	35.406	37.015	38.624	40.234	41.843	43.452	45.062	46.671
30	48.280	49.890	51.499	53.108	54.718	56.327	57.936	59.546	61.155	62.764
40	64.374	65.983	67.592	69.202	70.811	72.421	74.030	75.639	77.249	78.858
50	80.467	82.077	83.686	85.295	86.905	88.514	90.123	91.733	93.342	94.951
60	96.561	98.170	99.779	101.39	103.00	104.61	106.22	107.83	109.44	111.05
70	112.65	114.26	115.87	117.48	119.09	120.70	122.31	123.92	125.53	127.14
80	128.75	130.36	131.97	133.58	135.19	136.79	138.40	140.01	141.62	143.23
90	144.84	146.45	148.06	149.67	151.28	152.89	154.50	156.11	157.72	159.33
100	160.93	162.54	164.15	165.76	167.37	168.98	170.59	172.20	173.81	175.42

Table C–44 Kilometers into Miles

Km	0	1	2	3	4	5	6	7	8	9
	Miles	Miles	Miles	Miles	Miles	Miles	Miles	Miles	Miles	Miles
0	—	0.621	1.243	1.864	2.486	3.107	3.728	4.350	4.971	5.592
10	6.214	6.835	7.457	8.078	8.699	9.321	9.942	10.563	11.185	11.806
20	12.427	13.049	13.670	14.292	14.913	15.534	16.156	16.777	17.398	18.020
30	18.641	19.263	19.884	20.505	21.127	21.748	22.369	22.991	23.612	24.234
40	24.855	25.476	26.098	26.719	27.340	27.962	28.583	29.204	29.826	30.447
50	31.069	31.690	32.311	32.933	33.554	34.175	34.797	35.418	36.040	36.661
60	37.282	37.904	38.525	39.146	39.768	40.389	41.011	41.632	42.253	42.875
70	43.496	44.117	44.739	45.370	45.982	46.603	47.224	47.846	48.467	49.088
80	49.710	50.331	50.952	51.574	52.195	52.817	53.438	54.059	54.681	55.302
90	55.923	56.545	57.166	57.788	58.409	59.030	59.652	60.273	60.894	61.516
100	62.137	62.759	63.380	64.001	64.623	65.244	65.865	66.487	67.108	67.730

Table C–45 Units of Power

1 watt	0.73756 foot pound per second
1 foot pound per second	1.35582 watts
1 watt	0.056884 BTU per minute
1 BTU per minute	17.580 watts
1 watt	0.001341 U.S. horsepower
1 U.S. horsepower	745.7 watts
1 watt	0.01433 kilogram-calorie per minute
1 kilogram-calorie per minute	69.767 watts
1 watt	1×10^7 ergs per second
1 lumen	0.001496 watt

Table C-46 Pounds per Square Inch into Kilograms per Square Centimeter

Lb per Sq. In.	0	1	2	3	4	5	6	7	8	9
					Kg per Sq. Cm					
0	—	0.0703	0.1406	0.2109	0.2812	0.3515	0.4218	0.4922	0.5625	0.6328
10	0.7031	0.7734	0.8437	0.9140	0.9843	1.0546	1.1249	1.1952	1.2655	1.3358
20	1.4061	1.4765	1.5468	1.6171	1.6874	1.7577	1.8280	1.8983	1.9686	2.0389
30	2.1092	2.1795	2.2498	2.3201	2.3904	2.4607	2.5311	2.6014	2.6717	2.7420
40	2.8123	2.8826	2.9529	3.0232	3.0935	3.1638	3.2341	3.3044	3.3747	3.4450
50	3.5154	3.5857	3.6560	3.7263	3.7966	3.8669	3.9372	4.0075	4.0778	4.1481
60	4.2184	4.2887	4.3590	4.4293	4.4997	4.5700	4.6403	4.7106	4.7809	4.8512
70	4.9215	4.9918	5.0621	5.1324	5.2027	5.2730	5.3433	5.4136	5.4839	5.5543
80	5.6246	5.6949	5.7652	5.8355	5.9058	5.9761	6.0464	6.1167	6.1870	6.2573
90	6.3276	6.3980	6.4682	6.5386	6.6089	6.6792	6.7495	6.8198	6.8901	6.9604
100	7.0307	7.1010	7.1713	7.2416	7.3120	7.3822	7.4525	7.5228	7.5932	7.6635

Table C-47 Kilograms per Square Centimeter into Pounds per Square Inch

Kg per Sq. Cm	0	1	2	3	4	5	6	7	8	9
					Lb per Sq. In.					
0	—	14.22	28.45	42.67	56.89	71.12	85.34	99.56	113.79	128.01
10	142.23	156.46	170.68	184.90	199.13	213.35	227.57	241.80	256.02	270.24
20	284.47	298.69	312.91	327.14	341.36	355.58	369.81	384.03	398.25	412.48
30	426.70	440.92	455.15	469.37	483.59	497.82	512.04	526.26	540.49	554.71
40	568.93	583.16	597.38	611.60	625.83	640.05	654.27	668.50	682.72	696.94
50	711.17	725.39	739.61	753.84	768.06	782.28	796.51	810.73	824.95	839.18
60	853.40	867.62	881.85	896.07	910.29	924.52	938.74	952.96	967.19	981.41
70	995.63	1009.9	1024.1	1038.3	1052.5	1066.8	1081.0	1095.2	1109.4	1123.6
80	1137.9	1152.1	1166.3	1180.5	1194.8	1209.0	1223.2	1237.4	1251.7	1265.9
90	1280.1	1294.3	1308.6	1322.8	1337.0	1351.2	1365.4	1379.7	1393.9	1408.1
100	1422.3	1436.6	1450.8	1465.0	1479.2	1493.4	1507.7	1521.9	1536.1	1550.3

Table C–48 Pounds per Square Foot into Kilograms per Square Meter

Lb per Sq. Ft	0	1	2	3	4	5	6	7	8	9
					Kg per Sq. M					
0	....	4.88	9.77	14.65	19.53	24.41	29.30	34.18	39.06	43.94
10	48.82	53.70	58.59	63.47	68.35	73.23	78.12	83.00	87.88	92.76
20	97.65	102.53	107.42	112.30	117.18	122.06	126.95	131.83	136.71	141.59
30	146.47	151.35	156.24	161.12	166.00	170.88	175.77	180.65	185.53	190.41
40	195.30	200.18	205.07	209.95	214.83	219.71	224.60	229.48	234.36	239.24
50	244.12	249.00	253.89	258.77	263.65	268.53	273.42	278.30	283.18	288.06
60	292.95	297.83	302.72	307.60	312.48	317.36	322.25	327.13	332.01	336.89
70	341.77	346.65	351.54	356.42	361.30	366.18	371.07	375.95	380.83	385.71
80	390.59	395.47	400.36	405.24	410.12	415.00	419.89	424.77	429.65	434.53
90	439.43	444.30	449.19	454.07	458.95	463.83	468.72	473.60	478.48	483.36
100	488.24	493.12	498.01	502.89	507.77	512.65	517.54	522.42	527.30	532.18

Table C–49 Kilograms per Square Meter into Pounds per Square Foot

Kg per Sq. M	0	1	2	3	4	5	6	7	8	9
					Lb per Sq. Ft					
0	....	0.2048	0.4096	0.6144	0.8193	1.0241	1.2289	1.4337	1.6385	1.8433
10	2.0481	2.2530	2.4578	2.6626	2.8674	3.0722	3.2771	3.4819	3.6867	3.8915
20	4.0963	4.3011	4.5060	4.7108	4.9156	5.1204	5.3252	5.5300	5.7349	5.9397
30	6.1445	6.3493	6.5541	6.7589	6.9638	7.1686	7.3734	7.5782	7.7830	7.9878
40	8.1927	8.3975	8.6023	8.8071	9.0119	9.2167	9.4215	9.6262	9.8310	10.036
50	10.241	10.446	10.650	10.855	11.060	11.265	11.470	11.675	11.879	12.084
60	12.289	12.494	12.698	12.903	13.108	13.313	13.518	13.723	13.927	14.132
70	14.337	14.542	14.747	14.952	15.156	15.361	15.566	15.771	15.976	16.181
80	16.385	16.590	16.795	17.000	17.205	17.409	17.614	17.819	18.024	18.229
90	18.434	18.638	18.843	19.048	19.253	19.458	19.662	19.867	20.072	20.277
100	20.482	20.686	20.891	21.096	21.301	21.506	21.711	21.915	22.120	22.325

Table C–50 Water Heads and Equivalent Pressures

(weight of water at 62.4 lb per cubic ft)

Head in Feet	Pressure in Lb Sq. In.	Pressure in Lb Sq. Ft
5	2.17	312
10	4.33	624
15	6.50	936
20	8.66	1,248
30	12.99	1,872
40	17.32	2,496
50	21.65	3,120
60	25.98	3,744
70	30.31	4,368
80	34.64	4,992
90	38.97	5,616
100	43.30	6,240
125	54.13	7,800
150	64.95	9,360
175	75.78	10,920
200	86.60	12,480

Table C–51 Inches Vacuum into Feet Suction

In. Vacuum	Ft Suction
.5	.56
1.0	1.13
1.5	1.70
2.0	2.27
2.5	2.84
3.0	3.41
4.0	4.54
5.0	5.67
6.0	6.80
7.0	7.94
8.0	9.07
9.0	10.21
10.0	11.34
15.0	17.01
20.0	22.68
25.0	28.35
30.0	34.02

Table C–52 Friction of Water in Pipes

(approximate loss of heat in ft due to friction per 100 ft of pipeline)

U.S. Gal. per Min.	1 In. I.D.	1½ In. I.D.	2 In. I.D.	4 In. I.D.
10	11.7	1.4	0.5	
20	42.0	5.2	1.8	
30	89.0	11.0	3.8	
40	152.0	18.8	6.6	0.2
50		28.4	9.9	0.3
70		53.0	18.4	0.6
100		102.0	35.8	1.2
125			54.0	1.9
150			76.0	2.6
175			102.0	3.4
200			129.0	4.4
250				6.7

Table C–53 Friction in Fittings

(reduced to equivalent feet of pipeline loss)

Type of Fitting	1 In. I.D.	1½ In. I.D.	2 In. I.D.	4 In. I.D.
90° Elbow	2.8	4.3	5.5	11.0
45° Elbow	1.3	2.0	2.6	5.0
Tee Side Outlet	5.6	9.1	12.0	22.0
Close Return Bend	6.3	10.2	13.0	24.0
Gate Valve	.6	.9	1.2	2.3
Check Valve	10.5	15.8	21.1	42.3
Ball Valve (ported)	27.0	43.0	55.0	115.0

Examples:

A 2-in. inside diameter pipeline loses 29.7 ft of head pumping at 50 gal. per min for 300 ft of straight length. (9.9 per hundred ft × 3 hundreds)

If the pipeline in the above example also contained, say, three 90° elbows, three 45° elbows, two close return bends, two check valves and one ported ball valve, we could add to the loss of 29.7 ft of head as follows:

Three 90° elbows   = 3 × 5.5 = 16.5 additional pipeline feet of friction
Three 45° elbows   = 3 × 2.6 = 7.8 additional pipeline feet of friction
Two close return ends = 2 × 13.0 = 26.0 additional pipeline feet of friction
Two check valves   = 2 × 21.1 = 42.2 additional pipeline feet of friction
One ported ball valve = 1 × 55.0 = 55.0 additional pipeline feet of friction
Total fittings   = 147.5 additional pipeline feet of friction

Then 1.475 (hundreds of feet) × 9.9 (per 100 at 2 in. I.D.) = 14.6 additional ft of head loss

Grand total pipeline (300 straight ft plus 11 fittings) = 44.3 ft of head loss

44.3 ft of head = approximately 19.2 lb pressure per sq. in.

Table C–54 Temperature Conversions

°F	°C	°F	°C	°F	°C	°F	°C
−40.	−40.	2.	−16.67	44.60	7.	87.	30.56
−39.	−39.44	3.	−16.11	45.	7.22	87.80	31.
−38.20	−39.	3.20	−16.	46.	7.78	88.	31.11
−38.	−38.89	4.	−15.56	46.40	8.	89.	31.67
−37.	−38.33	5.	−15.	47.	8.33	89.60	32.
−36.40	−38.	6.	−14.44	48.	8.89	90.	32.22
−36.	−37.78	6.80	−14.	48.20	9.	91.	32.78
−35.	−37.22	7.	−13.89	49.	9.44	91.40	33.
−34.60	−37.	8.	−13.33	50.	10.	92.	33.33
−34.	−36.67	8.60	−13.	51.	10.56	93.	33.89
−33.	−36.11	9.	−12.78	51.80	11.	93.20	34.
−32.80	−36.	10.	−12.22	52.	11.11	94.	34.44
−32.	−35.56	10.40	−12.	53.	11.67	95.	35.
−31.	−35.	11.	−11.67	53.60	12.	96.	35.56
−30.	−34.44	12.	−11.11	54.	12.22	96.80	36.
−29.20	−34.	12.20	−11.	55.	12.78	97.	36.11
−29.	−33.89	13.	−10.56	55.40	13.	98.	36.67
−28.	−33.33	14.	−10.	56.	13.33	98.60	37.
−27.40	−33.	15.	− 9.44	57.	13.89	99.	37.22
−27.	−32.78	15.80	− 9.	57.20	14.	100.	37.38
−26.	−32.22	16.	− 8.89	58.	14.44	100.40	38.
−25.60	−32.	17.	− 8.33	59.	15.	101.	38.33
−25.	−31.67	17.60	− 8.	60.	15.56	102.	38.89
−24.	−31.11	18.	− 7.78	60.80	16.	102.20	39.
−23.80	−31.	19.	− 7.22	61.	16.11	103.	39.44
−23.	−30.56	19.40	− 7.	62.	16.67	104.	40.
−22.	−30.	20.	− 6.67	62.60	17.	105.	40.56
−21.	−29.44	21.	− 6.11	63.	17.22	105.80	41.
−20.20	−29.	21.20	− 6.	64.	17.78	106.	41.11
−20.	−28.89	22.	− 5.50	64.40	18.	107.	41.67
−19.	−28.33	23.	− 5.	65.	18.33	107.60	42.
−18.40	−28.	24.	− 4.44	66.	18.89	108.	42.22
−18.	−27.78	24.80	− 4.	66.20	19.	109.	42.78
−17.	−27.22	25.	− 3.89	67.	19.44	109.40	43.
−16.60	−27.	26.	− 3.33	68.	20.	110.	43.33
−16.	−26.67	26.60	− 3.	69.	20.56	111.	43.89
−15.	−26.11	27.	− 2.78	69.80	21.	111.20	44.
−14.80	−26.	28.	− 2.22	70.	21.11	112.	44.44
−14.	−25.56	28.40	− 2.	71.	21.67	113.	45.
−13.	−25.	29.	− 1.67	71.60	22.	114.	45.56
−12.	−24.44	30.	− 1.11	72.	22.22	114.80	46.
−11.20	−24.	30.20	− 1.	73.	22.78	115.	46.11
−11.	−23.89	31.	− 0.56	73.40	23.	116.	46.67
−10.	−23.33	32.	0.	74.	23.33	116.60	47.
− 9.40	−23.	33.	0.56	75.	23.89	117.	47.22
− 9.	−22.78	33.80	1.	75.20	24.	118.	47.78
− 8.	−22.22	34.	1.11	76.	24.44	118.40	48.
− 7.60	−22.	35.	1.67	77.	25.	119.	48.33
− 7.	−21.67	35.60	2.	78.	25.56	120.	48.89
− 6.	−21.11	36.	2.22	78.80	26.	120.20	49.
− 5.80	−21.	37.	2.78	79.	26.11	121.	49.44
− 5.	−20.56	37.40	3.	80.	26.67	122.	50.
− 4.	−20.	38.	3.33	80.60	27.	123.	50.56
− 3.	−19.44	39.	3.89	81.	27.22	123.80	51.
− 2.20	−19.	39.20	4.	82.	27.78	124.	51.11
− 2.	−18.89	40.	4.44	82.40	28.	125.	51.67
− 1.	−18.33	41.	5.	83.	28.33	125.60	52.
− 0.40	−18.	42.	5.56	84.	28.89	126.	52.22
0.	−17.78	42.80	6.00	84.20	29.	127.	52.78
1.	−17.22	43.	6.11	85.	29.44	127.40	53.
1.40	−17.	44.	6.67	86.	30.	128.	53.33

Table C–54 (*continued*)

°F	°C	°F	°C	°F	°C	°F	°C
129.	53.89	159.	70.56	188.60	87.	217.40	103.
129.20	54.	159.80	71.	189.	87.22	218.	103.33
130.	54.44	160.	71.11	190.	87.78	219.	103.89
131.	55.	161.	71.67	190.40	88.	219.20	104.
132.	55.56	161.60	72.	191.	88.33	220.	104.44
132.80	56.	162.	72.22	192.	88.89	221.	105.
133.	56.11	163.	72.78	192.20	89.	222.	105.56
134.	56.67	163.40	73.	193.	89.44	222.80	106.
134.60	57.	164.	73.33	194.	90.	223.	106.11
135.	57.22	165.	73.89	195.	90.56	224.	106.67
136.	57.78	165.20	74.	195.80	91.	224.60	107.
136.40	58.	166.	74.44	196.	91.11	225.	107.22
137.	58.33	167.	75.	197.	91.67	226.	107.78
138.	58.89	168.	75.56	197.60	92.	226.40	108.
138.20	59.	168.80	76.	198.	92.22	227.	108.33
139.	59.44	169.	76.11	199.	92.78	228.	108.89
140.	60.	170.	76.67	199.40	93.	228.20	109.
141.	60.56	170.60	77.	200.	93.33	229.	109.44
141.80	61.	171.	77.22	201.	93.89	230.	110.
142.	61.11	172.	77.78	201.20	94.	231.	110.56
143.	61.67	172.40	78.	202.	94.44	231.80	111.
143.60	62.	173.	78.33	203.	95	232.	111.11
144.	62.22	174.	78.89	204.	95.56	233.	111.67
145.	62.78	174.20	79.	204.80	96.	233.60	112.
145.40	63.	175.	79.44	205.	96.11	234.	112.22
146.	63.33	176.	80.	206.	96.67	235.	112.78
147.	63.89	177.	80.56	206.60	97.	235.40	113.
147.20	64.	177.80	81.	207.	97.22	236.	113.33
148.	64.44	178.	81.11	208.	97.78	237.	113.89
149.	65.	179.	81.67	208.40	98.	237.20	114.
150.	65.56	179.60	82.	209.	98.33	238.	114.44
150.80	66.	180.	82.22	210.	98.89	239.	114
151.	66.11	181.	82.78	210.20	99.	240.	115.56
152.	66.67	181.40	83.	211.	99.44	240.80	116.
152.60	67.	182.	83.33	212.	100.	241.	116.11
153.	67.22	183.	83.89	213.	100.56	242.	116.67
154.	67.78	183.20	84.	213.80	101.	242.60	117.
154.40	68.	184.	84.44	214.	101.11	243.	117.32
155.	68.33	185.	85.	215.	101.67	244.	117.78
156.	68.89	186.	85.56	215.60	102.	244.40	118.
156.20	69.	186.80	86.	216.	102.22	245.	118.33
157.	69.44	187.	86.11	217.	102.78	246.	118.89
158.	70.	188.	86.67				

Table C–55 Basic Conversions: Units of Weight

1 grain	.0029 ounces
1 grain	.000143 pounds
1 grain	.064799 grams
1 dram	60 grains
1 dram	.1371 ounces
1 dram	.008571 pounds
1 dram	3.88794 grams
1 ounce	437.5 grains
1 ounce	28.3495 grams
1 ounce	.0625 pounds
1 pound	453.592 grams
1 pound	16 ounces
1 ton	2000 pounds
1 ton	907.18581 kilograms
1 gram	.03527 ounces
1 gram	.002205 pounds
1 gram	15.432 grains
1 gram	.001 milligram
1 kilogram	1000 grams
1 kilogram	35.2739 ounces
1 kilogram	2.20462 pounds

Table C–56 Weight Conversion Formulas

To Convert	Into	Multiply By
grains	grams	0.0648
drams	grams	1.7718
ounces	grams	28.3495
pounds	grams	435.5924
tons	kilograms	907.18581
grams	grains	15.4324
grams	drams	0.5644
grams	ounces	0.0353
kilograms	pounds	2.2046

Table C–57 Weight Fractions

Oz	Fraction of a Lb	Decimal of a Lb	G	Oz	Fraction of a Lb	Decimal of a Lb	G
1/4	1/64	.0156	7.09	8-1/4	33/65	.5156	233.88
1/2	1/32	.0313	14.17	8-1/2	17/32	.5313	240.97
3/4	3/64	.0469	21.26	8-3/4	35/64	.5469	248.06
1	1/16	.0625	28.35	9	9/16	.5625	255.15
1-1/4	5/64	.0781	35.44	9-1/4	37/64	.5781	262.23
1-1/2	3/32	.0938	42.52	9-1/2	19/32	.5938	269.30
1-3/4	7/64	.1094	49.61	9-3/4	39/64	.6094	276.41
2	1/8	.125	56.70	10	5/8	.625	283.50
2-1/4	9/64	.1406	63.79	10-1/4	41/64	.6406	290.58
2-1/2	5/32	.1563	70.87	10-1/2	21/32	.6563	297.67
2-3/4	11/64	.1719	77.96	10-3/4	43/64	.6719	304.76
3	3/16	.1875	85.05	11	11/16	.6875	311.84
3-1/4	13/64	.2031	92.14	11-1/4	45/64	.7031	318.93
3-1/2	7/32	.2188	99.22	11-1/2	23/32	.7188	326.02
3-3/4	15/64	.2344	106.31	11-3/4	47/64	.7344	333.11
4	1/4	.250	113.40	12	3/4	.750	340.19
4-1/4	17/64	.2656	120.49	12-1/4	49/64	.7656	347.28
4-1/2	9/32	.2813	127.57	12-1/2	25/32	.7813	354.37
4-3/4	19/64	.2969	134.66	12-3/4	51/64	.7969	361.46
5	5/16	.3125	141.75	13	13/16	.8125	368.54
5-1/4	21/64	.3281	148.84	13-1/4	53/64	.8281	375.63
5-1/2	11/32	.3438	155.92	13-1/2	27/32	.8438	382.72
5-3/4	23/64	.3594	163.01	13-3/4	55/64	.8594	389.81
6	3/8	.375	170.10	14	7/8	.875	396.89
6-1/4	25/64	.3906	177.18	14-1/4	57/64	.8906	403.93
6-1/2	13/32	.4063	184.27	14-1/2	29/32	.9063	411.07
6-3/4	27/64	.4219	191.36	14-3/4	59/64	.9219	418.16
7	7/16	.4375	198.45	15	15/16	.9375	425.24
7-1/4	29/64	.4531	205.53	15-1/4	61/64	.9531	432.33
7-1/2	15/32	.4688	212.62	15-1/2	31/32	.9688	439.42
7-3/4	31/64	.4844	219.71	15-3/4	63/64	.9844	446.50
8	1/2	.500	226.80	16	1	1.000	453.59

Table C-58 Ounces into Grams

Oz	0	1	2	3	4	5	6	7	8	9
					G					
0	—	28.350	56.699	85.049	113.40	141.75	170.10	198.45	226.80	255.15
10	283.50	311.84	340.19	368.54	396.89	425.24	453.59	481.94	510.29	538.64
20	566.99	595.34	623.69	652.04	680.39	708.74	737.09	765.44	793.79	822.14
30	850.49	878.84	907.19	935.53	963.88	992.23	1020.6	1048.9	1077.3	1105.6
40	1134.0	1162.3	1190.7	1219.0	1247.4	1275.7	1304.1	1332.4	1360.8	1389.1
50	1417.5	1445.8	1474.2	1502.5	1530.9	1559.2	1587.6	1615.9	1644.3	1672.6
60	1701.0	1729.3	1757.7	1786.0	1814.4	1842.7	1871.1	1899.4	1927.8	1956.1
70	1984.5	2012.8	2041.2	2069.5	2097.9	2126.2	2154.6	2182.9	2211.3	2239.6
80	2268.0	2296.3	2324.7	2353.0	2381.4	2409.7	2438.1	2466.4	2494.8	2523.1
90	2551.5	2579.8	2608.2	2636.5	2664.9	2693.2	2721.6	2749.9	2778.3	2806.6
100	2835.0	2863.3	2891.6	2920.0	2948.3	2976.7	3005.0	3033.4	3061.7	3090.1

Table C-59 Grams into Ounces

G	0	1	2	3	4	5	6	7	8	9
					Oz					
0	—	0.035274	0.070548	0.10582	0.14110	0.17637	0.21164	0.24692	0.28219	0.31747
10	0.35274	0.38801	0.42329	0.45856	0.49384	0.52911	0.56438	0.59966	0.63493	0.67021
20	0.70548	0.74075	0.77603	0.81130	0.84658	0.88185	0.91712	0.95240	0.98767	1.0229
30	1.0582	1.0935	1.1288	1.1640	1.1993	1.2346	1.2699	1.3051	1.3404	1.3757
40	1.4110	1.4462	1.4815	1.5168	1.5521	1.5873	1.6226	1.6579	1.6932	1.7284
50	1.7637	1.7990	1.8348	1.8695	1.9048	1.9401	1.9753	2.0106	2.0459	2.0812
60	2.1164	2.1517	2.1870	2.2223	2.2575	2.2928	2.3281	2.3634	2.3986	2.4339
70	2.4692	2.5045	2.5397	2.5750	2.6103	2.6456	2.6808	2.7161	2.7514	2.7866
80	2.8219	2.8572	2.8925	2.9277	2.9630	2.9983	3.0336	3.0688	3.1041	3.1394
90	3.1747	3.2099	3.2452	3.2805	3.3158	3.3510	3.3863	3.4216	3.4569	3.4921
100	3.5274	3.5627	3.5979	3.6332	3.6685	3.7038	3.7390	3.7743	3.8096	3.8449

Table C–60 Pounds into Kilograms

Lb	0	1	2	3	4	5	6	7	8	9
					Kg					
0	—	0.454	0.907	1.361	1.814	2.268	2.722	3.175	3.629	4.082
10	4.536	4.990	5.443	5.897	6.350	6.804	7.257	7.711	8.165	8.618
20	9.072	9.525	9.979	10.433	10.886	11.340	11.793	12.247	12.701	13.154
30	13.608	14.061	14.515	14.969	15.422	15.876	16.329	16.783	17.237	17.690
40	18.144	18.597	19.051	19.504	19.958	20.412	20.865	21.319	21.772	22.226
50	22.680	23.133	23.587	24.040	24.494	24.948	25.401	25.855	26.308	26.762
60	27.216	27.669	28.123	28.576	29.030	29.484	29.937	30.391	30.844	31.298
70	31.752	32.205	32.659	33.112	33.566	34.019	34.473	34.927	35.380	35.834
80	36.287	36.741	37.195	37.648	38.102	38.555	39.009	39.463	39.916	40.370
90	40.823	41.277	41.731	42.184	42.638	43.091	43.545	43.999	44.452	44.906
100	45.359	45.813	46.266	46.720	47.174	47.627	48.081	48.534	48.988	49.442

Table C–61 Kilograms into Pounds

Kg	0	1	2	3	4	5	6	7	8	9
					Lb					
0	—	2.205	4.409	6.614	8.819	11.023	13.228	15.432	17.637	19.842
10	22.046	24.251	26.456	28.660	30.865	33.069	35.274	37.479	39.683	41.888
20	44.093	46.297	48.502	50.706	52.911	55.116	57.320	59.525	61.729	63.934
30	66.139	68.343	70.548	72.753	74.957	77.162	79.366	81.571	83.776	85.980
40	88.185	90.390	92.594	94.799	97.003	99.208	101.41	103.62	105.82	108.03
50	110.23	112.44	114.64	116.85	119.05	121.25	123.46	125.66	127.87	130.07
60	132.28	134.48	136.69	138.89	141.10	143.30	145.51	147.71	149.91	152.12
70	154.32	156.53	158.73	160.94	163.14	165.35	167.55	169.76	171.96	174.17
80	176.37	178.57	180.78	182.98	185.19	187.39	189.60	191.80	194.01	196.21
90	198.42	200.62	202.83	205.03	207.24	209.44	211.64	213.85	216.05	218.26
100	220.46	222.67	224.87	227.08	229.28	231.49	233.69	235.90	238.10	240.30

APPENDIX D

GLOSSARY

Amabile   Italian term referring to sweetness

Amaro   Italian term referring to bitterness

Amelioration   Dilution—the addition of water and/or sugar to juice, must, or wine

Anaerobic   Not requiring oxygen—microorganisms which do not require free oxygen in order to grow, such as wine yeasts

Anthocyanins   Phenolic color pigments found in the skins of grapes

AOC   Acronym for the French *Appellation d'Origine Controlee,* official regulatory system for wine origin and quality control

Aperitif   Appetizer wine—wine containing added flavoring from herbs, roots, seeds, and spices, such as Vermouth

Appearance   Sensory term relating to the entire visual aspect of wine evaluation, principally color and clarity

Appellation   Geographic source—the specific location where a given wine was grown

Aroma   Fruit fragrance—that portion of wine bouquet contributed by the fruit

Astringency   An aspirin-like sensory response on the palate generally due to tannins in wine

Austere   Sensory term relating to wines which have excessive acidity and/or astringency

Bacteria One-celled microorganisms which do not contain chlorophyll, such as bacilli (rod shaped) and cocci (spherical)

Balance Sensory term relating to the ratio of dryness and acidity

Balling Analytical measurement of wine viscosity

Barrique Wine barrel containing approximately 60 U.S. gallons

Baume Scale for measuring dissolved solids

Bentonite A montmorillonite clay compound used as a clarification agent in wine processing

Bianco Italian term relating to white wine

Big Sensory term used in reference to wines which exhibit an abundance of positive qualities

Bite Same as *Austere*

Blanc de Blancs French term relating to white wines made from white grapes

Blanc de Noirs French term relating to white wines made from black grapes

Bland Sensory term used to describe wines deficient in acidity and or tannin

Blush Very light pink wines, generally made from a brief fermentation of red must just prior to pressing

Body Sensory evaluation of wine viscosity—relates to lightness or heaviness of mouth feel

Bonded An ATF term relating to a premises on which wines may be legally stored prior to the payment of excise taxes

Bottle Fermentation Secondary fermentation of wine in special bottles designed for high pressures—captures natural carbon dioxide to make wine effervesce, or "sparkle"—the traditional Champagne bottle fermentation method is called "methode champenoise"

Bottle Shock Refers to wines suffering a temporary reduction in bouquet and flavor due to addition of preservatives, final filtration, and bottling—sometimes called "bottle sickness," recovery generally appears within a few weeks

Botrytis Cinerea Same as *Noble Mold*

Bouquet Sensory term relating to the entire fragrance of a wine; comprised chiefly by fruit aroma, fermentation flavors from yeast and/or bacteria, wood essences, and other components

Brandy The distillate or "spirits" of wine

Breathing The practice of uncorking a wine bottle well in advance of serving in order to allow the wine to expel headspace gases

Brilliant Sensory term relating to wines which have flawless clarity

Brix Analytical measurement of total dissolved solids in juice or must, most of which are sugars

Bulk Process Same as *Charmat*

Bung Stopper, usually made of wood or glass, which is used to seal a keg, barrel, cask, or some other bulk wine storage vessel

Bung Hole Opening at the top of a keg, barrel, cask, or some other bulk wine storage vessel in which a bung is inserted

Cantina Italian for "cellar"

Capsule The decorative finishing closure over a cork or cap on a wine bottle, usually made of plastic or nonlead metal alloys

Carbohydrate Chemical term relating to a food energy compound, such as sugar and ethyl alcohol, comprised of carbon, hydrogen, and oxygen

Carbon Dioxide The tasteless and odorless gas produced by fermentation

Carbonic Maceration Method by which whole grapes are fermented in a closed vessel during which the carbon dioxide gas generated from the fermentation permeates the grapes to release greater intensity of color and flavor

Carboy A narrow-mouthed glass container, usually 3 or 5 gal in capacity

Casa Vinicola Italian for "winery"

Cask Wooden vessel for bulk wine storage and aging, usually made of some species of white oak and containing at least 200 U.S. gal

Caskiness Sensory term relating to wines which may have been aged or stored in casks which were not cleaned or treated properly beforehand—much the same as "tanky"

Casse Sensory term describe wine haziness, generally due to high levels of copper and/or iron

Cava Spanish for "sparkling wine"

Cave French for "cellar"

Cepage French for "variety" or "cultivar"

Chai French for wine cellar—a term most often identifying Bordeaux wine cellars

Chambrer French for adjusting wine temperature to that in a cool room, or about 65°F

Chaptalization The addition of sugar to must or juice before fermentation into wine

Character Sensory term describing the entire profile of a wine including color, clarity, bouquet, balance, body, aftertaste, and overall impression

Charmat The bulk, or tank, method of making ordinary sparkling wines

Chateau French for "castle" or "mansion"—the wine estate comprising vineyards and chai in France, usually in Bordeaux

Chiaretto Italian for "light red"

Claret Term used by the English for red wines grown in the Bordeaux region of France

Clarify To make a wine brilliantly clear—same as *Fining*

Classified Growth A particular vineyard, domaine, estate, or other geographical locale which is officially recognized by a national government as a controlled appellation of wine origin

Clean Sensory term relating to a wine having no musty or other "off" elements in character

Clos French for "wall"—usually around a vineyard or estate

Cloudy Sensory term relating to a wine exhibiting heavy suspended solids

Cloying Sensory term for excessively sweet

Coarse   Sensory term relating to a wine having abnormally high elements of character, especially acidity and/or tannins

Colloid   The suspension material, usually protein, which causes wines to be cloudy

Color   Relates both to the hue and density of wine pigments

Complex   Sensory term relating to an interrelationship of extensive bouquet and/or flavor components

Concentrate   Dehydrated grape juice, usually at a Brix of 60° or more, which is used for sweetening juice, musts, or wines deficient in natural sugar—sometimes reconstituted with water and fermented to make low-grade wines

Cooked   Sensory used to describe a "baked" character or an oxidized bouquet and flavor

Cooperage   Cellar term relating to wooden containers, but sometimes loosely used to also include vessels constructed of glass, stainless steel, and other materials

Cooperative   Winery owned by a group of member grape growers

Cork   Bark of the cork oak, *Quercus suber,* largely grown in Portugal and Spain, from which wine bottle closures are made

Corkiness   Sensory term relating to the mold, *Penicillium expansum,* which manifests a foul mildew-like compound called trichloroanisole (TCA) in wine bouquet

Cradle   Device in which old bottles with sediment are placed for cork removal—often used for holding bottles while being poured

Cream of Tartar   Potassium bitartrate crystals precipitated by new wines, especially in colder storage temperatures

Cremant   Sparkling wine which has a reduced level of effervescence, not to be confused with "Cramant," which is an important winegrowing commune in the Champagne region of France

Cru   French for a specified "growth" or "vineyard"

Crust   Sediment which has collected and solidified on the inside surface of a bottle, often in association with old Port wines

Cuvée   A French enology term relating to a special blend, such as that which is designed for secondary fermentation into sparkling wine

Cultivar   Viticultural term relating to a man-made hybrid, set apart from the term "variety," which usually refers to natural progeny

Decant   Operation of delicately transferring wine from a bottle to a decanter so as to separate any sediment which have formed in the bottle during aging

Dégorgement   The technique of removing the plug of sediment from a bottle of sparkling wine after the *remuage,* or "riddling"

Delicate   Sensory term used in reference to wine with subtle bouquet and flavor

Demi-Sec   French for "half-dry" or "near dry"

Density   Sensory term relating to the amount of certain character constituents, usually color

Dessert Wine   A sweet wine generally consumed with, or instead of, dessert—made from sweet table wine having been "fortified" up to about 20% alcohol content with brandy

Dinner Wine Wine less that 14% alcohol generally consumed with meals.
Dirty Sensory term used in reference to wines tainted with musty odors and/or flavor
DOC Acronym for the Italian Denominazione di Origine Controllata official regulatory system for wine quality and origin control
Dolce Italian for "sweet"
Domaine French for "vineyard estate" or "wine estate," most commonly used in the Burgundy region of France
Dosage Addition of wine to sparkling wines directly after dégorgement in order to replace the expelled "ice plug"
Doux French for "sweet"
Dry Sensory term used in reference to an absence of sweetness
Earthy Sensory term used to describe soil flavors such as "chalky," "flinty," or "stony"—but not "dirty"
Edelfaule German for "noble mold"
Effervescent Sensory term for "sparkling"—the discharge of carbon dioxide bubbles in wine
Enology The scientific study of making wine
Enzymes Complex organic compounds which trigger chemical reactions
Estate-Bottled Label term referring to grapes having been grown on the estate of the vintner or vineyards under estate control
Esters Complex organic acid–alcohol flavor compounds
Ethyl Alcohol Essential wine compound produced from fermentation of grape juice or must—same as ethanol
Expressive Sensory term which relates to wine which has an abundance of character—the opposite of flat
Extract Analytical term referring to total dissolved solids in wine—the same as Brix in juice or must
Facultative Microorganisms which can grow either with or without an oxygen source
Fattoria Italian for "vintner"
Fermentation Usually relates to the transformation of sugar to alcohol, carbon dioxide, and energy in response to enzymes produced by the growth of yeasts—bacterial fermentations also exist in which malic acid is transformed to lactic acid, or in which ethyl alcohol is transformed to acetic acid
Fermentation Lock Cellar device used on fermenting vessels which allow CO_2 gas to escape, but disallow air to enter
Filtration Cellar treatment in which wine is pressured through media in which suspended solids or microorganisms are removed
Fining Cellar treatment in which agents such as bentonite are used to clarify wine
Finish Sensory term relating to the last impression of a wine just before, during, and after swallowing, including aftertaste
Firm Sensory term relating to wines which have ample acidity and/or tannins

Flat Sensory term relating to wines which have a deficiency of character—the opposite of expressive

Flavor Sensory term relating to olfactory values of taste such as "berry," "earthy," "vegetative," "fruity," "yeasty," and so forth

Flinty Sensory term relating to "stony" or "rock-like" flavors

Flor Refers to the surface-growing yeasts which synthesize acetaldehyde, the "nut-like" flavor compound common to Sherry wines

Floral Sensory term relating to a "flowery" or "blossom" value in bouquet

Flowers Cellar term for white film on the surface of wine which may indicate the growth of spoilage bacteria or wild yeasts

Fortified Wine Wine which has been blended with brandy to increase percentage of alcohol

Foxiness Sensory term loosely used to describe the aroma of wines made from native *Vitis labrusca* grapes such as Concord and Niagara

Free Run Juice or wine which flows from a press without the exertion of pressure

Frizzante Italian for "petillant" or "lightly sparkling"

Fruity Sensory term relating to a "fruit-like" value in bouquet and/or flavor

Full Sensory term relating to a heavy density of one or another element of wine character

Fume Sensory term relating to wines having a pronounced floral bouquet, usually in reference to Sauvignon Blanc white wines

Generic Label term relating to wine geography or type, such as "Burgundy" (wine grown in the Burgundy region of France) or "Vermouth" (wine type in which herb and spice essences are added)

Green Sensory term relating to wines made from premature grapes, or wines not fully developed in aging

Gross Lees Cellar term referring to the large amount of initial sedimentation which precipitates before and during fermentation

Harsh Sensory term relating to wine which is excessively high in total acidity and/or tannin

Heady Sensory term relating to table wines with a high alcohol content—the same as "strong"

Heavy Sensory term relating to high density of one or another character constituent, usually body

Hock English term for German white table wines

Hogshead Small wine cask

Hue Sensory term relating to the value of color, such as amber, straw–gold, brick red, crimson, ruby, and so forth.

Hybrid Viticultural term relating to the sexual crossing of one grape variety with another in order to create a new cultivar, such as the French-American hybrids Seyval Blanc and Chambourcin

Hydrometer Analytical device which responds to the specific gravity, or relative density, of a liquid—such as used to measure Brix and Balling

Indicator Analytical compound which changes color when a solution has been neutralized, such as phenolphthalein used to indicate when a total acidity titration is completed

Jeroboam Large wine bottle, usually with a capacity of 3 L

Keg Very small wooden wine vessel, usually with a capacity of less than 30 U.S. gal

KMS Acronym for potassium meta-bisulfite, used as a preservative in winemaking

Labrusca Viticultural term relating to *Vitis labrusca,* the native grapes of the northeastern United States, such as Concord and Niagara

Lactic Acid A "cheese-like"-flavored compound produced from malic acid in a malolactic fermentation by special bacteria

Late Harvest Wines made from grapes purposely left past peak ripeness in order to dehydrate berries, which concentrates sugar and flavor

Lees Production term relating to wine sedimentation, either from fermentation, clarification, or instability of acid salts, color pigments, and other constituents

Legs Sensory term relating to the reflux condensation of volatile wine components inside the glass—long "legs" widely believed to be the mark of superior wines—but there is no correlation, as condensation rates are determined by the relative temperatures of the wine and the glass

Light Sensory term relating to low density of one or another character constituents, usually color or body

Long Sensory term used in reference to wine flavors which linger in the mouth—extended aftertaste

Maderization The oxidation of ethyl alcohol and acetic acid into aldehydes—the process employed in the making of Madeira, Sherry, and Marsala wines

Magnum A double bottle containing 1.5 L

Malolactic Bacteria which transform malic acid to lactic acid, resulting in a "buttery" (diacetyl) flavor component

Marque French for "mark," usually in reference to "Grand Marque" Champagnes, the highest quality level

Methuselah Very large wine bottle, usually containing 6 L, often used in the Bordeaux and Champagne regions of France

Micron Filtration term relating to one-millionth of a meter, or about 0.00004 in.—most yeasts are removed at about 1.0 μ of filter porosity; most bacteria at about 0.5 μ

Millesime French for "vintage"

Moelleux French for "mellow"

Moldy Sensory term referring to an "off" value of bouquet and flavor, usually given to wines which have been made from moldy grapes and/or from wines stored in tankage which has harbored mold

Monopole French for "monopoly"—generally refers to an entire cru (vineyard) under a single ownership

Mousseux French for "foaming"—a term used to describe wines which are effervescent, or sparkling

Must Cellar term relating to crushed grapes

Musty Same as *Moldy*

Mutage The process of adding brandy to fermenting juice or must in order to arrest fermentation and retain remaining grape sugar as residual sweetness in the resulting wine—often used in reference to Port wines

Negociant French for "negotiant"—bulk wine buyers who age, blend, process, bottle, and market wines

Nero Italian for "dark red"

Noble Mold Viticultural term relating to *Botrytis cinerea,* a mold which permeates grape skins, allowing berry water to evaporate and remaining color, sugar, and flavor components to be concentrated

Nose Sensory term relating to the evaluation of aroma and bouquet

Nouveau French for "new"—generally relates to young red wines, such as "Beaujolais Nouveau," which are made and marketed soon after each vintage

Nutty Sensory term relating to the acetaldehyde "nut-like" flavor characteristic in Madeira and Sherry wines

Oenology English for "enology"

Oily Sensory term used in reference to wines which have a "fatty" consistency on the palate

Organoleptic Having to do with sensory evaluation

Oxidation Chemical term relating to the reaction of wine constituents with oxygen, such as the aging of wine or the browning of color

Pasteurization Cellar term relating to the inhibition of spoilage microorganisms by the controlled application of heat, generally between 160 and 180°F for less than a minute—rarely used in modern winemaking

Petillant Sensory term relating to wines which are just slightly effervescent

pH Analytical term relating to the entire range of the most intense acidity at pH 0.0 to the most intense base at pH 14.0, with neutrality at 7.0—often used as one of several indicators of optimal grape ripeness

Phylloxera The grapevine root louse, more precisely known as *Phylloxera vastatrix*

Pierce's Disease A bacterial disease which blocks the passage of fluids in a grapevine

Piquant French for "tart"

Pomace Cellar term referring to the seeds and skins remaining in the press

Pop Wine Wine containing added flavoring(s) from other berries, melons, and fruits

Potassium Bitartrate Same as *cream of tartar*

ppm Abbreviation for the chemical measurement expression "parts per million," which is roughly equivalent to milligrams per liter

Press Juice or Press Wine Cellar terms relating to the juice or wine extracted under pressure—usually more dense in color and excessively astringent from higher levels of phenols

Puncheon British wine vessel measure of about 70 U.S. gal

Punt Indentation in the bottom of a wine bottle originally intended to provide added strength to the container

Racking Cellar term relating to the decanting operation in separating juice or wine from lees

Rancio Sensory term used in reference the bouquet of some maderized wines—has no connection with "rancid"

Recolte French for "harvest"

Remuage The process of placing unfinished bottle-fermented sparkling wines into a table or rack in order to "work" the sediment into the neck of the bottle prior to dégorgement—the same as *Riddling*

Rehoboam A large wine bottle, about 4.5 liters in capacity, often used in the Bordeaux and Champagne regions of France

Rich Sensory term having the same general meaning as *Heavy*

Round Sensory term referring to good proportions of constituency

Rosé Pink wines, generally made from a brief fermentation of red grape must just prior to pressing—typically more pigmented than "blush" wines

Rotten Eggs Sensory term relating to wines which have degenerated because of the formation of hydrogen sulfide (H_2S)

Round Sensory term relating to wines which have a good balance and harmony of character and constituency

Salamanazar Very large wine bottle, containing 9 L, now rarely found in use commercially

Sec French for "dry"

Secco Italian for "dry"

Sekt German for "sparkling wine"

Soft Sensory term referring to wines having lower intensities of total acidity and astringency

Sorbic Acid Preservative used to inhibit yeast growth, ineffective against bacteria

Sour Sensory term referring to spoilage—often misused in reference to wines having higher intensities of total acidity or astringency

Stemmy Sensory term referring to wines which express a rather astringent vegetative character often related to excessive contact with the grape stems during fermentation

Strong Sensory term referring to table wines expressing a burn of alcohol in the bouquet

Sulfite Cellar term generally referring to KMS, or potassium meta-bisulfite—a source of sulfur dioxide used as preservative and antioxidant in wines

Sur Lies Cellar term referring to holding new wines on the gross fermentation lees in order to promote malolactic fermentation and extract complex flavors

Tanky Same as *Caskiness*

Tannic Sensory term relating to phenolic astringency in wines with a rather "leather-like" flavor

Tannin   A group of phenolic compounds found in grape seeds, skins, and stems, as well as in wood used for aging—tannins are antioxidants and thus serve to slow the wine-aging process

Tart   Sensory term relating to wines having higher intensities of total acidity

Tartaric Acid   Principal organic acid in grapes—combines with potassium to form potassium bitartrate salts, or cream of tartar

Tastevin   A small shallow cup, usually made of silver, typically related to the Burgundy region of France

Tawny   Sensory term referring to amber–red color, usually as a result of wines aged for long periods of time

Teinturier   A red grape which has particularly high density in color, often used in blending to enhance color-deficient wines

Tenuta   Italian for "estate"

Tirage   Cellar term referring to the laying of bottles horizontally, tier upon tier

Tonneau   Wine cask with a capacity of about 238 U.S. gal, traditional to the Bordeaux region in France

Total Acidity   The aggregate of all acids measured in a given wine, the sum of fixed and volatile acidities

Triage   The removal of defective clusters and/or berries, usually on a moving belt inspection line

Ullage   Air space in a bottle or cellar container resulting from assimilation, evaporation, and seepage

Varietal Wine   A wine labeled for the variety of vine from which it was predominantly made

Variety   An individual natural strain of vine recognized with a group of other closely related strains—for example, Chardonnay, Trebbiano, Johannisberg Riesling, Pinot Noir, Cabernet Sauvignon, and Nebbiolo, are each a 'variety' of the *Vitis* genus and *vinifera* species

Vendange   French for "grape harvest"

Vendemmia   Italian for "grape harvest"

Veraison   Viticultural term referring to the point when grape berries start to ripen, softening and changing color

Vert   French for "green"—often used in reference to wines made from underripened grapes, or for tart young white wines

Vigne   French for "vine"

Vignoble   French for "vineyard"

Vin   French for "wine"

Vino   Italian for "wine"

Vintage   Label term relating to the year in which the grapes were grown—sometimes used in reference to a particularly good year of growth

Vintner   A wine producer

Viticulture That part of horticulture which to do with the cultivation of grape vines

Vitis Vinifera Viticultural term referring to the genus and species of "Old World" vines—the native wine-grape vines of Europe such as Chardonnay and Cabernet Sauvignon

Volatile Acidity Analytical term referring to the acidity produced by spoilage microorganisms

Woody Sensory term referring to wine which has the bouquet and/or taste of excessive wood aging

Yeast Single-cell plant organisms that produce enzymes which convert simple sugars into ethyl alcohol, carbon dioxide gas, and energy

Yeasty Sensory term referring to wine which has the bouquet and/or taste of excessive exposure to yeast in "sur lies"

Zymase The enzymes produced by wine yeasts which convert sugars to ethyl alcohol, carbon dioxide gas, and heat energy during fermentation

BIBLIOGRAPHY

———. 1994. *Marketing Without Money!* Lincolnwood, IL: NTC Business Books.

———. 1983. *The Dartnell Marketing Manager's Handbook.* Chicago: Dartnell.

ADAMS, L.D. 1985. *The Wines of America,* 3rd ed., New York: McGraw-Hill.

BARCLAY, V. 1995. Aftermarketing: How to Keep Lifelong Customers. *Vineyard & Winery Management* 21(6): 10–11.

BARCLAY, V. 1995. How To Find And Keep A Good Distributor. *Vineyard & Winery Management* 21(7): 12–14.

BARCLAY, V. 1995. How To Manage Successful Programs. *Vineyard & Winery Management* 21(5): 10–11.

BARCLAY, V. 1994. Planning For Success: How To Formulate Your Marketing Plan. *Vineyard & Winery Management* 20(6): 12–14.

BARR, R. 1981. Financing Winery Operations. *The Economics of Small Wineries.* Berkeley: University of California Press.

BARSBY, S.L., STEVEN L. BARSBY and ASSOCIATES, INC. 1989. *The Economic Contributions of the Wine Industry to the States' Economies.* Washington, DC: The National Wine Coalition.

BORDELON, B. 1995. *Growing Grapes in Indiana,* Purdue University Cooperative Extension Service Bulletin HO-45, IN: West Lafayette, IN.

BOULTON, R.B., V.L. SINGLETON, L.F. BISSON and R.E. KUNKEE 1995. *Principles and Practices of Winemaking.* New York: Chapman and Hall.

BROOME, Jr., T. 1995. How and Where to Borrow Money. Part 1—SBA, *Vineyard & Winery Management* 21(6): 13–15.

CLARKE, OZ. 1995. *Wine Atlas.* New York: Little Brown.

CLANCY, K.J., SHULMAN, R.S. 1991. *The Marketing Revolution: A Radical Manifesto for Dominating the Marketplace.* New York: Harper Business.

DAVIDSON, J. 1989. *The Marketing Sourcebook For Small Business.* New York: Wiley.

DAVIDSON, J. 1988. *Marketing On A Shoestring.* New York: Wiley.

FURLONG, C. 1993. *Marketing For Keeps.* New York: Wiley.

HARKNESS, E. 1994. *Identification & Prevention of Common Wine Defects.* Kentucky Vineyard Society, Bardstown, KY.

JORJORIAN, P. 1994. Get Together With Your Marketing. *Vineyard & Winery Management* 20(2): 38–40.

LEDGERWOOD, L.D. 1981. Financing the New On-Farm Winery. *Eastern Grape Grower and Winery News.* 20–23.

MCKEE, L.J. 1995. It Can Pay to Help Restaurants Sell Wine. *Wine East* 23(4): 16–17, 38.

NOBLE, A.C, ARNOLD, R.A., BUECHENSTEIN, J., LEACH, E.J., SCHMIDT, J.O., and STERN, P.M. 1987. Research note: modification of a standardized system of aroma terminology. *Am. J. Enol. Vitic.* 38(2): 143–146.

NOBLE, A.C., ARNOLD, R.A., MASUDA, B.M., PECORE, S.D., SCHMIDT, J.O., and STERN, P.M. 1984. Progress towards a standardized system of wine aroma terminology. *Am. J. Enol. Vitic.* 35: 107–109.

NOBLE, A.C. 1990.The Aroma Wheel. *http://www.wineserver.edu.*

RIEGER, T. 1995. Market, Merchandise and Make Friends! *Vineyard & Winery Management* 21(4): 32–36.

SEXTON, J.D. 1995. The Tasting Room As Profit Center. *Wines Vines* 76(8): 21–23.

THOMAS, J.W. 1995. Advertising: How Best To Use It. *Wines Vines* 76(3): 31–33.

VADEN, D.H. and T.K. WOLF. 1994. The Cost of Growing Winegrapes in Virginia. *Virginia Cooperative Extension Publication.* 463: 006.

VAN DE WATER, L. 1990. *Evaluating and Treating Wine Defects.* Business Bulletin, Napa, CA.

VINE, R.P. 1996. *Wine Appreciation.* New York: Wiley.

VINE, R.P. (ed.). 1992. *Fruit and Berry Wine Symposium Proceedings.* West Lafayette, IN: Purdue University.

VINE, R.P. 1983. Keys to Successful Winery Establishment. *Vinifera Wine Growers Association Proceedings,* Middleburg, VA.

VINE, R.P. 1983. Concepts and Approaches on Wine Marketing. *Texas Grape Growers Association Proceedings,* Lubbock, TX.

VINE, R.P. 1981. *Commercial Winemaking,* Westport, CT: AVI Publishing Co.

ZOECKLEIN, B.W., K.C. FUGELSANG, B.H. GUMP AND F.S. NURY. 1995. *Wine Analysis and Production.* New York: Chapman and Hall.

ZOECKLEIN, B.W., K.C. FUGELSANG, B.H. GUMP AND F.S. NURY. 1990. *Production Wine Analysis.* Westport, CT AVI/New York: Van Nostrand.

INDEX

Acebacter, 80
Acetic acid, 101
Acidity. *See also* Total acidity
  of apple varieties, 262
  of bramble berries, 263
Acids, and winemaking, 101–102, 219, 240
Adams, Leon, 17
Adlum, John, 5
Advertising, and marketing, 289–94
Aftertaste, and wine judging, 358
Aging. *See also* Barrel aging; Oak aging
  of blush wines, 259
  as controlled process of oxidation, 102–103
  of red wines, 248–49, 250
  of white wines, 228
Agrobacterium tumefaciens, 72
Alcohol Beverage Commission (ABC), and marketing, 298
Alcohol burner, 85
Alcoholic content
  ATF labeling regulations on, 142, 150
  determination of by ebulliometer, 336–38

Alexander, John, 4
Altimira, Father Jose, 10
Amelioration, and addition of sugar, 215–16, 218–19, 239
American oak, compared to French oak, 126
American Society of Enologists, 20
American Vine Dresser's Guide, The (Dufour, 1826), 5
American Vintners Association (AVA), 160, 173
American wine, ATF labeling regulations for, 141–42
American Wine Society, 188
Amino acids, and wine production, 107–108
Amis du Vin, Les (wine society), 188
Analytical instrumentation and procedures, for winemaking
  alcohol by ebulliometer, 336–38
  bottle sterility, 315–17
  Brix-balling by hydrometer, 320
  Brix by refractometer, 321–22
  extract by nomograph, 338, 339
  gram stain, 312–13

428 *Index*

malolactic fermentation determination by
   paper chromatography, 314–15
  microscopy, 309–12
  pH meter, 319–20
  sensory evaluation, 348–61
  sulfur dioxide levels, 344–48
  volatile acidity by cash still, 342–44
  wine production and, 184–88
  winery sanitation and, 317–19
Anthocyanins, 106
Aperitif wines, and classification of wines, 152
Appellation
  ATF blending regulations on, 116–17
  value of grape varieties and, 47
Apple wines, 262–63
Aroma identification kits, 109
Aromas
  of apple varieties, 262
  grape flavors and, 109–10
  sensory evaluation of wine and, 352
Aroma wheel, 353
Ascorbic acid, 205
Aspirator filter pump, 85
Astringency, as gustatory term, 357
ATF. *See* Bureau of Alcohol, Tobacco, and Firearms

Bacteria, and microbiology of winemaking, 80–83
Balanced pruning, 65
Balancing
  of blush wines, 259
  of red wines, 249
  of white wines, 228–29
Balling. *See also* Brix
  analytical instrumentation for, 185–86, 320
  definition of, 99
Baltimore, Lord, 3
Barrel aging, and wine production, 120–31, 248–49. *See also* Oak aging
Batonage, 121, 213
B-Cap closure system, 136–37
Bench grafting, 42, 43
Bentonite, as fining agent
  blueberry wines and, 264
  preparation of, 205
  white wines and, 223
Berry, grape
  diagram of, 31
  fermentation of whole, 213

Berry wines. *See* Fruit and berry wines
Bilateral Cordon training system, 62
Billboards, 292–93
Biological turbidity, 227
Bitterness, and phenolic compounds, 357
Black rot, 67, 68
Blending
  of blush table wines, 258
  of red table wines, 246–47
  of white table wines, 222–23
  wine production and, 111–19
Blueberry wines, 264–65
Blush wines
  aging, balancing, and preservation of, 259
  blending of, 112, 258
  bottling and corking of, 260
  clarification and stabilization of, 259
  crushing and destemming for, 256–57
  filtration of, 259–60
  grape varieties for, 255–56
  pectinase treatment and, 97
  pressing and fermentation of, 257–58
  racking of, 258
Body, and wine judging, 358
Bonds, ATF and excise taxes, 171–72
Bootleggers, during Prohibition, 16
Bostwick, Father William, 7
Botrytis cinerea. *See* Noble mold
Bottle aging, of red wines, 250
Bottled wine, microbial spoilage of, 91–93
Bottled wine record, 179, 183
Bottles and bottling
  analytical procedure for determining sterility of, 315–17
  of blush wines, 260
  packaging of wine and, 134, 206–208
  of red wines, 249
  of white wines, 231, 232
Bottle shock, 231, 260
Bottling line operation, 232
Bouquet, and sensory evaluation, 352
Boutique wineries, and wine boom of 1970s–1980s, 21–22
Bramble berry wines, 263–64
Brandy, addition of to wines, 151
Brettanomyces yeasts, 77, 78, 90, 104
Brix
  analytical instrumentation for, 185–86, 320, 321–22
  blueberry wines and, 265

definition of, 99
hydrometers and, 320
raisin wines and, 268
red wines and, 238–40
refractometers and, 321–22
of soluble solids at twenty degrees Celsius, 322–36
white wines and, 215–19
Brochures, advertising, 290–91
Brown rot, of cherries, 265
Budget
  for marketing, 273
  vineyard establishment and, 49–55
Buds, of grapevine, 28–30
Buena Vista Vinicultural Society, 11
Bulk wine, microbial spoilage of, 89–91
Bulk wine book inventory card, 234
Bulk wine record, 179, 182
Bull, Ephraim, 8
Bunch grape, definition of, 24
Bunch rot, 67, 68
Bureau of Alcohol, Tobacco, and Firearms (ATF)
  advertising and, 289
  amelioration and, 218–19
  blending and, 112–17
  chaptalization and, 216–18
  definition of wine, 95–96
  dessert wines and, 151
  dried fruit wines and, 268–69
  European generic names on U.S. wines, 149
  honey wine and, 269
  labeling regulations of, 140–45, 209–10
  marketing and, 271
  permit application process, 169–73
  record keeping regulations of, 177–78
  sparkling wines and, 151
  storage tank regulations of, 195–96
  table wines and, 150
Business plan, for winery, 157–59

Cabernet Franc, as cultivar of *Vitis vinifera*, 37
Cabernet Sauvignon, as cultivar of *Vitis vinifera*, 35
Cactus wine, 1
Calcium sulfate sediment, 226
Calcium tartare sediment, 226
California, history of winemaking in, 9–14, 16
Candida, 77, 78, 90
Capsules, and bottle closures, 136–37, 209

Cardinal scales, for wine evaluation, 361
Carpinteria vine (California), 11
Cartridge filtration, 132–33, 198, 199, 230
Cased goods book inventory form, 235
Cases, for wine bottles, 210
Cash-flow financing, 157, 159
Cash Volatile Acid distillation apparatus, 187, 342–44
Catawba grape, 5, 38
Cellulose pad filters, 225
Certificate of label approval, and ATF labeling regulations, 144–45
Chambourcin, as cultivar of French-American hybrid, 44–45
Champagne, definition of, 150
Champlain, Samuel de, 3
Champlin, Charles, 7
Chancellor, as cultivar of French-American hybrid, 45
Chapman, Joseph, 11
Chaptalization, and addition of sugar, 215–16, 216–18, 239
Chardonnay grape, as cultivar of *Vitis vinifera*, 32
Charles II (King of England), 3
Chemical agents, for wine preservation, 89
Cherry wines, 265–66
Chlorine treatment, of wooden barrels, 130
Churches, and history of winemaking in America, 6–7, 9–11
Citric acid
  peach wines and, 267
  treatment of wooden barrels with, 130–31
  as wine additive, 94, 102, 205
Clarification, of blush wines, 259
Classification, of wines
  aperitif wines and, 152
  dessert wines and, 151
  pop wines and, 152
  sparkling wines and, 150–51
  table wines and, 148–50
Clay, Henry, 5
Climate, and site selection for viticulture, 26–27
Clones, of grapevine, 30
Closures, for wine bottles, 135–37
Cluster thinning, and pruning, 65
Cold Duck, and wine boom of 1950s and 1960s, 21
Cold stabilization, 223
Cold storage, of bramble berry wines, 264. *See also* Refrigeration

Color
  of blush wines, 255
  of bottles, 134
  of labels, 138–39
  pressing of red wines and, 244
  sensory evaluation of wine and, 350–52
  wine production and, 106–107
Commercial starter cultures, 87–88
Compiled mailing lists, 291
Computer technology
  advertising and, 293
  sources of software, 307
Concord grape, 8, 39
Conditioning, of oak storage barrels, 126, 128
Constitution, eighteenth amendment (1920), 14, 17
Consumers, direct sales to, 301–302
Contingency management, 154
Contour planting, 59
Cooperative Extension Service, 26
Copper sulfite turbidity, 226
Cordon training systems, 62
Corks
  for blush wines, 260
  closure of wine bottles and, 135–36, 208–209
  for red wines, 249
Corkers, as winery equipment, 201, 202
Corporations, and signature authority, 173
Cost
  of barrel aging, 121–26
  blending of wines and, 119
  example of inventory cost analysis, 156
  of vineyard establishment, 46–60
Cream of tartar, as wine additive, 223, 224
Crown gall, 71, 72
Crushers and crushing
  apple wines and, 263
  blush wines and, 256–57
  cherry wines and, 265–66
  equipment requirements, 189–90
  red table wines and, 238
  strawberry wine and, 267
  sulfur dioxide treatment at, 104–105
  white table wines and, 214–15
Cultivars, of grapes
  definition of, 42
  selection of, 32
  site selection and, 26–27
Customer service, 298
Cutin, of grape berry, 73

Database marketing, 292
Delaware, as cultivar of *Vitis labrusca*, 38
Delaware, Lord, 3
Design, of wineries
  case studies of, 163–68
  motifs for, 163
Dessert wines, and classification of wines, 151
Destemmers and destemming
  blush wines and, 256–57
  equipment for, 189–90
  red table wines and, 238
  white table wines and, 214–15
Detartration
  maintaining temperatures for, 223
  wine production and, 119–20
Diammonium phosphate (DAP)
  blueberry wines and, 265
  peach wines and, 266
Diatomaceous earth (DE), and wine filtration, 131–32, 200, 225
Direct mail, and advertising, 291
Disaccharides, 98
Diseases, control of and vineyard management, 67–70
Downy mildew, 67, 68
Drainage, of soil for viticulture, 27, 55, 59
Dried fruit wines, 268–69
Dry red wines, blending parameters for, 112
Dry white wines, blending parameters for, 112
Dufour, Jean Jacques, 4, 5

Ebulliometer, 336–38
Economics. *See* Budget; Cost
Education, and marketing, 277–78
Edelfaule wines, 83
Egg whites, fining of red wines with, 205, 247–48
El Camino Real (California), 10
Electronic scales, 189
Embden-Meyerhof reaction, 96
Enology. *See also* Blush wines; Red wines; White wines; Wines and winemaking
  acids, acidity, and pH, 101–102
  barrel aging and, 120–31
  definition of, 95
  detartration and, 119–20
  filtration and, 131–33
  fining and, 119
  flavors of grapes and, 108–11
  grape and wine components, 96, 97

late-harvest grapes and, 99–101
malolactic fermentation, 108
oxygen and oxidation, 102–103
packaging and, 134–45
pectic enzymes and, 96–97
perils and pitfalls in, 145–47
phenols, phenolics, and polyphenols, 105–107
proteins and, 107–108
sugars and sweetness, 98–99
sulfur dioxide, 103–105
wine blending and, 111–19
Environmental impact statements, and ATF permit process, 172
Equipment
  for microbiological laboratory in winery, 85–86, 184–88, 305
  sources for, 305–306
  vineyard establishment and costs of, 47–48, 50
  for winery, 188–202
Erikson, Leif, 1
Estate bottled, definition of, 117
Ethyl acetate, 110
European Economic Community (EEC), regulations on sparkling wines, 150–51
Euvitis, description of as subgenus, 24
Evaluation, of wine, 188
Extract. *See also* Brix
  definition of in winemaking, 99
  determination of by nomograph, 338, 339

Fay, Elijah, 6
Feasibility studies, for new winery, 157–59
Federal excise tax, 150, 151, 171–72
Fermentation
  of apple wines, 263
  of blueberry wines, 265
  of blush wines, 257–58
  of cherry wines, 266
  of fruit and berry wines, 261–62, 263–64
  of mead, 269
  of peach wines, 266–67
  of red table wines, 240–44
  of sparkling wines, 150, 151
  of strawberry wine, 267–68
  of white table wines, 220
  of whole clusters and berries, 213
  yeasts and process of, 73–77
Fermentation control cards, examples of, 182, 253

Fermentation locks, 196–97
Ferric phosphate turbidity, 227
55-plus beverage monitor, 230
Fillers, and winery equipment, 200
Filters and filtration
  blush table wines and, 259–60
  bramble berry wines and, 264
  red table wines and, 249
  white table wines and, 229–31
  wine production and, 131–33
  winery equipment and, 198–200
Financing, of new winery, 157–59
Finger Lakes region (New York), history of winemaking in, 6–7, 8
Fining
  of red table wines, 247
  of white table wines, 223
  wine production and, 119, 204–205
First racking. *See also* Racking
  sulfur dioxide treatment during, 105
  of white table wines, 221–22
Fixed costs, and vineyard establishment, 49
Flavonoids, 106
Flavors, of wines. *See also* Fruit flavor
  oak barrel aging and, 120
  wine production and, 108–11
Floral aromas, 109, 352–53
Florida, history of winemaking in, 2
Flor yeasts, 76–77
Flower clusters, of grapevine, 30
Fournier, Charles, 19–20
Frank, Dr. Konstantin, 20
Free ammonium nitrogen (FAN), 100
Free-run red wine, 245
Free sulfur dioxide, 344–47
Freezing, of wines, 223
French-American grape hybrids
  Brix and, 238
  development and introduction of, 18–19
  phenolic color profile of, 107
  viticulture and, 43–45
French oak, compared to American oak, 126
Frost-free sites, 27
Fruit aromas, 109–10, 353–54
Fruit and berry wines
  apple wines, 262–63
  blueberry wines and, 264–65
  bramble berry wines, 263–64
  cherry wines and, 265–66
  definition of wine and, 95–96

dried fruit wines, 268–69
fermentation of, 261–62
mead, 269
peach wines, 266–67
strawberry wines, 267–68
wine boom of 1950s and 1960s and, 21
Fruit clusters, of grapevine, 30
Fruit flavor
  in red wines, 237
  in white wines, 212–13
Fruiting buds, and pruning, 65

Gallo, Ernest & Julio, 20
Galvez, General Jose de, 9
Gamay, as cultivar of *Vitis vinifera*, 36
Gay-Lussac, Joseph-Louis, 74
Gelatin, and fining of wines, 205, 223
Generic wines, 149
Geneva Double Curtain vine training system, 62, 63
Gewürztraminer, as cultivar of *Vitis vinifera*, 34
Gift shop, at winery, 298–300
Glucosides, 106
Gold Rush (California, 1849), 11
Gold Seal Vineyards (Hammondsport, New York), 20
Government. *See* Alcohol Beverage Commission; Bureau of Alcohol, Tobacco, and Firearms; States
Grafting, of grapevines, 42
Gram stain procedure, 312–13
Grape berry moth, 70
Grape flea beetle, 70
Grapes. *See also* Viticulture; *Vitis labrusca*; *Vitis vinifera*; Wine and winemaking; specific varieties
  blush wines and varieties of, 255–56
  composition of, 96, 97
  cultivars of, 26–27, 32, 42
  diseases of, 67–70
  flavors of, 108–11
  grafting of, 42
  insects and, 70–72
  planting of, 59
  pruning of, 28, 64–67
  red wines and varieties of, 236–37
  sources for, 306–307
  structure of, 24, 30, 31
  trellising of, 29, 61–64
  white wines and varieties of, 211–12

Great Depression, 17
Greeley, Horace, 9
Gross lees, 221
Group tours, of winery, 275–76
Growing season, length of, 27
Guides, for winery tours, 296
Gustatory mode, of sensory evaluation, 355–58

Half-bottles, as bottle type, 208
Hansenula, 90
Haraszthy, Agoston, 11–12
Harvesting
  equipment for, 63–64
  red wines and, 238
  white wines and, 214
Hawkins, Sir John, 2
Hearst, George, 13
Herbicide injury, to grapevines, 64
Hilgard, Eugene Waldemar, 12–13
Histamine, in wines, 81, 83
History, of winemaking in America
  in early twentieth century, 17–20
  Eastern America and, 2–9
  from late 1940s to present, 20–23
  Native Americans and, 1
  Norse explorers and wild grapevines, 1–2
  Prohibition and, 14–17
  Western America and, 9–11
Honey. *See* Mead
Hoses, as winery equipment, 192–93
House mailing lists, 291
Humor, on wine labels, 139
Hybrid corks, 209
Hybrid grapevines, and viticulture, 42–45
Hydrogen sulfide, 103
Hydrolysis, of sugar, 98
Hydrometers, 185, 186, 320
Hydroxybenzene, 106
Hypochlorite solutions, and winery sanitation, 93

Imperial, as bottle type, 208
Indiana
  history of winemaking in America and, 4
  state regulation of winemaking, 174–75
Information for New Wineries (ATF), 169–70
Insects, vineyard management and control of, 70–72
Instability, determination, causes, and recovery of positive, 226–27

Insurance, for winery operations, 159–60
Internet, and marketing, 293
Interviews, with media, 281–82
Inventory, examples of forms for
  bulk wine form, 234
  cased goods form, 235
  cost analysis, 156
Inversion, of sugar, 98
Ion-exchange process, and detartration, 120
Iron-tannin turbidity, 227
Ives, as cultivar of *Vitis labrusca*, 40
Ives, Henry, 40

Jacket, and fermenter temperature, 243
Jacques, Jean, 6
Japanese beetle, 70
Jefferson, Thomas, 4, 5, 16
Jeroboam, as bottle type, 208
Johannisberg Riesling, as cultivar of *Vitis vinifera*, 33–34
Journalism, and marketing, 278–79
Judging, of wines by sensory evaluation, 348–61
Jug wines, 149–50
Juice and juice concentrates
  microbial spoilage in, 88–89
  sources for purchase of, 306–307

Kansas, prohibition movement in, 14
Keuka Low Renewal cane training system, 62
Kieselsol, and fining of wines, 205, 223
Killer yeasts, 104
Kloeckera, 77, 79

Labelers, as winery equipment, 201–202, 203
Labeling
  ATF regulations and, 115–17, 173
  blending of wines and, 115–17
  conversion to varietal in U.S., 21
  of fruit wines, 96
  packaging of wine and, 137–45, 209–10
Laboratory, and microbiology of winemaking, 83–86, 184–88, 305, 307. *See also* Analytical instrumentation and procedures
Laboratory analysis record, 179, 180
Labor costs, and vineyard establishment, 47
Lactic acid, 101
Lactobacillus brevis, 81, 82
Late-harvest wines, production of, 99–101, 212
Layout, of vineyard, 59

Legal requirements. *See* Bureau of Alcohol, Tobacco, and Firearms; States
Legaux, Pierre, 6
Lenoir, Jean, 109
Leuconostoc oenos, 80, 81
Lincoln, Mary Todd, 8
Location, of winery, 160–62
London Company, of Virginia, 3
Longfellow, Henry Wadsworth, 6
Longworth, Nicholas, 6
Lot origination forms, examples of, 179, 181, 252
Lyon, Abraham de, 3

Maceration, of fermented red wine, 244
Maceration carbonique process, 237
Machine harvesting, 63–64
Madeira, as fortified wine, 151
Magazines, and marketing, 281
Magnum, as bottle type, 208
Mailing lists, 291
Maine, prohibition movement in, 14
Maintenance, of oak storage barrels, 129
Malic acid, 101, 263
Malolactic fermentation, 81, 108, 247, 314–15
Malvidin, 106
Management, of winery, 154
Marechal Foch, as cultivar of French-American hybrid, 45
Market evaluation, 302
Marketing
  advertising and, 289–94
  budget for, 273
  feedback, 302–304
  plan for, 271–73
  public relations and, 273–84
  rationale for, 270–71
  of varieties of grapes, 47
Marketing plan, 271–73
Marsala, as fortified wine, 151
Mason, John, 3
Massachusetts, prohibition in, 14
Masson, Jules, 7
Materials, for wine production, 202–10
Mead, 269
Media kits, and marketing, 283–84
Media relations, and marketing, 278–84
Meiosis, 42
Membrane filtration, 230–31
Meritage blending, 117–18

Merlot, as cultivar of *Vitis vinifera*, 35
Microbiology, of wine. *See also* Analytical instrumentation and procedures
  bacteria and, 80–83
  commercial starter cultures, 87–88
  equipment sources for, 305
  factors affecting microbial growth, 87
  laboratory and, 83–86
  molds and, 83
  spoilage of bottled wine, 91–93
  spoilage of bulk wine, 89–91
  spoilage of juice concentrates, 89
  spoilage in juice and must, 88–89
  yeasts and, 73–79
Microfiltration, 133
Microscope and microscopy, 86, 309–12
Missions, and early winemaking in California, 9–11
Mississippi, and prohibition movement, 17
Missouri, history of winemaking in, 8–9
Modified 20-point cardinal scale, for wine evaluation, 359, 360, 361
Moisture, and microbial growth, 87
Molds, and wine microbiology, 83
Monosaccharides, 98
Montmorency cherries, 265
Mourvedre, as cultivar of *Vitis vinifera*, 38
Muscadines (*Muscadinia*)
  as cultivars of *Vitis rotundifolia*, 41
  French Huguenots and early winemaking in Florida, 2
  as subgenus, 24
Muscat de Frontignan, as cultivar of *Vitis vinifera*, 34
Must pumps, 189–90
Mute juice, 213

Name, of winery, 272
Napa Valley (California), history of winemaking in, 12
Nation, Carrie, 14
Native Americans, and winemaking, 1
Nebbiolo, as cultivar of *Vitis vinifera*, 37
Networking, and marketing, 277
Newsletters, and marketing, 288–89
Newspapers, and marketing, 280–81
News releases, and marketing, 279–80
New York, history of winemaking in, 6–7, 8
Nez du Vin, Le (The Nose of Wine) aroma identification kits, 109
Niagara, as cultivar of *Vitis labrusca*, 38–39

Niche sales, 296
Noble mold, and production of late harvest wines, 83, 100, 212
Nomograph, 338, 339
Nonflavonoids, 106
Nutrients, and microbial growth, 87

Oak aging
  blueberry wines and, 265
  red wines and, 248–49
  white wines and, 228
  wine production and, 120–31
Olfactory mode, of sensory evaluation of wine, 352–55
Operating costs, of vineyards, 46
Opportunity cost, 123
Organic copper turbidity, 226
Overall impression, and sensory evaluation of wines, 358, 361
Overripened grapes, wines made from, 212
Oxidation, and wine production, 102–103
Oxidative yeasts, 77
Oxygen, and wine production, 102–103

Packaging
  of red wines, 252, 254
  of white wines, 231, 233
  wine production and, 134–43
Pad filtration, 132–33
Papago Indians, 1
Paper chromatography, 86, 314–15
Paris World Exposition of 1889, 14
Pasteur, Louis, 74
Pasteurization, of wine, 89
Peach wines, 266–67
Pectinic acid, 96–97
Pectins, and winemaking, 96–97, 204
Pectolytic enzyme, 263
Pediococcus cerevisiae, 81, 82, 83
Penicillium expansum, 83
Penn, William, 3, 5
Pennsylvania Vine Company, 6
Pericarp, of fruit cluster, 30
Periodicals, on winemaking, 307–308
Permits
  ATF for commercial wine premises, 169–73
  for special events at wineries, 286
Personnel
  customer service and, 298
  sales and, 295–96

ATF permit process and questionnaire for, 172–73
Pest control, and vineyard management, 67–72
Pesticides, responsible use of, 67, 70
Petite Sirah, as cultivar of *Vitis vinifera*, 38
pH. *See also* pH meter
  blueberry wines and, 265
  microbial growth and, 87
  red wines and, 236–37
  of soils for viticulture, 55
  sulfur dioxide treatment and, 105
  white wines and, 211, 212
  winemaking and, 101–102
Phenols, and wine production, 105–107
Phloem, and grafting, 42
pH meter, 184–85, 319–20
Photographs, and marketing, 282–83
Photosynthesis, 28
Phylloxera root louse
  grafting and, 42
  history of winemaking in California and, 12–13, 72
Pichia, 77, 79, 90
Pierce's disease, 13, 70
Pigment-tannin sediment, 227
Pinot Blanc, as cultivar of *Vitis vinifera*, 34
Pinot Noir, as cultivar of *Vitis vinifera*, 35
Pinto Grigio, as cultivar of *Vitis vinifera*, 35
Piston-type basket press, 190–91
Plantagenet, Beauchamp, 2
Planting, and vineyard establishment, 59
Plastic corks, 209
Plate-and-frame filter, 198, 200
Pleasant Valley Wine Company, 7
Pneumatic basket press, 191
Polyphenols, and wine production, 105–107
Polysaccharides, 98–99
Polyvinylpolypyrrolidone (PVPP), 266
Pomace, handling of, 215
Pop wines
  classification of wines and, 152
  wine boom of 1950s and 1960s and, 21
Portola, Gaspar de, 9
Potassium bitartrate (KHT), as wine additive, 119, 223, 224
Potassium meta-bisulfite (KMS), as wine additive, 94, 105, 203–204, 246, 263
Pourriture noble wines, 83
Powdery mildew, 67, 68
Practical Winery & Vineyard (magazine), 273
Prasta, Charles C., 1–2

Preservation
  of blush wines, 259
  of red wines, 249
  of white wines, 228–29
  wine production and, 88–89
Presses and pressing
  apple wines and, 263
  blush wines and, 257–58
  bramble berry wines and, 264
  cherry wines and, 266
  red wines and, 244, 245
  of whole clusters, 213
  winery equipment and, 190–92
Prisse de Mousse yeast, 263
Pro forma income statement, for winery, 157, 158–59
Prohibited practices, in ATF labeling regulations, 143–44
Prohibition, and history of winemaking, 14–17
Propagation, of grapevines, 30, 32
Proprietary blending, 118
Proprietary wines, definition of, 118
Proteins, and wine production, 107–108
Pruning
  fundamental formula for, 67
  grape quality and, 28
  vineyard management and, 64–67
Public relations, and marketing, 273–84
Pump over, and fermentation of wine, 242
Pumps, as winery equipment, 192
Punching, and fermentation of wine, 242

Quality and quality control
  of berries for bramble berry wines, 263
  bottles and, 134
  establishment of program for, 183–84
Quercus suber, 208

Racking. *See also* First racking; Second racking
  of blush wines, 258
  of red wines, 245–46
Radio, and marketing, 281
Raisin wines, 268–69
Raspberry wines, 263
Record keeping, for wine production, 176–83
Reducing sugars, 98
Red varieties of grapes
  of French-American hybrids, 44–45
  of *Vitis labrusca*, 39–40
  of *Vitis vinifera*, 35–38

Red wines
  acid additions to, 240
  balancing and preservation of, 249
  blending of, 112, 246–47
  bottle aging of, 250
  bottling and corking of, 249
  Brix adjustments for, 238–40
  crushing-destemming for, 238
  fermentation of, 240–44
  filtration of, 249
  fining of, 205, 247–48
  free-run and press wine, 245
  fruit flavor intensity in, 237
  grape varieties for, 236–37
  harvesting of grapes for, 238
  maceration and, 244
  malolactic fermentation of, 247
  oak-aging of, 248–49
  packaging of, 252, 254
  pressing of, 244
  racking of, 245–46
  stabilization of, 247–48
  suggested bottles for types of, 207
  yeast inoculation of, 240
Referrals, and marketing, 304
Refractometer, 186, 321–22
Refrigeration. *See also* Cold storage
  stabilization of wines by, 223
  winery equipment and, 197–98
Regulation, of winemaking. *See also* Bureau of Alcohol, Tobacco, and Firearms
  Alcohol Beverage Commission and marketing, 298
  for special events at wineries, 286
  by state of Indiana, 174–75
Reselling, and marketing, 302–303
Residual sugar, 186
Residual sweetness, 99
Response mailing lists, 291
Resveratrol, 106
Retail sales
  as marketing channel, 301
  shops at wineries and, 296–98
Retsina, 152
Rhizopus molds, 83
Rhode Island, prohibition movement in, 14
Ripper method, for sulfur dioxide, 187
Roosevelt, Franklin Delano, 17
Rootstock, and grafting, 42
Rosé wines. *See* Blush wines
Row spacing, 59

Saccharides, 98
Saccharomyces apiculata, 74
Saccharomyces beticus, 76–77
Saccharomyces cerevisiae, 75–76, 77, 90
Saguaro cactus, 1
Saignée system, of must concentration, 107, 237, 257
St. James Episcopal Church (Hammondsport, New York), 6–7
Sales, and marketing, 294–302
San Gabriel Mission (California), 10
Sangiovese, as cultivar of *Vitis vinifera*, 36–37
Sangria, 21
Sanitation, in wine production
  analytical procedures for, 317–19
  importance of, 183, 237
  molds and, 83
  winery and, 93–94
Sauvignon Blanc, as cultivar of *Vitis vinifera*, 32–33
Scales, for measurement of grapes and fruit, 188–89
Schoonmaker, Frank, 21
Schwann, Theodor, 74
Scion, and grafting, 42
Scuppernongs, and *Vitis rotundifolia*, 41
Second racking, of white wines, 222. *See also* Racking
Semillon, as cultivar of *Vitis vinifera*, 34
Sensory evaluation, of wine
  analytical procedures for, 348–61
  equipment for, 187–88
Serra, Padre Junipero, 9–10
Seyval Blanc, as cultivar of French-American hybrid, 43
Shapes, of labels, 138
Sherry, as fortified wine, 151
Shoots, of grapevine, 30
Shoulders, of fruit cluster, 30
Side-arm vacuum flask, 85
Signage, for wineries, 293
Signature authority, and ATF permit process, 173
Site preparation, and vineyard establishment, 55, 59
Site selection, for viticulture, 26–27
Small Business Administration (SBA), 159
Smith, Adam, 270
Smith, Captain John, 2
Soda ash treatment, of wooden barrels, 129–30

Sodium hypochlorite, 93
Soils, and viticulture, 27, 55
Solids, and flavor in white wines, 213
Sonoma (California), history of winemaking in, 11–13
Sorbic acid, and control of spoilage, 93, 206
Sparkling wines, classification of, 150–51
Speakeasy clubs, during Prohibition, 16
Special events, and marketing, 284–88
Splits, as bottle type, 207–208
Spoilage, microbial
  of bottled wine, 91–93
  of bulk wine, 89–91
  determination of causes of, 92
  flavors and, 110–11
  in juice and must, 88–89
Stabilization
  of blush wines, 259
  of red wines, 247–48
  of white wines, 223
Stainless steel tanks, aging of wine and cost of, 122–25
Stamens, of grape flower, 30
Stanford, Leland, 13
Starter cultures, 87–88
States, regulation of winemaking by, 169, 174–75
Sterile filtration, 88, 93
Sterility, analytical procedure for bottle, 315–17
Stevenson, Robert Louis, 12
Strawberry wine, 267–68
Stuck fermentation, 220–21, 243, 269
Styrofoam incubator, 86
Sucrose, 98
Sugars, and winemaking, 98–99, 263, 267. *See also* Brix
Sulfite
  blueberry wines and, 265
  cherry wines and, 266
  citric acid sanitizing and, 94
Sulfur dioxide
  analytical instrumentation and levels of, 187, 344–48
  control of spoilage in bottled wines and, 93
  wine production and, 103–105
Sulfurous acid, 94
Sur lies, 221–22
Swab test, for winery sanitation, 317–19
Sweetness, and winemaking, 98–99. *See also* Brix
Sweet red wines, blending parameters for, 112

Sweet white wines, blending parameters for, 112
Sylvaner, as cultivar of *Vitis vinifera*, 34–35
Synthetic corks, 136
Syrah, as cultivar of *Vitis vinifera*, 37

Table wines, and classification of wines, 148–50. *See also* Blush wines; Red wines; White wines
Talk shows, television, 281
Tank presses, 191–92
Tanks, for wine storage, 122–25, 194–96, 306
Tannins, and wine production, 106, 238, 244, 357
Tartaric acid, as wine additive, 101, 219
Tasting, and sensory evaluation of wines, 348–61
Tasting room, at winery, 296–98
Taxation. *See* Federal excise tax
Taxonomy
  of grapevine, 28
  of yeasts, 75
Taylor, Greyton, 18
Taylor, Walter, 8
Teiser, Ruth, 11
Television, and marketing, 280
Temperance movement, 14
Temperature, and microbial growth, 87
Tendrils, of grapevine, 28, 30
Testimonials, and marketing, 303–304
Thermolabile protein turbidity, 227
Total acidity (TA), 101, 186–87, 240, 265, 338, 340–41
Tourism, and marketing, 274–76
Tours, of winery, 296–98
Trade associations, 277
Training systems, for grapevines, 61–64
Trellising, and viticulture, 29, 61–64
Trichoroanisole (TCA), 83
2, 4-D herbicide, 64

Ugarte, Father Juan, 9
Ultrafiltration, 133
Umbrella Kniffin cane training system, 61–62
Unfiltered wines, 133, 249
Universities, and technological advances in wine production, 22

Vacuum aspiration method, for sulfur dioxide, 187
Value, of grapes in wine at various prices per ton, 47

Variable costs, and vineyard establishment, 48–49, 50
Varietal wines, 148–49
Vegetal aromas, 110, 354
Vermont, prohibition movement in, 14
Verrazano, Giovanni da, 2
Veraison, 27, 98
Vidal Blanc, as cultivar of French-American hybrid, 43–44
Video, of winery, 275
Vignoles, as cultivar of French-American hybrid, 44
Vine, of grape
  characteristics and description of, 27–32
  training of during vineyard establishment, 59–60
  trellising of, 29, 61–64
Vinegar bacteria, 80
Vineyard & Winery Management (magazine), 273
Vineyards
  establishment of, 46–60
  management of, 64–72
Vintage wine, definition of, 117
Virginia, history of winemaking in, 3
Visual aids, and marketing, 282–83
Visual mode, of sensory evaluation of wine, 350–52
Viticulture
  cultivar selection, 32
  definition of, 24
  economic incentives for in colonial America, 3
  grafting and, 42
  grapevine structure and, 27–32
  hybrids and, 42–45
  site preparation for, 55, 59
  site selection for, 26–27
  soils and, 27
  trellising and, 61–64
  vineyard establishment and, 46–60
  vineyard management and, 64–72
  Vitis labrusca and, 38–40
  Vitis riparia and, 40
  Vitis vinifera and, 32–38
  wild American species of importance to, 40–41
Vitis, subgenera of, 24
Vitis aestivalis, 41
Vitis berlandieri, 41

Vitis labrusca
  Brix and, 215, 238
  characteristics of, 24–25
  history of winemaking in Eastern America and, 2–3, 4
  phenolic color profile of, 106
  viticulture and, 38–40
Vitis riparia, 40
Vitis rotundifolia
  French Huguenots and early wine production in Florida, 2
  phenolic color profile of, 106
  viticulture and, 41
Vitis rupestris, 41
Vitis vinifera
  Brix and, 215, 238
  characteristics of, 24
  history of winemaking in California and, 9, 11–14
  history of winemaking in Eastern America and, 3
  phenolic color profile of, 106
  viticulture and, 32–38
Volatile acidity, 187, 342–44
Volstead Act (1920), 14, 16–17

Wagner, Philip, 17–18
Web pages, and advertising, 293
Weed control, 55, 64
White varieties of grapes
  of French-American hybrids, 43–44
  of *Vitis labrusca*, 38–39
  of *Vitis vinifera*, 32–35
  white table wines and, 211–12
White wines
  acid additions to, 219
  aging of, 228
  balancing and preservation of, 228–29
  blending of, 112, 222–23
  bottling of, 231, 232
  Brix adjustment and, 215–19
  crushing and destemming, 214–15
  fermentation of, 220
  filtration of, 229–31
  fining and stabilization of, 223
  first racking of, 221–22
  flavor intensity and, 212–13
  grape varieties and, 211–12
  harvesting and, 214
  packaging of, 231, 233

pectic enzyme and, 97
pressing and, 215
second racking of, 222
stuck fermentation and, 220–21
suggested bottles for types of, 207
yeast inoculation of, 219–20
Whole clusters and/or berry, pressing of grapes in form of, 213
Wholesale channels, and marketing, 300–301
Wilder, Marshall, 7
Wine East (magazine), 273
Wine Institute, 17
Winery. *See also* Marketing
  case studies of designs for, 163–68
  equipment for, 188–202
  feasibility studies and financing of, 157–59
  first commercial in America, 6
  insurance and, 159–60
  location of, 160–62
  management of, 154
  name of, 272
  sanitation in, 93–94, 183, 317–19
  size and scale of, 162–63
Winestone sediment, 226
Wines and Vines (magazine), 273
Wine trails, 276
Wine and winemaking. *See also* Blush wines; Enology; Red wines; Viticulture; White wines
  analytical instrumentation and procedures for, 184–88, 309–61
  blending of wines, 111–19
  classification of wines, 148–52
  composition of wine, 96, 97
  definition of wine, 95–96
  design of winery and, 153–68

history of in America, 1–23
marketing and, 270–304
materials for, 202–10
microbiology of, 73–94
quality control and, 183–84
record keeping and, 176–83
regulation of, 169–75
sources of equipment for and information on, 305–308
viticulture and, 24–72
white table wines and, 211–35
Winthrop, John, 3
Wood aromas, 110, 354
Wood storage containers. *See also* Barrel aging; Oak aging
  loss of wine contents from, 248
  oxidation and, 102–103, 228
Word of mouth, advertising by, 294
Work orders, examples of, 155

Xylem, and grafting, 42

Yeast and mold swab test kits, 85
Yeasts
  blueberry wines and, 265
  as material in winemaking, 204
  microbiology of winemaking and, 73–79, 90
  oxygen and, 102
  red wines and, 240
  sulfur dioxide and, 104
  white wines and, 219–20
Yields, of grape varieties and vineyard profitability, 47

Zinfandel, as cultivar of *Vitis vinifera*, 37
Zytorrhysis, 99